Biomedical Engineering

Biomedical Engineering

Materials, Technology, and Applications

Hossein Hosseinkhani

WILEY-VCH

Author

Dr. Hossein Hosseinkhani
Matrix, Inc.
Innovation Center for Advanced Technology
1740 Broadway 15th floor
NY 10019
New York
United States

Cover Image: © ArtemisDiana/Shutterstock

Library of Congress Card No.: applied for

British Library Cataloguing-in-Publication Data
A catalogue record for this book is available from the British Library.

Bibliographic information published by the Deutsche Nationalbibliothek
The Deutsche Nationalbibliothek lists this publication in the Deutsche Nationalbibliografie; detailed bibliographic data are available on the Internet at <http://dnb.d-nb.de>.

Print ISBN: 978-3-527-34746-9
ePDF ISBN: 978-3-527-82669-8
ePub ISBN: 978-3-527-82668-1
oBook ISBN: 978-3-527-82667-4

Typesetting Straive, Chennai, India
Printing and Binding CPI Group (UK) Ltd, Croydon, CR0 4YY

C103423_221022

Contents

Preface

Biomedical engineering aims to develop a set of simple, inexpensive, portable, and robust tools that could be commercialized and used in various fields of biomedical sciences such as drug discovery, diagnostic tools, and therapeutic approaches in regenerative medicine. Specifically, biomaterials technology aims to achieve this goal by developing cheap and widely applicable technologies that can be easily used by researchers in academia and industry in various biological, medical, and engineering disciplines. Biomaterials technology is very interdisciplinary in nature and brings together the fields of polymer chemistry, pharmaceutical science, biology, and basic and clinical medicines. The elucidation of using biomaterials technology will open many doors to significantly improve the quality of biological tools and lead identification as well as therapeutic approaches. Biomaterials described here are suitable for testing specifically differentiation capability of numerous types of stem cells. Especially in the case of human stem cells, it may be of clinical relevance for future cell-based therapeutic applications. Also, it will provide attractive combinational strategy of tissue engineering principles with materials engineering to accelerate and enhance tissue regeneration. It covers wide range of applications including drug discovery, micro- and nano-engineering, cellular microenvironment, and high-throughput technologies. Whichever is the final goal for experimental biology and clinical medicine, the first key issue to be dealt is to engineer cellular microenvironment in biomaterials as efficiently as possible and to facilitate the *in vivo*-like condition. The wide-range applications of materials science and engineering involve nanotechnology, biomaterials, and drug delivery systems (DDSs), as well as in microscale technology. Based on basic and scientific knowledge, the ability of such a technology is applicable to produce various kinds of products applicable in medicine, pharmaceutical drugs (e.g. cytokines, growth factors, and antibodies), and stem cells technology. In other words, combinational technology of materials science and engineering with biology is expected to enhance the quality of valuable biological, pharmaceutical, and medical devices. However, the idea underlying biomaterials, that cellular microenvironment might be mimicked by combinational technology of materials science and biology, rather than by conventional technology, has yet to make its mark in clinical medicine. Although the concept may appear to be elegantly straightforward, the most direct application of biomaterials technology must be in biological field. Recent researches have indicated that successful implementation of biomaterials in clinic will require the coordinated development of a variety of new technologies and the establishment of unique interactions between investigators from divergent medical and basic science disciplines.

This book is intended to bring together the various experimental aspects of biomaterials of interest to particularly medical physicians and to show how the subject works. The book starts with brief introduction to biomaterials followed by chapters dealing with the synthesis, structure, and properties of various types of biomaterials.

We trust that beginners, teachers, and practitioners of the subject will find the book useful and instructive. The book could profitably be used as the basis of a university course in the subject.

New York, USA
July 2022

Dr. Hossein Hosseinkhani

1

Concepts of Biomedical Engineering

1.1 Introduction

Biomaterials technology is one newly emerging biomedical form to create new device and induce the regeneration of detective and injured body tissues and organs as well as to substitute the biological functions of damaged organs. To this end, the cells of high proliferation and differentiation potentials are being combined with some cell scaffolds and biological signals of growth factor and gene. Since there are some cases in which cells are genetically innovated to produce the growth factors inducible angiogenesis and tissue regeneration, the technology of gene delivery is also necessary for tissue engineering. Current developments in the technological fields of biomedical and tissue engineering, bioengineering, biomechanics, microfabrication, and microfluidics have led to highly complex and pertinent new tools for *in vitro* and *in vivo* applications. The purpose of biomaterials technology is to mimic organ tissues *in vitro* in order to partially reduce the amount of *in vivo* testing. These types of systems can enhance functionality of cells by mimicking the tissue architecture complexities when compared to *in vitro* analysis and at the same time present a more rapid and simple process when compared to *in vivo* testing procedures. The development of new technology for analysis of engineered tissues can be achieved through the combination of these research domains. Combining these advanced research domains, we then present a new area of technology that allows analysis *in vitro* on engineered tissues. An extension of the biomaterials technology has also allowed tissue and organ development, which can be considered as a first step toward the replacement of animal testing using a combined organ model.

1.2 Frontiers in Biomedical Engineering

Nowadays, pharmaceutical technology and research for new drugs and formulations is of great importance, as scientists attempt to discover even more revolutionary and efficient methods to treat various diseases. At the same time the correct dosing and the side targeting are equally important for clinical success. Research in sustained drug-release systems is very promising toward such direction, while it offers advantages and potential rewards contrary to the traditional therapy. The

Biomedical Engineering: Materials, Technology, and Applications, First Edition. Hossein Hosseinkhani.

drug plasma concentrations remain inside the therapeutic range for a longer time period compared to the conventional formulations. In addition, sustained-release formulations may increase the likelihood for the patient to respond to therapy, since drug formulations are generally characterized as once-daily given dose. The rate at which a drug is released from a sustained-release formulation depends on many factors, while excipients play the most important role. Most of sustained-release formulations are based on biodegradable polymers in the form of a drug-encapsulating matrix or membrane. Examples range from monolithic devices, polymer-coated capsules and implant devices, hydrogels, to injectable systems based on suspensions of microspheres, nanospheres, or polymer solutions. The advantages and disadvantages of these different formulation systems are being extensively discussed in this book.

Administration forms, capable of sustained-drug release, became an important part of medication in terms of improved treatment effect, reduction of side effects, and patients' convenience. Sustained release of active drug provides many therapeutic advantages, the most important of which is that the drug blood levels can be maintained for a long time with minimal fluctuation. The problem with each dose of an immediate-release drug is that the concentration of drug available to the body immediately peaks and then declines rapidly. When the drug concentration reaches very high levels, it contributes to adverse side effects, while when it remains at lower levels, it is not possible to provide therapeutic benefit. Thus, it is desirable to release drugs at a constant rate, therefore maintaining drug concentration within the therapeutic range and eliminating the need for frequent dosages. Other advantages of sustained-release devices can include delivery to the required site, reduced dangers of overdose or side effects, and economic advantages by virtue of more efficient dosage, at the expense of possibly more complicated fabrication.

Sustained-release devices with sophisticated design, along with the used polymers, are considered important for the efficient control of the drug release. Modern sustained-release dosage forms require reliable excipients to ensure a release rate of the active drug, which is reproducible in a narrow range. Since a limited number of polymers fulfill this requirement, there is an interest in either modulating drug release via addition of common excipients or developing new polymers, designed for the specific application. Frequently, the approaches used to achieve adequate control of drug release include hydrophilic and lipophilic matrix systems, in which the mechanism of drug release is based on a combination of diffusion and erosion processes. Waxes or other hydrophobic materials, swellable hydrophilic excipients, and lipid combination of hydrophobic and hydrophilic excipients are the most common applications in order to achieve the desired release rate. Most of them are insoluble and permeable biocompatible polymers. For sensitive drugs, such as proteins, excipients are required for drug stabilization during manufacturing process, storage, and release. In general, controlled-release preparations appear mainly in three forms: a dosage having an internal matrix typically based on polymers, a dosage wherein active drug-containing particles are coated (mainly tablets or microparticles), and a dosage relying on osmotic pressure. From those three types, the matrix form tablet is the one that has achieved the greatest popularity, mainly due to its ease of manufacturing process. From all these formulations, the delivery mechanism should control the rate of release.

The ideal release mechanism should be at a constant rate (zero order) and can be broadly classified into physical and chemical mechanisms.

The physical mechanisms include diffusion of drug molecules through a polymer layer, dissolution or degradation of polymer matrix controlling the drug-release rate, osmotic pressure for drug release (solvent-activated release), and use of ion exchange for ionized drugs. One of the main advantages of using physical mechanisms is that the drug-release kinetics can be controlled by the drug-delivery system itself. Each drug-delivery system has predetermined drug-release kinetics that can be adjusted by varying simple parameters, e.g. the type of a polymer used, the thickness of the polymer membrane, and the surface area. In a diffusion system, the drug is either encapsulated in a polymer membrane or suspended within a polymer matrix. The procedure of the release mechanism from such a system can be described as follows: water diffuses into the membrane or matrix, the drug dissolves, and finally the dissolved drug diffuses out of the polymer. In a membrane system, diffusion of water through the polymer is the rate-limiting step. Furthermore, the rate of release remains constant, and it is proportional to the concentration of the initially appeared drug. Solvent-activated systems use several mechanisms. The most common one employs a semipermeable membrane that contains a small, laser-drilled hole. Within the membrane there is a high concentration of an osmotic agent, either the drug itself or a salt, which causes water to enter through the membrane. The drug is then forced out through the hole because of the increased pressure (osmotic pumps). Polymeric degradation is perhaps the most interesting method of drug release. As with the diffusion method, the drug is contained within a polymer membrane or matrix. The polymer is designed to degrade and release the drug at a specific location in the body. As the polymer degrades, the drug is freed and it is then available to the body.

The chemical mechanisms are based on breaking the covalent bonds that connect drug molecules to a delivery vehicle, such as polymer chains, through the implementation of either chemical or enzymatic degradation. The main disadvantage of using the chemical mechanisms is that drug molecules have to be chemically modified for grafting to the delivery vehicle. Such a situation leads to the development of new chemical entities, which are called prodrugs. For this reason, the physical mechanisms have been widely implemented.

Drug control release can be accomplished in numerous ways: intravenously, transdermally, or orally. The present review will focus first on the different sustained-release formulations administered orally, second on the materials used for their preparation, and third on the mechanisms that have been developed and patented during the past years in order to control the drug release. The examples mostly used, focus on innovative formulations and are being investigated during the last 5–7 years.

From the viewpoint of designing a biomaterial, understanding materials' characteristics is critical for predicting what function the material can exhibit as a biomaterial. For example, the design of a ceramic or metal implant emphasizes the mechanical functions, such as the geometric shape, surface, and mechanical properties, to bear and transmit loading. The metal stent serves as mechanical supports to prevent localized flow constriction by its rigid structure. The contact lenses made by polymer provide optical function to allow visible light to precisely transmit and refract into pupil. Cochlear implant using electrical impulses stimulates cochlear nerves that can create a perception of sound. The electric function such as cardiac pacemaker stimulates muscle to induce rhythmical pulses. Also, piezoelectric implant transfers mechanical stress into electric current to improve bone growth.

From the viewpoint of safety, in general, toxic reaction is directly associated with chemical or electrochemical properties of biomaterials interacting with body fluid. The passive characteristic is the basic and important requirement for invaded biomaterials. Accordingly, the physical interactions between biomaterials and the living body rather than chemical interactions should be particularly concerned. From several clinical cases, failures also occur when utilizing biomaterials with unfavorable physical properties. For instance, comparing with bone, a relatively higher elastic modulus of titanium alloy that served as an artificial hip joint may induce the stress-shielding effect and may result in implant loosening after a long-term implantation. The main reason for stress-shielding effect is an unsymmetrical loading-stress transfer from implant to neighboring bone due to their mismatch stiffness.

According to above-mentioned two viewpoints of safety and design, understanding physical properties of materials is an important step to improve or further develop novel biomedical devices. The physical properties of material including electric, thermal, and optical properties will be introduced in this book while deducing how to apply these specific properties to make a material to replace functions of living tissues.

Electronic biomaterials can be divided into two classifications. The first is using nonbiological electronic materials to serve as biomedical devices due to their particular electronic properties, such as conducting wire, lithium battery, and pulse generator, contained cardiac pacemaker to provide electronic properties. The second is utilizing biologicals or biologically derived materials as templates. For instance, a template made by DNA is used to link electrodes with nanoparticles. This book highlights the electronic conductivity and resistance of these materials.

Recently, the optical properties of materials have been widely applied in biomedical field, including medical image and biological spectrum to obtain image of biological body by combining many techniques such as radiology, thermography, nuclear medicine, medical photography, and microscopy . The medical images consisted of X-ray radiography, ultrasound, computed topography, nuclear medicine, magnetic resonance image, and positive emission tomography created by different radiation. On the other hand, the biological spectrum techniques, used to detect the biochemical, structural, and functional changes of biological molecules, such as atomic absorption spectroscopy, X-ray diffraction spectroscopy, fluorescence spectroscopy and nuclear medicine, magnetic resonance spectroscopy, will be introduced in this book.

The thermal properties for polymers are relatively important for biomedical applications as compared with metals and ceramics because their thermal properties directly affect their mechanical properties. For example, biodegradable synthetic polymers show that the lower the glass transition temperature, the weaker the mechanical properties. Hence, this book introduces three kinds of thermal properties, i.e. thermal expansion, conductivity, and capacity of materials.

A variety of materials, such as metals, ceramics, polymers, and composites, are used in bioengineering. The understanding and measurement of the mechanical properties and failure processes of these biomaterials are quite important for their successful and safe applications. Mechanical behavior of a material reflects its response to the applied load or force. It is necessary to know the mechanical behavior of materials to have a suitable design for their applications and to prevent the occurrence of fracture when in use. In this chapter, the fundamentals in mechanics of materials, methods of mechanical testing, and failure processes are demonstrated. The basic knowledge in this chapter is applicable to various materials, certainly including most of the biomaterials.

The basic mechanics of materials are introduced in this book. All solid materials can be deformed when subjected to external loads. The deformation may be elastic, viscoelastic, and/or plastic ones. The mechanics of materials mainly ascertain the relationship between the responsive deformation and applied load. A variety of vocabularies and several important phenomena related to the mechanics of materials, such as engineering stress–strain, true stress–strain, modulus of elasticity, shear modulus, yielding, necking, and dislocation slip, have to be clearly demonstrated in discussing the mechanical properties of materials.

A number of mechanical tests are demonstrated in this book. The tensile test is widely used to measure the strength of materials. An engineering stress–strain curve is constructed from the load–elongation measurements. Yield strength is the stress required to produce a small specified amount of plastic deformation. Hardness testing is a measure of a material's resistance to localized plastic deformation. A small indenter is forced into the surface of a tested material, under controlled conditions of load and rate of application. Thereafter, the depth or size of the resulting indentation is measured, which in turn is related to a hardness index number. The softer the material, the larger and deeper the indentation, and the lower the hardness index number is. Impact tests are established to ascertain the characteristics of material fracture at high loading rates. Charpy and Izod are two standardized impact tests to measure the impact energy or as called notch toughness. The toughness of a material is the ability to absorb energy in the plastic range without fracturing.

Several fracture behaviors of materials, such as ductile and brittle fracture, fatigue failure, and wear, are discussed in this book. Ductile materials exhibit substantial plastic deformation with high energy absorption before fracture. Brittle fracture normally has little or no plastic deformation with low energy. The material fracture includes two steps, crack initiation and propagation, in response to the applied stress. The fracture mode is highly dependent on the mechanism of crack propagation. Ductile fracture is characterized by extensive plastic deformation in the vicinity of an advancing crack. For brittle fracture, cracks may spread quite rapidly, with a little or even no accompanying plastic deformation. Fatigue is a failure that occurs for materials subjected to dynamic and fluctuating stresses. The failure of metals involves alternating stresses of which maximum stress is lower than the materials strength. A stress–cycle (S—N) curve is an important parameter that characterizes a material's fatigue behavior. The fatigue limit or endurance limit and fatigue strength are defined in the S—N curves for a variety of materials. The fatigue fracture is characterized by three distinct steps: crack initiation, crack propagation, and final fracture. The initiation of fatigue crack always occurs on the surface of a component at some sites with stress concentration. The macroscopic fatigue failure surface is divided into two areas with different appearances. The polished or burnished area decorated with clamshell marking or beach marks is associated with slow crack growth when the cracks rub each other as the specimen is deformed back and forth through each stress cycle. The rough or granular area fractured as a result of overload when the crack propagates rapidly. The fatigue behavior of engineering materials is highly sensitive to a number of variables, such as the mean stress level, geometrical design, metallurgical variables, surface effects, and environment. Commercially, the processes of shot peening, carburizing, or nitriding are often executed on the surfaces to improve the material's fatigue properties. Wear is an important factor in the deterioration of components that move over each other. This phenomenon often limits both the using life and the performance of these components. Wear loss of material from the surface is mainly by transfer to another surface or the creation of wear debris. Besides the wear loss, some surface defects may be created in the surface. These surface defects may exacerbate other types of damage, such as fatigue and stress corrosion. Adhesive wear and abrasive wear are two primary wear mechanisms. These two wear mechanisms are mainly ascribed to the sliding wear. Sliding wear could be affected by various conditions, such as lubrications, materials, loading, sliding speed, hardness, and surface roughness.

Metallic materials, ceramic materials, and polymeric materials represent a versatile class of biomaterials being extensively applied in multitude of biomedical applications. An in-depth understanding and valid modification of physical, chemical, biological, and engineering properties is highly relevant to the performance and development of medical devices. Surface properties of biomaterials play a major role in determining biocompatibility. They have a significant influence on biological response and also determine the long-term performance. The main goal in designing biomaterials is therefore to ensure that they exhibit appropriate surface properties as well as desired physical and mechanical characteristics, which would enable them to function properly in the biological environment. It is hard to satisfy all of these characteristics. In order to improve surface properties, the surface treatment techniques should be applied. However, it is still a highly challenging task to modify surface properties – producing hemocompatible surfaces of an ECMO film, for example. Biological response to biomaterials is very complex and still not fully investigated. For the ECMO, the surface of the biomaterial is responsible for initiating the primary interaction with blood fluids due to the mechanism of blood coagulation. Therefore, the effective control of the surface structure and properties of biomaterials would be very beneficial for getting insights into physicochemical properties and their corresponding applications.

The surface treatments of biomaterials could be simply separated into interchange of surface species and directly cover compounds. It is of vital importance to ensure that the surface of the materials is suitably conditioned to ensure an appropriate biocompatibility. For that reason, if the surface structures of the materials can be properly modified to improve their biological reaction properties with keeping their original natural functions, the surface treatment is viable. This definitely requires advanced instrumentation and characterization techniques to understand surface structure properties and functional performance relationships of various materials and further to open the way toward more efficient medical devices, therapies, and other biomedical applications.

The surface treatments of materials by chemical methods, electrochemical methods, plasma methods, and ion beam implantation are principally explained and recommended in this book. Those methods can provide an improvement of surface properties to greatly expand the biomedical applications of materials. For an evaluation of the surface properties of materials, image technologies such as scanning electron microscope (SEM), atomic force microscope (AFM), near-field scanning optical microscope (NSOM), and tip-enhanced Raman scattering (TERS) imaging and spectroscopy methods such as surface-enhanced Raman scattering (SERS), X-ray absorption near-edge spectroscopy (XANES), and extended X-ray absorption fine structure (EXAFS) are also interpreted.

Cell therapy using biomaterials technology will only be a topic of the future rather than the present and is based on combinational technology of materials science, stem cells technology, and reconstructive surgery that aim to regenerate natural tissues as well as to create biological substitutes for defective or lost organs and tissues. The design of materials that can regulate cell behavior such as proliferation and differentiation is a key component for the fabrication of tissue engineering scaffolds. From the viewpoint of immune system response of the body, the implanted biomaterials should mimic the structure and biological function of native extracellular matrix (ECM), both in terms of chemical composition and physical properties. Therefore, in order to mimic the biological function of ECM proteins, the scaffold materials used in tissue engineering need to be chemically functionalized to provide appropriate niches for cell proliferation and differentiation and ultimately tissue regeneration as ECM does. Fabricated tissue-engineered scaffolds used in regenerative medicine have micron dimensions that fail to mimic the structure of natural ECM. Nanotechnology and the use of new nanomaterials underpin the race for tissue-engineered products because of the strong biological activity that current researches appear to indicate for these materials. It is no doubt that tissue engineering and regenerative medicine using nanomaterials will be the novel innovation of the future. Remarkably, the recent identification of nanotechnology with enhanced ability to mimic natural ECM proteins such as collagen structure has led to the discovery of a class of nanomaterials with specific properties for tissue engineering applications and regenerative medicine therapy. One of the common approaches is to produce fibers similar to ECM proteins is self-assembly of collagen. This technique is sufficient to cause its rapid transduction into a variety of different tissues *in vitro* as well as *in vivo*. Moreover, this novel technique for proteins and peptides appears to circumvent many problems associated with cells and drug-based methods. This review will consider the self-assembled systems and the recent developments for their potential applications in regenerative medicine. The self-assembling proteins and peptides for designing novel biomimetic nanomaterials and their potential applications in regenerative medicine and biomedical applications will be also discussed in detail from the viewpoint of their biological applications.

Many three dimensional (3D) models currently in practice require expensive equipment, large sample volumes, long incubation times, and/or extensive expertise, and their greatest disadvantage is that they are too far from the nature of human organs. Because of the above problems, research and development on drug discovery, regenerative medicine, biotech, and pharmaceutical industries is very costly and takes several years to bring a single drug/product to the market. The goal of biomedical engineering is to merge biomaterials science, nanotechnology, and biological principles to generate 3D *in vitro* living organs to mimic organ/tissues in order to partially reduce the amount of *in vitro* and *in vivo* animal testing and clinical trials, and to solve the above problems; to sum up, the final goal is to jump from the lab bench to the market. It is proposed to do all the above costly and timely tests in a rapid and cost-affordable way. At the nanoscale, chemistry and materials help us to fabricate novel type of hydrogels that are similar to human organs, infusing the cell-laden hydrogels with ECM molecules and gradients of signaling molecules to influence cell development and aggregation. At microscale, fabrication technologies adopted from the semiconductor industry, such as photolithography, will help us to mass produce identical building blocks in a variety of shapes and sizes. These products will have to mimic the physical, chemical, and biological properties of natural organs and tissues at different scales, from molecules to cells to building blocks to organized clusters to reach the final device. It is envisioned that the proposed method can be used to generate vascularized organs and tissues with controlled cell–cell and cell–matrix interactions that will be useful as *in vitro* diagnostics tools and drug screening applications as well as for transplantation.

We anticipate that elucidation of the preceding goals will open many doors and lead to significant improvements in biological tools, drug discovery process, lead identification as well as therapeutic approaches. The miniaturization of this approach allows one to perform many more experiments than previously possible in a simpler manner. Biomaterials technology aims to develop set of tools that are simple, inexpensive, portable, and robust that could be commercialized and used in various fields of biomedical sciences such as drug discovery, diagnostic tools, and therapeutic approaches in regenerative medicine. It will build up interdisciplinary critical and experimental research media aimed at overcoming the fundamental limits to information processing. It will enhance graduate education in the local universities and produce high-quality researchers among local students and encourage them to pursue research careers by creating new knowledge in selected areas of focus. Also, it will consist of a top academic and high-tech objective to fabricate, investigate, and implement novel advanced microstructures, nanostructures, and superstructures based on ordered highly functionalized materials to meet the demands of maximum efficient, active hydrogel materials of high and sustained reactivity as well as long-term stability. It will also enable new fundamental research and development for the next generation of biomedical materials as well as exploit such novel structures to develop novel biomedical devices, and transfer the knowledge to academia and industry for a future implementation of novel knowledge and technology to the world. In turn, it should increase the international competitiveness of the world into a knowledge-intensive micro- and bioengineering organ-based biosensors by focusing on high-impact research to generate new breakthroughs aimed at solving significant practical problems of biomedical sciences while seeking to extend the boundaries of understanding. It aims to raise research profile of bioengineering as a vibrant center for medical and technological applications through a bottom-up approach that embraces both elements of basic and applied research to enhance the competencies in existing technologies, to seek out promising new areas, and to develop an integrated, cutting-edge research program by growing a pool of top research talent and developing the platforms on which local universities could create research breakthroughs of importance to the world. Since advancement in device development technologies is a significant indicator of the developed societies, rapidly growing market for biomedical devices provides competitive advantage in R&D and commercialization for this field. Such advancement can be reached by means of setting up and helping local universities to establish their own facilities using our technology created in local universities.

1.3 Impact of Biomedical Engineering

Biodegradable materials such as collagen, gelatin, dextran, and carbohydrate derivatives are traditionally used as coating biomaterials for the clinical preparation of nanoparticles. Cross-linked iron oxide nanoparticles have also been developed as a result of the amination or carboxylation of dextran; they have improved conjugation capacity for more complex, integrative applications. The polymer-encapsulated method is very flexible, since it allows the preparation of iron oxide nanoparticles that can carry a wide variety of stabilizers. Various proteins, antibodies, peptides, and aptamers have been attached to the surface of the iron oxide nanoparticle. These act as targeting agents for specific biomarkers on target cells, providing a more exciting avenue to modulate nanotoxicity *in vivo*. The surface charge also plays a role in nanotoxicity, as it influences the adsorption of molecules and ions that may change organism/cellular responses toward nanoparticles. The surface charge is a major determinant of nanoparticle colloidal behavior that influences the organism response by changing the shape and size of nanoparticle through aggregate/agglomerate formation. Generally, neutral surfaces are the most biocompatible, whereas cationic surfaces are more likely to induce hemolysis and platelet aggregation. This may be as a result of the affinity of cationic nanoparticles to the negative phospholipid head groups or protein domains on cellular membranes. In addition, the surface charge can influence plasma protein binding. Hence, it affects the *in vivo* biodistribution and the clearance of the iron oxide nanoparticles from the circulation. The surface roughness is one of the main factors involved in the nonspecific binding forces that promote cellular uptake (similar to hydrophobicity and cationic charge). Small radii surface coarseness greatly minimizes electrostatic or hydrophobic–hydrophilic repulsive interactions, therefore promoting cell–nanoparticle interactions. However, significant nanotoxicities can also be caused by the enhancement of certain physicochemical properties of nanoparticles.

1.3.1 Target Drug Delivery

It is becoming clear that due to its instability and degradability, naked drug is rarely applied in systemic delivery; accordingly, this section will deal primarily with drug-loaded carriers, such as nanospheres, nanocapsules, liposomes, micelles, microemulsions, conjugates, and other nanoparticles. Polymeric biomaterials are classically biodegradable and positively charged (e.g. cationic-cell-penetrating peptides, cationic polymers, dendrimers, and cationic lipids) that are widely used as drug (gene, growth factor) carrier [1–20]. Conjugation of drugs with a variety of small molecules (e.g. cholesterol, bile acids, and lipids), polymers, peptides, proteins (e.g. antibodies), as well as aptamers (e.g. RNAs), and encapsulating drugs in nanoparticulate formulations improves the stability, cellular internalization, or cell-specific active targeted delivery synthetic polymers that have been widely investigated for drug delivery. These synthetic polymers may enhance intracellular delivery by facilitating endosomal escape and inducing lysosomal disruption, endosomal release, and drug protection from lysosomal degradation by way of buffering the endosomes [21–30]. Biodegradable polymers have been reported to undergo hydrolytic degradation and yielding nontoxic and neutral pH degradation products, thereby providing sustained gene delivery [22, 31–44].

1.3.2 Early Stage Detection

The initial interaction between cells and solid substrates is important for designing biomaterials to a variety of biological applications such as tissue engineering and cell-based biochip. Microfluidic devices have been recently used to fabricate vascularized tissue-engineered constructs and miniaturized bioreactors. Microscale technologies, which were traditionally used in the microelectronics industry to fabricate computer chips, have recently emerged as a useful approach to control the various aspects of the cellular microenvironment. In particular, the development of techniques such as soft lithography that can be cheaply and easily used to fabricate micro- and nano-devices, without the need for microfabrication facilities, has greatly enhanced the widespread application of microscale technologies in drug discovery. Microscale technologies have emerged as a powerful tool to pattern cells on substrates. In this approach, techniques such as microcontact printing and micromolding are used to generate adhesive micropatterns on substrates. These techniques have been used to create arrays of cells as well as to control the shape of individual cells. Furthermore, by using these techniques novel biological insights have been gained regarding the effects of cell shape on apoptosis and differentiation. Micropatterning techniques are a powerful method in standardizing *in vitro* drug discovery assays since they can be easily incorporated within the microwell systems used currently for drug screening.

1.3.3 Personalized Medicine

Therapies based on drugs are entering clinics, especially for diseases requiring locoregional treatments, including age-related macular degeneration, diabetic macular edema, respiratory virus infection, pachyonychia congenital, hepatitis, human immunodeficiency virus infection, and cancer. There are a number of obstacles and concerns that should be overcome before personalized medicine will be used as a new therapeutic technique.

1.4 General Applications of Biomedical Engineering

The initial interaction between cells and solid substrates is important for designing biomaterials to a variety of biological applications such as tissue engineering and cell-based biochip. On the contrary, nanoparticles have many potential applications ranging from imaging to drug-delivery systems. Therefore, design of an advanced *in vitro* culture system that can analyze the interaction of particles with cells may be beneficial.

1.4.1 Pharmaceutic

One of the future approaches with stem cells is the combinational therapies with gene. The cells can not only function as the carrier vehicle to target genes to the site of action but can also positively participate in the process of tissue repairing. Therefore, several researches have been investigated on the therapy of tissue regeneration by genetically engineered stem cells. Over the last decade gene therapy has captured the scientific and public interest with the promise to deliver genes and proteins to specific tissues or to replace deficient host cell populations. In planning gene therapy strategies for tissue engineering, the aim is to deliver a therapeutic gene of a growth factor or cytokine, into the target tissue. However, considering the immunological and safety issues of viral vectors, necessity in the development of nonviral vector systems has been increasingly magnified. Better understanding of molecular mechanism on the differentiation of stem cells will eventually allow us to properly manipulate stem cells both *in vivo* and *in vitro* for the regeneration of tissues and organs. This paradigm provided a proof of principle for the use of tissue engineering techniques as a means to improve methods of gene transduction. Tissue-engineered cell construct can be potentially used as therapeutic cell-based gene delivery or as an *in vitro* model system for testing of genetic manipulations in order to understand the effects of gene expression on tissue development.

1.4.2 Medicine

Materials technology aims to develop a technique that draws from microscale engineering, novel biomaterials, and biological principles to overcome the limitations of the current approaches to generate 3D *in vitro* living organs. These include the inability to generate 3D constructs that mimic the complexity of native tissue structure as well as to generate vascularized structures within 3D tissue culture system. The success of hydrogels as tissue implants or biomedical devices is strongly dependent on their bulk properties. The ability to directly seed cells within macroporous hydrogels will be important for overcoming challenges associated with uniform cell-seeding density and vascularization. In general, hydrogels from natural sources can be derived from polymers such as collagen, hyaluronic acid (HA), fibrin, alginate, agarose, or chitosan. Depending on their origin and composition, various natural polymers have specific utilities and properties. Many natural polymers, such as collagen, HA, and fibrin, are derived from various components of the mammalian ECM. The advantages of natural polymers include low toxicity and biocompatibility.

1.4.3 Consumer Goods

Metallic materials, ceramic materials, and polymeric materials represent versatile class of biomaterials being extensively applied in multitude of biomedical applications. An in-depth understanding and valid modification of physical, chemical, biological, and engineering properties is highly relevant to the performance and development of medical devices. Surface properties of biomaterials play a major role in determining biocompatibility. They have a significant influence on biological response, and they also determine the long-term performance. The main goal in designing biomaterials is therefore to ensure that they exhibit appropriate surface properties as well as desired physical and mechanical characteristics, which would enable them to function properly in the biological environment. It is hard to satisfy all of these characteristics. In order to improve surface properties, the surface treatment techniques should be applied. Biological response to biomaterials is very complex and still not fully investigated. Therefore, the effective control of the surface structure and properties of biomaterials would be very beneficial for getting insights into physicochemical properties and their corresponding applications.

1.5 Summary and Challenges

The surface treatments of biomaterials could be simply separated into interchange of surface species and directly cover compounds. It is of vital importance to ensure that the surface of the materials is suitably conditioned to ensure an appropriate biocompatibility. For that reason, if the surface structures of the materials can be properly modified to improve their biological reaction properties while keeping their original natural functions, the surface treatment is viable [4, 12, 14–18, 22, 31, 37, 38, 45–78]. This definitely requires advanced instrumentation and characterization techniques to understand surface structure properties and functional performance relationships of various materials and further to open the way toward more efficient medical devices, therapies, and other biomedical applications. Cell therapy using biomaterials technology will only be a topic of the future rather than the present and is based on combinational technology of materials science, stem cells technology, and reconstructive surgery that aim to regenerate natural tissues as well as to create biological substitutes for defective or lost organs and tissues [13, 19–21, 39–42, 78–100]. The design of materials that can regulate cell behavior such as proliferation and differentiation is a key component for the fabrication of tissue engineering scaffolds [9, 27, 28, 65, 101–126]. From the viewpoint of immune system response of the body, the implanted biomaterials should mimic the structure and biological function of native ECM, in terms of both chemical composition and physical properties [127–144]. Therefore, in order to mimic the biological function of ECM proteins, the scaffold materials used in tissue engineering need to be chemically functionalized to provide appropriate niches for cell proliferation and differentiation and ultimately tissue regeneration as ECM does.

References

1 Hall, A.H., Wan, J., Shaughnessy, E.E. et al. (2004). RNA interference using boranophosphate siRNAs: structure–activity relationships. *Nucleic Acids Res.* 32: 5991–6000.

2 Dowler, T., Bergeron, D., Tedeschi, A.L. et al. (2006). Improvements in siRNA properties mediated by 2′-deoxy-2′-fluoro-beta-D-arabinonucleic acid (FANA). *Nucleic Acids Res.* 34: 1669–1675.

3 Mahmoudi, M., Hosseinkhani, H., Hosseinkhani, M. et al. (2010). Magnetic resonance imaging tracking of stem cells *in vivo* using iron oxide nanoparticles as a tool for the advancement of clinical regenerative medicine. *Chem. Rev.* 111: 253–280.

4 Mottaghitalab, F., Rastegari, A., Farokhi, M. et al. (2017). Prospects of siRNA applications in regenerative medicine. *Int. J. Pharm.* 524: 312–329.

5 Abedini, F., Hosseinkhani, H., Ismail, M. et al. (2011). *In vitro* intracellular trafficking of biodegradable nanoparticles dextran-spermine in cancer cell lines. *Int. J. Nanotechnol.* 8: 712–723.

6 Abedini, F., Ismail, M., Hosseinkhani, H. et al. (2011). Effects of CXCR4 siRNA/dextran-spermine nanoparticles on CXCR4 expression and serum LDH levels in a mouse model of colorectal cancer metastasis to the liver. *Cancer Manage. Res.* 3: 301.

7 Abedini, F., Ismail, M., Hosseinkhani, H. et al. (2010). Toxicity evaluation of dextran-spermine polycation as a tool for genetherapy *in vitro*. *J. Cell Anim. Biol.* 4: 170–176.

8 Azzam, T., Eliyahu, H., Shapira, L. et al. (2002). Polysaccharide-oligoamine based conjugates for gene delivery. *J. Med. Chem.* 45: 1817–1824.

9 Hosseinkhani, H. (2006). DNA nanoparticles for gene delivery to cells and tissue. *Int. J. Nanotechnol.* 3: 416–461.
10 Hosseinkhani, H., Azzam, T., Kobayashi, H. et al. (2006). Combination of 3D tissue engineered scaffold and non-viral gene carrier enhance *in vitro* DNA expression of mesenchymal stem cells. *Biomaterials* 27: 4269–4278.
11 Hosseinkhani, H., Azzam, T., Tabata, Y., and Domb, A. (2004). Dextran–spermine polycation: an efficient nonviral vector for *in vitro* and *in vivo* gene transfection. *Gene Ther.* 11: 194–203.
12 Alibolandi, M., Abnous, K., Sadeghi, F. et al. (2016). Folate receptor-targeted multimodal polymersomes for delivery of quantum dots and doxorubicin to breast adenocarcinoma: *in vitro* and *in vivo* evaluation. *Int. J. Pharm.* 500: 162–178.
13 Abedini, F., Hosseinkhani, H., Ismail, M. et al. (2012). Cationized dextran nanoparticles-encapsulated CXCR4-siRNA enhanced correlation between CXCR4 expression and serum ALP in colorectal cancer. *Int. J. Nanomed.* 7: 4159–4168.
14 Mottaghitalab, F., Shokrghozar, M.A., Farokhi, M., and Hosseinkhani, H. (2015). Silk fibrin nanoparticles as novel drug delivery systems. *J. Controlled Release* 206: 161–176.
15 Yeo, W.Y., Hosseinkhani, H., Rahman, S.A. et al. (2014). Safety profile of dextran-spermine gene delivery vector in mouse lungs. *J. Nanosci. Nanotechnol.* 14: 3328–3336.
16 Hosseinkhani, H., Abedini, F., Ou, K.L., and Domb, A.J. (2015). Polymers in gene therapy technology. *Polym. Adv. Technol.* 26: 198–211.
17 He, W.J., Hosseinkhani, H., Hong, P.D. et al. (2013). Magnetic nanoparticles for imaging technology. *Int. J. Nanotechnol.* 10: 930–944.
18 Alibolandi, M., Abnous, K., Ramezani, M. et al. (2014). Synthesis of AS1411-aptamer-conjugated CdTe quantum dots with high fluorescence strength for probe labeling tumor cells. *J. Fluoresc.* 24: 1519–1529.
19 Hosseinkhani, H. and Hosseinkhani, M. (2009). Biodegradable polymer-metal complexes for gene and drug delivery. *Curr. Drug Safety* 4: 79–83.
20 Hosseinkhani, H., Hosseinkhani, M., Chen, Y.-R., and Subramani, K. (2011). Innovative technology of engineering magnetic DNA nanoparticles for gene therapy. *Int. J. Nanotechnol.* 8: 724–735.
21 Hosseinkhani, M., Hosseinkhani, H., Chen, Y.-R., and Subramani, K. (2011). *In vitro* physicochemical evaluation of DNA nanoparticles. *Int. J. Nanotechnol.* 8: 736–748.
22 Amini, R., Azizi Jalilian, F., Abdullah, S. et al. (2013). Dynamics of PEGylated-dextran-spermine nanoparticles for gene delivery to leukemic cells. *Appl. Biochem. Biotechnol.* 170: 841–853.
23 Hosseinkhani, H., Chen, Y.R., He, W. et al. (2013). Engineering of magnetic DNA nanoparticles for tumor-targeted therapy. *J. Nanopart. Res.* 15: 1–10.
24 Hosseinkhani, H., He, W.J., Chiang, C.H. et al. (2013). Biodegradable nanoparticles for gene therapy technology. *J. Nanopart. Res.* 15: 1–15.

25 Hosseinkhani, H., Hosseinkhani, M., Gabrielson, N.P. et al. (2008). DNA nanoparticles encapsulated in 3D tissue-engineered scaffolds enhance osteogenic differentiation of mesenchymal stem cells. *J. Biomed. Mater. Res. A* 85 (1): 47–60.

26 Hosseinkhani, H., Inatsugu, Y., Hiraoka, Y. et al. (2005). Impregnation of plasmid DNA into three-dimensional scaffolds and medium perfusion enhance *in vitro* DNA expression of mesenchymal stem cells. *Tissue Eng.* 11 (9–10): 1459–1475.

27 Hosseinkhani, H. and Tabata, Y. (2006). Self assembly of DNA nanoparticles with polycations for the delivery of genetic materials into cells. *J. Nanosci. Nanotechnol.* 6 (8): 2320–2328.

28 Hosseinkhani, H., Aoyama, T., Ogawa, O., and Tabata, Y. (2002). Liver targeting of plasmid DNA by pullulan conjugation based on metal coordination. *J. Controlled Release* 83 (2): 287–302.

29 Hosseinkhani, H. (2019). *Nanomaterials in Advanced Medicine*. New York: Wiley, 224 p.

30 Saberianpour, S., MoHeidarzadeh, M., Geranmayeh, M.H. et al. (2018). Tissue engineering strategies for the induction of angiogenesis using biomaterials. *J. Biol. Eng.* 12 (1): 38.

31 Abedini, F., Ebrahimi, M., and Hosseinkhani, H. (2018). Technology of RNA interference in advanced medicine. *MicroRNA* 7 (2): 74–84.

32 Abedini, F., Ebrahimi, M., Roozbehani, A.H. et al. (2018). Overview on natural hydrophilic polysaccharide polymers in drug delivery. *Polym. Adv. Technol.* 29 (10): 2564–2573.

33 Abdullah, S., Wendy-Yeo, W.Y., Hosseinkhani, H. et al. (2010). DNA complexes gene transfer into the lung by nanopartide dextran-spermine/plasmid. *J. Biomed. Biotechnol.* 2010 (10): 284840.

34 Ghadiri, M., Vasheghani-Farahani, E., Atyabi, F. et al. (2017). *In-vitro* assessment of magnetic dextran-spermine nanoparticles for capecitabine delivery to cancerous cells. *Iran. J. Pharm. Res.* 16 (4): 1320–1334.

35 Ghadiri, M., Vasheghani-Farahani, E., Atyabi, F. et al. (2017). Transferrin-conjugated magnetic dextran-spermine nanoparticles for targeted drug transport across blood-brain barrier. *J. Biomed. Mater. Res. A* 105 (10): 2851–2864.

36 Jain, A., Hosseinkhani, H., Domb, A.J., and Khan, W. (2015). Cationic polymers for the delivery of therapeutic nucleotides. In: *Polysaccharides* (ed. K. Ramawat and J.M. Mérillon), 1969–1990. Switzerland: Springer International Publishing.

37 He, W., Hosseinkhani, H., Mohammadinejad, R. et al. (2014). Polymeric nanoparticles for therapy and imaging. *Polym. Adv. Technol.* 25 (11): 1216–1225.

38 Hosseinkhani, H., Hong, P.D., and Yu, D.S. (2013). Self-assembled proteins and peptides for regenerative medicine. *Chem. Rev.* 113 (7): 4837–4861.

39 Khan, W., Hosseinkhani, H., Ickowicz, D. et al. (2012). Polysaccharide gene transfection agents. *Acta Biomater.* 8 (12): 4224–4232.

40 Taheri, M.M., Vasheghani-Farahani, E., Hosseinkhani, H. et al. (2012). Fabrication and characterization of a new MRI contrast agent based on a magnetic dextran-spermine nanoparticle system. *Iran. Polym. J.* 21: 239–251.

41 Amini, R., Hosseinkhani, H., Abdulamir, A. et al. (2012). Engineered smart biomaterials for gene delivery. *Gene Ther. Mol. Biol.* 14: 72–86.

42 Hosseinkhani, H., Hong, P.D., Yu, D.S. et al. (2012). Development of 3D *in vitro* platform technology to engineer mesenchymal stem cells. *Int. J. Nanomed.* 7: 3035–3043.

43 Sharifzadeh, G. and Hosseinkhani, H. (2017). Biomolecule-responsive hydrogels in medicine. *Adv. Healthcare Mater.* 6 (24): 1700801.

44 Hosseinkhani, H. and Domb, A.J. (2019). Biodegradable polymers in gene-silencing technology. *Polym. Adv. Technol.* 30 (10): 2647–2655.

45 Wang, H., Hosseinkhani, H., Chaswal, V. et al. (2016). *The World Scientific Encyclopedia of Nanomedicine and Bioengineering I: Nanotechnology For Translational Medicine: Tissue Engineering, Biological Sensing, Medical Imaging, and Therapeutics (A 5-Volume Set).* World Scientific Publishing, Inc. ISBN: ISBN-13: 978-9814667654.

46 Hosseinkhani, H. and Ou, K.L. (2014). *The Nanoscience in Translational Medicine.* Wu-Nan Book Inc. ISBN: ISBN: 978-957-11-7839-4.

47 Hosseinkhani, H. and Ou, K.L. (2013). *Advanced Biomaterials for Biomedical Engineering.* Wu-Nan Book Inc. ISBN: ISBN: 978–957–11-7169-2.

48 Khalaji, S., Ebrahimi, N.G., and Hosseinkhani, H. (2020). Enhancement of biocompatibility of PVA/HTCC blend polymer with collagen for skin care application. *Int. J. Polym. Mater. Polym. Biomater.* 70: 459–468.

49 Hosseini, V., Maroufi, N.F., Saghati, S. et al. (2019). Current progress in hepatic tissue regeneration by tissue engineering. *J. Transl. Med.* 17 (1): 383.

50 Toosi, S., Naderi-Meshkin, H., Kalalinia, F. et al. (2019). Bone defect healing is induced by collagen sponge/polyglycolic acid. *J. Mater. Sci. - Mater. Med.* 30 (3): 33.

51 Saberianpour, S., Heidarzadeh, M., Geranmayeh, M.H. et al. (2018). Tissue engineering strategies for the induction of angiogenesis using biomaterials. *J. Biol. Eng.* 12: 1–15.

52 Hosseinkhani, H. (2018). 3-D culture systems for cell culture technology. *J. Cell Res.* 1: 1–4.

53 Shahrezaei, M., Habibzadeh, S., Babaluo, A.A. et al. (2017). Study of synthesis parameters and photocatalytic activity of TiO_2 nanostructures. *J. Exp. Nanosci.* 12: 45–61.

54 Toosi, S., Naderi-Meshkin, H., Kalalinia, F. et al. (2017). Long bone mesenchymal stem cells (Lb-MSCs): clinically reliable cells for osteo-diseases. *Cell Tissue Bank* 18: 489–500.

55 Abassi, N., Hashemei, S.M., Salehi, M. et al. (2016). Influence of oriented nanofibrous PCL scaffolds on quantitative gene expression during neural differentiation of mouse embryonic stem cells. *J. Biomed. Mater. Res. A* 104: 155–164.

56 Farokhi, M., Mottaghitalab, F., Shokrghozar, M.A. et al. (2016). Importance of dual delivery systems for bone tissue engineering. *J. Controlled Release* 225: 152–169.

57 Gholivand, M.B., Mohammadi-Behzad, L., and Hosseinkhani, H. (2016). Application of Cu-chitosan/multi-walled carbon nanotube film modified electrode for the sensitive determination of rutin. *Anal. Biochem.* 493: 35–43.
58 Toosi, S., Naderi-Meshkin, H., Kalalinia, F. et al. (2016). PLGA-incorporated collagen: toward a biodegradable composite scaffold for bone-tissue engineering. *J. Biomed. Mater. Res. A* 104: 2020–2028.
59 Toosi, S., Naderi-Meshkin, H., Kalalinia, F. et al. (2017). Comparative characteristics of mesenchymal stem cells derived from reamer-irrigator-aspirator, iliac crest bone marrow, and adipose tissue. *Cell. Mol. Biol.* 62: 68–74.
60 Mottaghitalab, F., Hosseinkhani, H., Shokrghozar, M.A. et al. (2015). Silk as potential candidate for bone tissue engineering. *J. Controlled Release* 215: 112–128.
61 Seyedpour, S.M., Pachenari, M., Janmaleki, M. et al. (2015). Effects of antimitotic drug on mechanical behaviors of cytoskeleton in distinct grades of colon cancer cell lines. *J. Biomech.* 48: 1172–1178.
62 Han, H.C., Lo, H.C., Chen, K.H. et al. Nano-textured fluidic biochip as biological filter for selective survival of cells and biological microorganisms. *J. Biomed. Mater. Res. A* 103: 2015, 2015–2023.
63 Jahani, H., Azizi Jalilian, F., Kaviani, S. et al. (2015). Controlled surface morphology and hydrophilicity of polycaprolactone towards differentiation of mesenchymal stem cells into neural cells. *J. Biomed. Mater. Res. A* 103: 1875–1881.
64 Baheiraei, N., Azami, M., and Hosseinkhani, H. (2015). Investigation of magnesium incorporation within gelatin/calcium phosphate nanocomposite scaffold for bone tissue engineering. *Int. J. Appl. Ceram. Technol.* 12: 245–253.
65 Jain, A., Hosseinkhani, H., Domb, A.J., and Khan, W. (2015). Cationic polymers for the delivery of therapeutic nucleotides. *Polysaccharides: Bioactivity Biotechnol.* 1969–1990.
66 Pachenari, M., Seyedpour, M., Babazadeh Shayan, S. et al. (2014). Mechanical properties of cancer cytoskeleton depend on actin filaments to microtubules content: investigating different grades of colon cancer cell lines. *J. Biomech.* 47: 373–379.
67 Rahbarghazi, R., Nassiri, S.M., Ahmadi, S.H. et al. (2014). Dynamic induction of pro-angiogenic milieu after transplantation of marrow-derived mesenchymal stem cells in experimental myocardial infarction. *Int. J. Cardiol.* 173: 453–466.
68 Ghodsizadeh, A., Hosseinkhani, H., Piryaei, A. et al. (2014). Galactosylated collagen matrix enhanced *in vitro* maturation of human embryonic stem cell-derived hepatocyte-like cells. *Biotechnol. Lett* 36: 1095–1106.
69 Shi, D., Tatu, R., Liu, Q., and Hosseinkhani, H. (2014). Stem cells based tissue engineering for regenerative medicine. *Nano LIFE* 4 (2): 1–13.
70 Hu, C.S., Tang, S.L., Chiang, C.H. et al. (2014). Characterization and antitumor effects of chondroitin sulfate-chitosan nanoparticles delivery system. *J. Nanopart. Res.* 16: 1–15.
71 Ou, K.L. and Hosseinkhani, H. (2014). Development of 3D *in vitro* technology for medical applications. *Int. J. Mol. Sci.* 15: 17938–17962.

72 Ou, K.L., Chu, J.S., Hosseinkhani, H. et al. (2014). Biomedical nanostructured coating for minimally invasive surgery devices applications: characterization, cell cytotoxicity evaluation and an animal study in rat. *Surgical Endoscopy* 28: 2174–2188.

73 Huang, C.F., Huang, C.F., Chiang, H.J. et al. (2014). Comparison of cell response and surface characteristics on titanium implant with SLA and SLAffinity functionalization. *J. Electrochem. Soc.* 161: 15–20.

74 Hosseinkhani, H., Chen, Y.R., He, W. et al. (2013). Engineering of magnetic DNA nanoparticles for tumor-targeted therapy. *J. Nanopart. Res.* 15: 1345–1355.

75 Hosseinkhani, H., Hiraoka, Y., Li, C.H. et al. (2013). Engineering three-dimensional collagen-IKVAV matrix to mimic neural microenvironment. *ACS Chem. Neurosci.* 4: 1229–1235.

76 Hosseinkhani, H., He, W.J., Chiang, C.H. et al. (2013). Biodegradable nanoparticles for gene therapy technology. *J. Nanopart. Res.* 15: 1794–1809.

77 Chiang, C.-H., Hosseinkhani, H., Cheng, W.-S. et al. (2013). Improving drug loading efficiency and delivery performance of micro- and nanoparticle preparations through optimizing formulation variables. *Int. J. Nanotechnol.* 10: 996–1006.

78 Ou, S.-F., Chen, C.-S., Hosseinkhani, H. et al. (2013). Surface properties of nano-structural silicon-doped carbon films for biomedical applications. *Int. J. Nanotechnol.* 10: 945–958.

79 Hosseinkhani, H. and Chen, K.H. (2013). Editorial: nanotechnology research in Taiwan. *Int. J. Nanotechnol.* 10: 837–839.

80 Subramani, K., Pathak, S., and Hosseinkhani, H. (2012). Recent trend in diabetes treatment using nanotechnology. *Digest J. Nanomater. Biostruct.* 7: 85–95.

81 Hosseinkhani, H. (2012). 3D *in vitro* technology for drug discovery. *Curr. Drug Safety* 7: 37–43.

82 Mahmoudi, M., Hosseinkhani, H., Hosseinkhani, M. et al. (2011). Magnetic resonance imaging tracking of stem cells *in vivo* using iron oxide nanoparticles as a tool for the advancement of clinical regenerative medicine. *Chem. Rev.* 111: 253–280.

83 Hosseinkhani, H. (2011). Editorial: on nanomedicine. *Int. J. Nanotechnol.* 8: 615–617.

84 Abedini, F., Hosseinkhani, H., Ismail, M. et al. (2011). *In vitro* intracellular trafficking of biodegradable nanoparticles of dextran-spermine in cancer cell lines. *Int. J. Nanotechnol.* 8: 712–723.

85 Sarabi, R.S., Sadeghi, E., Hosseinkhani, H. et al. (2011). Polyrotaxane capped quantum dots as new candidates for cancer diagnosis and therapy. *J. Nanostruct. Polym. Nanocompos.* 7: 18–31.

86 Subramani, K., Mathew, R., Hosseinkhani, H., and Hosseinkhani, M. (2011). Bone regeneration around dental implants as a treatment for peri-implantitis: a review of the literature. *J. Biomimetics Biomater. Tissue Eng.* 11: 21–33.

87 Kalhor, H.R., Farzaneh Shahin, V., Fouani, M.H., and Hosseinkhani, H. (2011). Self-assembly of tissue transglutaminase into amyloid-like fibrils using physiological concentration of Ca^{2+}. *Langmuir* 27: 10766–10784.

88 Abedini, F., Ismail, M., Hosseinkhani, H. et al. (2011). Effects of CXCR4 siRNAs/dextran-spermine nanoparticles on CXCR4 expression and serum LDH levels in a mouse model of colorectal cancer metastasis to the liver. *Cancer Manage. Res.* 3: 301–309.

89 Hosseinkhani, H., Hosseinkhani, M., Hattori, S. et al. (2010). Micro and nanoscale *in vitro* 3D culture system for cardiac stem cells. *J. Biomed. Mater. Res. A* 94: 1–8.

90 Mohageri, S., Hosseinkhani, H., Ebrahimi, N.G. et al. (2010). Proliferation and differentiation of mesenchymal stem cell on collagen sponge reinforced with polypropylene/polyethylene terephathalate blend fibers. *Tissue Eng. Part A* 16: 3821–3830.

91 Lindstrom, S., Iles, A., Persson, J. et al. (2010). Nanoporous titania coating of microwell chips for stem cell culture and analysis. *J. Biomech. Sci. Eng.* 5: 272–279.

92 Abdullah, S., Yeo, W.Y., Hosseinkhani, H. et al. (2010). Gene transfer into the lung by nanoparticle dextran-spermine/plasmid DNA complexes. *BioMed Res. Int.* 2010: 1–10.

93 Abedini, F., Ismail, M., Hosseinkhani, H. et al. (2010). Toxicity evaluation of dextran-spermine polycation as a tool for gene therapy *in vitro*. *J. Cell Anim. Biol.* 4: 170–176.

94 Amini, R., Rosli, R., Abdullah, S. et al. (2010). Delivery of plasmid expressing green fluorescent protein by PEGylated dextran-spermine to acute myeloid leukemic cells. *Myeloid Leukemia* 65: 1–10.

95 Balbarini, A., Magera, A., Barsotti, M.C. et al. (2010). Self assembling peptide amphiphile nanofibers as a scaffold for endothelial progenitor cells. *J. Am. Coll. Cardiol.* 55: E1652.

96 Subramani, K., Hosseinkhani, H., Khraisat, A. et al. (2009). Targeting nanoparticles as drug delivery systems for cancer treatment. *Curr. Nanosci.* 5: 134–140.

97 Hosseinkhani, H., Hosseinkhani, M., Vasheghani, E., and Nekoomanesh, M. (2009). *In vitro* sustained release and degradation study of biodegradable poly (D,L-lactic acid) microspheres loading theophylline. *Adv. Sci. Lett.* 2: 70–77.

98 Di Stefano, R., Barsotti, M.C., Magera, A. et al. (2009). A biological self assembling amphiphilic peptide enhances endothelial progenitor cells growth and paracrine release. *Eur. Heart J.* 30: 177–177.

99 Tian, F., Hosseinkhani, H., Hosseinkhani, M. et al. (2008). Quantitative analytical of cell adhesion on aligned micro- and nanofibers. *J. Biomed. Mater. Res. A* 84: 291–299.

100 Hosseinkhani, H., Hosseinkhani, M., Khademhosseini, A. et al. (2008). DNA nanoparticles encapsulated in 3-D tissue engineered scaffold enhance osteogenic differentiation of mesenchymal stem cells. *J. Biomed. Mater. Res. A* 85: 47–60.

101 Hosseinkhani, H. and Hosseinkhani, M. (2008). Suppression effect of basic fibroblast growth factor on mesenchymal stem cell proliferation activity; part I: release characteristics. *Chem. Today* 26: 30–32.

102 Hosseinkhani, H. and Hosseinkhani, M. (2008). Suppression effect of basic fibroblast growth factor on mesenchymal stem cell proliferation activity; part II: biological characteristics. *Chem. Today* 26: 35–37.

103 Hosseinkhani, M., Hosseinkhani, H., Khademhosseini, A. et al. (2007). Bone morphogenetic protein-4 enhances cardiomyocytes differentiation of cynomolgus monkey ES cells in knockout serum replacement medium. *Stem Cells* 25: 571–580.

104 Furong, T., Hosseinkhani, H., Estrada, G., and Kobayashi, H. (2007). Quantitative method for the analysis of cell attachment using the aligned scaffold structure. *J. Phys.* 61: 587–590.

105 Hosseinkhani, H., Yamamoto, M., Inatsugu, Y. et al. (2006). Enhanced ectopic bone formation using combination of impregnation of plasmid DNA into 3-D scaffold and bioreactor perfusion culture. *Biomaterials* 27: 1387–1398.

106 Hosseinkhani, H., Hosseinkhani, M., Tian, F. et al. (2006). Osteogenic differentiation of mesenchymal stem cells in self-assembled-peptide amphiphile nanofibers. *Biomaterials* 27: 4079–4086.

107 Hosseinkhani, H., Azzam, T., Kobayashi, H. et al. (2006). Combination of 3-D tissue engineered scaffold and non-viral gene carrier enhance *in vitro* DNA expression of mesenchymal stem cells. *Biomaterials* 27: 4269–4278.

108 Hosseinkhani, H., Hosseinkhani, M., Tian, F. et al. (2006). Ectopic bone formation in collagen sponge-self assembled peptide amphiphile nanofibers hybrid scaffold in a perfusion culture bioreactor. *Biomaterials* 27: 5089–5098.

109 Hosseinkhani, H., Hosseinkhani, M., Khademhosseini, A. et al. (2006). Enhanced angiogenesis through controlled release of basic fibroblast growth factor from peptide amphiphile for tissue regeneration. *Biomaterials* 27: 5836–5844.

110 Hosseinkhani, H., Hosseinkhani, M., and Kobayashi, H. (2006). Proliferation and differentiation of mesenchymal stem cells by using self-assembly of peptide-amphiphile nanofibers. *Biomed. Mater.* 1: 8–15.

111 Hosseinkhani, H., Hosseinkhani, M., and Kobayashi, H. (2006). Design of tissue-engineered nanoscaffold through self-assembly of peptide amphiphile. *J. Bioact. Compat. Polym.* 21: 277–296.

112 Hosseinkhani, H., Hosseinkhani, M., and Khademhosseini, A. (2006). Emerging applications of hydrogels and microscale technologies in drug discovery. *Drug Discovery* 1: 32–34.

113 Hosseinkhani, H., Hosseinkhani, M., and Khademhosseini, A. (2006). Tissue regeneration through self-assembled peptide amphiphile nanofibers. *Yakhte Med. J.* 8: 204–209.

114 Hosseinkhani, H., Kobayashi, H., and Tabata, Y. (2006). Design of tissue-engineered nano-scaffold using peptide-amphiphile for regenerative medicine. *Pept. Sci.* 2005: 341–344.

115 Hosseinkhani, H., Kobayashi, H., and Tabata, Y. (2006). Selective differentiation cardiomyocyte cells by using peptide-amphiphile nanofibers. *Pept. Sci.* 2005: 63–66.

116 Konishi, M., Tabata, Y., Kariya, M. et al. (2005). In vivo anti-tumor effect of dual release of cisplatin and adriamycin from biodegradable gelatin hydrogel. *J. Controlled Release* 103: 7–19.

117 Hosseinkhani, H., Inatsugu, Y., Hiraoka, Y. et al. (2005). Impregnation of plasmid DNA into 3-D scaffold and medium perfusion enhance in vitro DNA expression of mesenchymal stem cells. *Tissue Eng.* 11: 1459–1475.

118 Hosseinkhani, H., Inatsugu, Y., Hiraoka, Y. et al. (2005). Perfusion culture enhances osteogenic differentiation of rat mesenchymal stem cells in collagen sponge reinforced with poly (glycolic acid) fiber. *Tissue Eng.* 11: 1476–1488.

119 Hosseinkhani, H., Azzam, T., Tabata, Y., and Domb, A.J. (2004). Dextran-spermine polycation: an efficient non-viral vector for *in vitro* and *in vivo* gene transfection. *Gene Ther.* 11: 194–203.

120 Hosseinkhani, H., Aoyama, T., Ogawa, O., and Tabata, Y. (2003). Ultrasound enhances the transfection of plasmid DNA by non-viral vector. *Curr. Pharm. Biotechnol.* 4: 109–122.

121 Hosseinkhani, H., Aoyama, T., Ogawa, O., and Tabata, Y. (2002). Ultrasound enhancement of *in vitro* transfection of plasmid DNA by a cationized gelatin. *J. Drug Targeting* 10: 193–204.

122 Hosseinkhani, H., Aoyama, T., Ogawa, O., and Tabata, Y. (2002). *In vitro* transfection of plasmid DNA by amine derivatives of gelatin accompanied with ultrasound irradiation. *Pharm. Res.* 19: 1469–1477.

123 Hosseinkhani, H., Aoyama, T., Ogawa, O., and Tabata, Y. (2001). *In vitro* transfection of plasmid DNA by different-cationized gelatin with or without ultrasound irradiation. *Proc. Jpn. Acad. Ser. B* 77: 161–166.

124 Hosseinkhani, H. (2013). Innovation technology to engineer 3D living organs as intelligent diagnostic tools. In: *Characterization and Development of Biosystems and Biomaterials* (ed. A. Öchsner, L.F.M. Silva and H. Altenbach), 183–192. Springer Publication.

125 Hosseinkhani, H. (2013). Controlled release systems for bone regeneration. In: *Polymeric Biomaterials, Third Edition, Volume II: Medicinal and Pharmaceutical Applications of Polymers and Technology* (ed. S. Dumitriu and V. Popa), 170–182. USA: CRC Press/Taylor and Francis.

126 Hosseinkhani, H., Hosseinkhani, M., and Subramani, K. (2012). Bone regeneration using self-assembled nanoparticles-based scaffolds. In: *Emerging Nanotechnologies in Dentistry* (ed. K. Subramani and W. Ahmed), 225–237. UK: Elsevier.

127 Hosseinkhani, H. and Hosseinkhani, M. (2009). Tissue engineered scaffolds for stem cells and regenerative medicine. In: *Trend in Stem Cells Biology and Technology* (ed. H. Baharvand), 367–387. USA: HUMANANA Press Inc.

128 Hosseinkhani, H., Hosseinkhani, M., Zhang, S., and Subramani, K. (2008). Self-assembly of nanomaterials for engineering cell microenvironment. In: *Micro and Nanoengineering of the cell microenvironment: Applications and Technologies* (ed. A. Khademhosseini), 275–290. USA: Artech House Publishers.

129 Hosseinkhani, H., Hosseinkhani, M., and Khademhosseini, A. (2007). Emerging technology of hydrogels in drug discovery. In: *Topics in Multifunctional Biomaterials and Devices* (ed. N. Ashammakhi), 120–129. London, UK.

130 Hosseinkhani, H., Aoyama, T., Ogawa, O., and Tabata, Y. (2005). Tumor targeting of plasmid DNA by dextran conjugation based on metal coordination. In: *Key Engineering Materials*, vol. 288–289, 109–112. Switzerland: Trans Tech Publications.

131 Yamamoto, M. and Hosseinkhani, H. (2004). *Liver Targeting of Plasmid DNA by Polymer-Metal Conjugation*, 119–132. Japan: Yodosha Tech Publications, Japanese version, Med Tech Publications.

132 Hosseinkhani, H. and Hosseinkhani, M. (2009). Design of 3D culture systems to enhance *in vitro* gene expression of mesenchymal stem cells. *Proc. Mol. Ther.* 17: S270–S270.

133 Hosseinkhani, H., Hosseinkhani, M., Khademhosseini, A. (2007). A new injectable tissue engineered scaffold induces angiogenesis. *Proceedings of the AIChE Annual Meeting*.

134 Hosseinkhani, M. and Hosseinkhani, H. (2007). Post-translational modification of GATA-4 involved in the differentiation of monkey ES cell into cardiac myocytes. *Circulation* 116 (16): 202–203.

135 Hosseinkhani, M. and Hosseinkhani, H. (2007). Bone morphogenetic protein-4 enhances cardiomyocyte differentiation of cynomolgus monkey ES cells in knockout serum replacement medium. *Eur. Heart J.* 28: 230–231.

136 Hosseinkhani, H., Kobayashi, H., and Tabata, Y. (2006). Design of a nano-vessel-like network for controlled proliferation and differentiation of mesenchymal stem cells for regenerative medicine. *Tissue Eng.* 12 (4): 993–994.

137 Hosseinkhani, H., Hosseinkhani, M., and Khademhosseini, A. (2006). A new injectable tissue engineered scaffold for regenerative medicine. *Proc. Int. Conf. Microtechnol. Med. Biol.* 4281294: 10–11.

138 Hosseinkhani, H. (2005). Selective differentiation cardiomyocyte cells by using peptide-amphiphile nanofibers. *Proceedings of the 42nd Japanese Peptide Symposium*. Tokyo, Japan 30, 25–30.

139 Hosseinkhani, H. (2005). Design of tissue-engineered nano-scaffold using peptide-amphiphile for regenerative medicine. *Proceedings of the 42nd Japanese Peptide Symposium* 30: 21–34. Osaka, Japan, The Japanese Peptide Society.

140 Hosseinkhani, H., Tabata, Y. (2004). PEGylation enhances tumor targeting of plasmid DNA by an artificial cationized protein with repeated RGD sequences, Pronectin® cationized. *Proceedings 7th World Biomaterials Congress*. Sydney, Australia, Australian Society for Biomaterials.

141 Hosseinkhani, H. and Tabata, Y. (2004). Tumor targeting of plasmid DNA by spermine derivative of dextran combined with ultrasound. *Polym. Preprints* 53 (2): 2PE179.

142 Hosseinkhani, H. and Tabata, Y. (2004). Ultrasound enhances expression level of plasmid DNA by PEGylation of cationized dextran in tumor. *Journal Code: X0225A* 19 (3): 299.

143 Hosseinkhani, H., Aoyama, T., Ogawa, O., and Tabata, Y. (2002). Tumor targeting of plasmid DNA by dextran conjugation based on metal coordination. *Drug Delivery* 17 (3): 260.

144 Vasheghani, E., Hosseinkhani, H., and Nekomanesh, M. (2001). "Effect of preparation conditions on theophylline release from biodegradable poly (DL-lactic acid) microspheres", proceedings of the JCR symposium. *J. Controlled Release* 72: 287–291.

2

Biomaterials

2.1 Introduction

Materials capable of exhibiting biorecognition events have begun to revolutionize medicine and biomedical sciences by providing tools for inducing or probing biomolecular and cellular interactions. Biomedical materials may mimic properties of the biological milieu and, in particular, are promising for creating bioactive materials for a variety of applications ranging from tissue engineering to microdevices. The most common strategy to produce bioactive materials has been to bind peptides, proteins, nucleic acids, or other biological molecules directly to a synthetic or inert surface, structure, or polymeric molecule. However, simple and robust methods of functionalizing these materials with multiple biomolecules would be a critical advance for widespread application of these materials. Therefore, it is necessary to synthesize and characterize robust multifunctional materials containing immobilized biomolecules that can be used to functionalize materials with a variety of biological molecules. Specifically, it is important to utilize the high binding affinity and specificity of the biomolecules bonding to conjugate functionally active biomolecules to biomaterials. The conjugation of biomolecules to materials can be used to generate multifunctional biomaterials that can, with high specificity, bind or "capture" a plethora of biotinylated biomolecules. In addition, it is important to test the utility of these materials by generating gradients of biomolecules within materials and analyze their ability to influence cell function. Biomedical materials are useful for many biomedical applications such as high-throughput drug screening and biosensing assays, as thin films for coating medical devices, or as scaffolds for tissue engineering.

In medical diagnostics and therapeutics, there is a continuous effort to enhance methods, materials, and devices to improve patient care. In recent years, the development of novel biomaterials and their application to medical problems have dramatically enhanced the treatment of many diseases [1, 2]. Biomaterials such as polymers, ceramics, and metals have been used for several decades for medical applications. In addition to the over 40 000 pharmaceutical preparations in use, it is estimated that currently there are over 8000 medical devices and 2500 diagnostic products that employ biomaterials [1]. Despite the widespread use of materials in medicine, many biomaterials lack the desired functional properties

Biomedical Engineering: Materials, Technology, and Applications, First Edition. Hossein Hosseinkhani.

for stimulating a specific biological response and have not been engineered for optimized performance. Therefore, there is an increasing need to develop new materials to address such problems in medicine and biology.

2.2 Biomedical Materials

Poly(ε-caprolactone) (PCL), an aliphatic polyester that is bioresorbable and biocompatible, is generally used in pharmaceutical products [3]. The experimental results have shown that, although synthetic biodegradable PCL supports cell growth, to proliferate more and encourage cell ingrowth for better integration between cells and the scaffold, the biologically inert PCL nanofibers need effective hybridization with bioactive molecules [4]. It has been reported that electrospinning of PCL with collagen gives encouraging results in improving the cell–scaffold interactions [4, 5]. Besides, polyethersulfone (PES) has many fascinating properties including favorable mechanical strength, thermal and chemical resistance, and excellent biocompatibility [6, 7]. Therefore, the polymer blend of PCL/collagen/PES can overcome the shortcomings of natural and synthetic polymers, resulting in a new biomaterial with good biocompatibility and improved mechanical, physical, and chemical properties.

Poly(glycolic acid) (PGA) is such kind of absorbable polymers that can possess high molecular weight (thus good processability and mechanical strength) and excellent degradation properties because of its extremely hydrolyzable backbones, which are of good properties for biomedical applications such as degradable and absorbable sutures and implants [8–10]. However, due to high hydrophilicity, PGA degrades very rapidly [11], limiting its clinical applications. To overcome this, it is necessary to modify the degradation rates of PGA. The use of poly(L-lactic acid) (PLLA) as a biomedical material has been widely studied and utilized in the areas of tissue engineering, bone healing materials, and drug delivery because of its good biodegradability, biocompatibility, and reasonably good mechanical properties [12]. Due to low hydrophilicity and high crystallinity, degradation of PLLA is extremely slow (complete degradation often takes years) [13]. We can combine PLLA with PGA in order to improve the hydrolytic stability of PGA.

Mass loss is monitored throughout degradation and was defined using Eq. (2.1) [12], where m_i is the initial dry mass and m_d is the dry mass after degradation:

$$\text{Mass loss} = (m_i - m_d)/m_i \times 100\% \tag{2.1}$$

It is evident that the pure PGA rods have lost nearly 35% of its mass at the degradation period of six weeks. However, the PLLA/PGA composite rods underwent less mass loss during the degradation process, especially in the case of the composite rods containing 50% PLLA, which just lost 22% of its mass. It was obvious that PLLA improved the hydrolytic stability of the composite rods due to its hydrophobic methyl group in the repeating unit. The supplemented PLLA can protect PGA from rapid degradation by decreasing its hydrophilicity, resulting in blockage of water entry into the rods; thus, the composite rods undergo minimal mass loss. Subsequent changes in pH value and SEM micrographs can also confirm this result. The pH value changes are defined by Eq. (2.2), where pH after and pH before were the pH values of degradation media after and before degradation, respectively:

$$\Delta pH = \mathrm{pH}_{\mathrm{after}} - \mathrm{pH}_{\mathrm{before}} \quad (2.2)$$

Large changes in pH values are observed for all tested materials even in the early degradation stage. Micro-sized degraded fragments are observed to be produced during the degradation, which may be the reason for the large pH value decline. The pH values of pure PGA samples dropped obviously in comparison with that of the other two composite materials. It is evident that 50% PLLA/PGA composite rods possess the least changes of pH values due to its higher PLLA content. These results confirm that the presence of PLLA can protect PGA from rapid degradation by decreasing its hydrophilicity and produce less acidic degradation products. The results showed that PLLA can weaken media acidity, thus inhibiting the autocatalysis degradation caused by acidic degradation products. As a result, the degradation rate will decrease [12]. This result indicated that it was possible to control the degradation rate of the blend by changing its composition; therefore, it could be proposed that this kind of composite material can be used as a potential applicable biomaterial. Addition of PLLA to PGA resulted in formation of a new kind of hydrolytic stabilized biomaterial. The mass and pH value of the pure PGA rods showed notable changes throughout almost the entire degradation period, but those of the composite rods underwent less change than pure PGA. PLLA can improve the hydrolytic stability of PGA effectively and will show some direction in modifying the degradation of PGA.

Hydrophilic polymers and especially their cross-linked forms known as hydrogels are a class of biomaterials with a wide spectrum of biomedical and biological applications such as for high-throughput drug screening and biosensing systems [14], as immunoisolation barriers for microencapsulation [15–17], as thin films for coating medical devices or tissues [18], as scaffolds for tissue engineering [19–22], or as micro-/nanoparticles for drug delivery. The reason for such a diverse set of applications is due to the high water content of hydrogels as well as our ability to engineer their properties and their ability to mimic natural tissues [23, 24]. Desired characteristics of hydrogel scaffolds include physical parameters such as mechanical strength and degradability, while biological properties include biocompatibility and the ability to provide a biologically relevant microenvironment. In addition to engineering hydrogels for specific *in vivo* performance, these materials may be modified to exhibit specific functionality *in vitro*. One of the most widely used and studied hydrogel materials for biomedical applications is poly(ethylene glycol) (PEG), which is nontoxic and nonimmunogenic and is approved by the FDA for various clinical uses. In many cases, PEG has been applied as a "stealth material" since it is inert to most biological molecules such as proteins. Some of the earliest work on the use of PEG and polyethylene oxide (PEO) as hydrophilic biomaterials was performed by Merrill et al. [25] who showed that adsorption of these materials onto glass surfaces prevented protein adsorption. Since then, many forms of PEG surface modification have been used in order to render a surface protein-resistant and to enhance surface biocompatibility [26]. PEG polymers can be covalently cross-linked using a variety of methods to form hydrogels. A particularly appealing method of cross-linking PEG chains is through photopolymerization using acrylate terminated PEG monomers [27]. Photopolymerizable hydrogels may be gelled *in situ*, enabling the polymer to conform to a particular shape or to an implantation site. Various parameters of these polymers can be controlled to modify cell behavior. For example, UV exposure, photoinitiator concentration, monomer chain length, and conjugation of various biological molecules can be used to modify gel properties. PEG hydrogels may be modified with peptide sequences to induce degradation [28] or modify cell adhesion [29]. In addition to chemical modification, block copolymers of PEG such as triblock copolymers of PEO and poly(propylene oxide), degradable PEO and poly(lactic acid) (PLA), and other similar methods can be used to add specific properties to the PEG hydrogels [30]. Also, acrylated PEG molecules can be photocross-linked with other materials such as hyaluronic acid (HA) to generate composite hydrogels with enhanced mechanical and degradation properties [31]. PEG hydrogels have been used within microdevices for various applications such as to fabricate high-throughput screening systems or to minimize protein adsorption on surfaces [14, 32, 33]. For example, PEG surface modifications have been used within microfluidic channel to pattern proteins and cells [34, 35], and PEG microstructures have been fabricated within channels to capture cells within specific regions for sensing, cell separation, and cell-based microreactors [36]. Also, environmentally responsive hydrogels, which can change properties in response to stimuli such as light, pH, or temperature, have been used as functional components of microdevices such as biosensors and valves. For example, functional valves have been fabricated from pH-sensitive photocross-linkable PEG hydrogels. These valves sense the pH of the solution and in response change their swelling, which results in actuation [37]. One attractive feature of PEG hydrogels is that they can be modified by incorporating specific peptide sequences to promote, sense, or elucidate cell–biomaterial interactions. Conjugation of biologically active peptide sequences can cause the normally inert and nonadhesive PEG-based hydrogel to become a bioactive material. For example, by immobilizing adhesive peptides such as RGD, the PEG gel may become cell adhesive. Also, proteins have also been conjugated to PEG molecules and have been subsequently incorporated within hydrogels [38]. However, the direct conjugation of proteins to hydrogels has been hindered by protein inactivation and the inability to easily conjugate various proteins to hydrogels. Although modest success has been achieved, the precise conditions for performing PEG–peptide or PEG–protein chemical conjugation must be determined individually for each peptide or protein. In addition, functional modification with peptides or proteins must be performed during the manufacture of the gel, which makes it laborious to examine multiple conditions such as peptide/protein densities that dictate cellular response and to produce hydrogels with multiple types of immobilized ligands. Therefore, it would be advantageous to develop a hydrogel-based system with the option of binding the ligand after formation of the gel. In such a system, hydrogels could be rapidly functionalized with a variety of ligands at multiple concentrations. Presently, there are no hydrogel systems that permit this type of functionalization scheme. Streptavidin is a tetrameric protein (i.e. four biotin binding sites) that is used in a wide range of FDA-approved diagnostic test kits due to its high affinity to biotin (a small molecule) and its robustness in relation to a wide pH range, detergents, and ionic strength. Streptavidin has become a nearly ubiquitous protein used in affinity separations, biomolecular assays, and clinical diagnostics [39]. Additionally, the streptavidin–biotin system is characterized by low nonspecific interactions resulting in reduced background signals. Its widespread use is due to its extremely high affinity for biotin ($K_d \sim 10^{-15}$), which it binds with great specificity [40]. Conjugation of streptavidin with a number of synthetic, responsive polymers has been met with abundant success. Other investigators have demonstrated, for example, that conjugation of stimuli-responsive polymers to streptavidin near the biotin binding site allows for control of biotin binding by external stimuli such as temperature and pH [41–43]. Also, many proteins have been conjugated with biotin, and the biotinylation of proteins has become a standard practice for which a number of commercially available kits exist. Therefore, streptavidin-conjugated gels are useful because they can be easily modified with peptides and proteins conjugated to biotin with little or no loss in activity. The functionalization of hydrogels with proteins can be useful for various analytical and therapeutic applications. For example, antibody conjugation to the hydrogels can be used to generate sensitive ELISA assays. Given that streptavidin can potentially bind to four biotinylated biomolecules within a small proximity, the signals generated could be amplified and greatly enhance the sensitivity of the assay. Similarly, the functionalized gels can be used to study the effects of proteins on cell behavior with applications in generating *in vitro* microenvironments for toxicology studies and tissue engineering applications. The proposed material has a number of capabilities that include (i) multifunctionality, (ii) ad hoc functionalization, and (iii) 3D patterning or construction. It is important to synthesize and characterize these materials and then to demonstrate their effectiveness in an example involving cell function on concentration gradients of adhesive ligands. These multifunctional hydrogels may find a number of applications including, but not limited to, biomolecular analysis, diagnostics, biosensors, molecular separations, cell sorting, migration assays, controlled drug release devices, and drug targeting (through immobilized antibodies or aptamers). These materials will form the basis for a new line of investigation, which it can carry out on a larger scale upon successful completion of these proposed exploratory experiments.

PEG together with acrylate-based chemistries (e.g. PEG diacrylate) represent the most common strategy for generating photocross-linkable protein and cell-repellant materials that may be modified with biomolecules to exhibit a specific biological response. By conjugation of streptavidin to photocross-linkable PEG hydrogels, we are able to capture a wide variety of biomolecules. After binding streptavidin to PEG, the resulting precursor molecule may be directly integrated into hydrogels by, for example, mixing with various other polymer precursors that contain acrylate groups such as PEG diacrylate or acrylated HA. The density of biomolecules within the hydrogel may be directly controlled by varying the density of streptavidin molecules within the hydrogel. To produce hydrogels containing immobilized streptavidin that are useful for a variety of biological applications, we can address the following questions:

1.1 How does the stoichiometric ratio of PEG to streptavidin during the chemical reaction affect the distribution of bound PEG molecules per streptavidin?
1.2 How does the number of conjugated PEGs per streptavidin affect biotin binding?
1.3 How does the stoichiometric ratio of PEG-streptavidin to PEG diacrylate affect hydrogel formation and biotin binding?

Smart polymers such as hydrogel–protein bioconjugates can be useful for various tissue engineering or cell biology applications. For example, they can be used to generate scaffolds that induce neural extension for axon regeneration or direct cell migration. The conjugation of streptavidin to hydrogels can be used to anchor any biotinylated biomolecule within hydrogels, which can then be used as a platform technology for many diagnostic and therapeutic applications. As a proof of concept, we can generate concentration gradients of adhesive molecules such as fibronectin and laminin within the hydrogels. This will be accomplished by using microfluidic gradient generators that are capable of producing microscale hydrogels with controlled spatial features. By combining microfluidics and photolithography, we are able to generate microgels from the materials designed earlier that can be subsequently functionalized with any biotinylated protein. We therefore are able to address the following questions:

2.1 Can hydrogels be synthesized that contain concentration gradients of immobilized streptavidin? If so, can various biotinylated proteins (such as fibronectin and laminin) be immobilized within the hydrogel gradients?
2.2 How does the concentration gradient of laminin or fibronectin influence cell adhesion to PEG hydrogels?

Table 2.1 shows examples of synthetic and natural polymers.

2.3 Biomaterials in Medicine

Chemical functional groups of biomaterials could influence the behaviors of biological properties of cells and tissues, such as methyl ($-CH_3$), hydroxyl (–OH), carboxyl (–COOH), and amino ($-NH_2$) groups that have been presented in biomaterials. Poly(acrylic acid)-modified surface of tissue culture plates promoted osteogenic differentiation but not chondrogenic differentiation of mesenchymal stem cells (MSCs) [44]. Curran et al. [45, 46] reported that the $-NH_2$ and –SH-modified surfaces of clean glass promoted and maintained osteogenesis both in the presence and absence of biological stimuli, but these surfaces did not support long-term chondrogenesis under any test conditions. In addition, the mechanical properties of substrate also play important roles for cells behaviors. Most cell types spread more, adhere better, and appear to survive better on stiffer matrices, and some cannot grow on very soft (<50 Pa) substrate [47]. Studying the relationship between physicochemical microenvironment and behaviors of MSCs

Table 2.1 Synthetic and natural polymers used as biomedical materials.

Synthetic polymers: Advantages: Control of the molecular weight during the synthesis and of the physical properties Disadvantages: Lack of intrinsic biological activity	
PGA (polyglycolic acid)	Relatively hydrophilic, degradable through hydrolysis at the
ester bond in aqueous solution or *in vivo* (2–4 weeks)	
PLA (polylactic acid)	More hydrophobic than PGA, chirality, degradable in
months or years	
PCL (poly ε-caprolactone)	Hydrophobic, degrade at a much slower rate than PGA and PLA
Natural polymers: Advantages: Possess intrinsic biological activity, enzymatically degradable Disadvantages: Limited control over parameters such as molecular weight, the potential for adverse immunological responses, variation in degradation rates due to difference in host enzyme levels, inferior mechanical properties	
Collagen	Triple helix structure, each polypeptide chain has general sequence of Gly–X–Y where X and Y are often Pro and hydroproline, adhesive functions
Chitosan	Deacetylated derivatives of chitin, structure similar to cellulose, stable, insoluble above pH 7, cationic
GAGs (chondroitin 6 sulfate), (hyaluronan)	Unbranched, highly negatively charged, strongly hydrophilic, covalently linked to proteins in the form of proteoglycans that form hydrated gels to resist compressive forces in mammalian ECM

could be important for bone tissue engineering. It is a challenge to build appropriate microenvironment for proliferation and differentiation of MSCs.

Natural polysaccharide and protein are widely used to build the microenvironment for cell growth. They have advantages in that they facilitate cell attachment and maintain differentiation of cells. These natural polymers include multiple functional groups, such as $-CH_3$, $-NH_2$, $-OH$, and COOH. Flexible and rigid chain segments of these polymers provide the possibility to modulate the mechanical performance of the microenvironment. Chitosan, gelatin, and pectin are selected to build the microenvironment for MSCs growth and directional differentiation due to their excellent physicochemical properties and biocompatibility.

2.3.1 Gelatin

Gelatin, the denatured type of collagen, is composed of a unique sequence of amino acids. Characteristic features of gelatin are the high content of the amino acids such

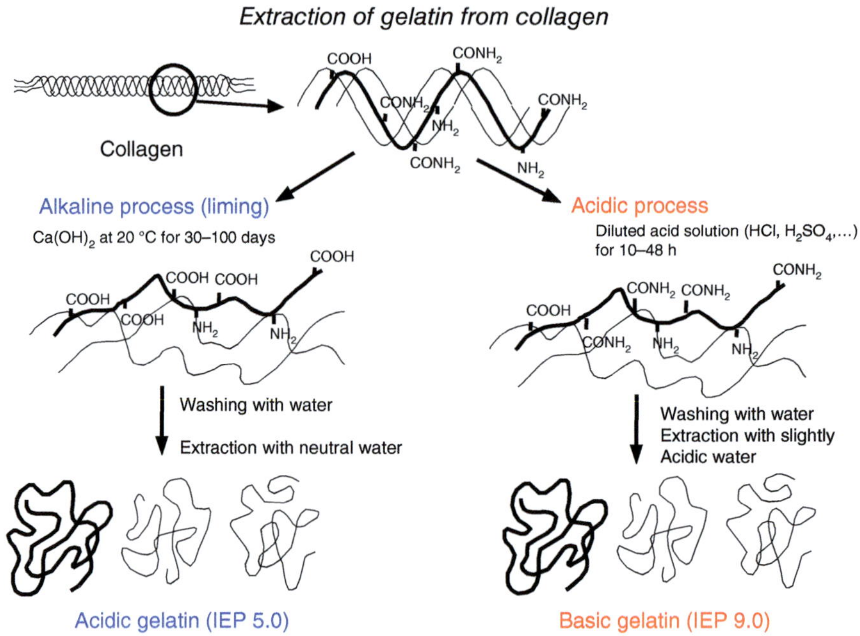

Figure 2.1 Extraction of gelatin from collagen.

as glycine, proline, and hydroxyproline. With a similar structure to GAGs and collagen, the chitosan/gelatin composites has been used as tissue engineering scaffolds, mimicking the natural extracellular matrix (ECM), for the repair of damaged bone and cartilage with promising results [48, 49]. Porous chitosan/gelatin network scaffolds have been developed via polyelectrolyte complex formation, freeze-drying, and post-cross-linking with glutaraldehyde. The bilaminar texture for skin tissue engineering has excellent cytocompatibility [50–52]. Figure 2.1 shows the extraction of gelatin from collagen.

2.3.2 Pectin

Pectins are polysaccharides enriched in galacturonic acid and galacturonic acid methyl ester units. Combined with proteins and other polysaccharides, pectins are skeletal tissues of plants and closely imitate the structure of polysaccharides found in the ECMs of mammals. They are chemically stable and physically strong. With high molecular weight and a polyanionic nature, pectin reacts to their environments through a continuum of physical states, ranging from dense gels to dilute solutions. These properties enable pectin polymers to carry signal molecules and support various biologically active substances. Moreover, pectin can form network structures with basic biopolyelectrolytes such as chitosan. Thus, their properties and general availability make pectin viable to consider when engineering new biomedical materials. The combination of pectin and a second polymer into a composite may modulate the degree of swelling and mechanical properties [53–55].

Morra et al. [56] modified polystyrene tissue culture plate by the covalent linking of two different enzymatically modified hairy regions (HRs) from pectin. It was found that depending on the HR preparation, cells shifted from spread morphology and adhesion behavior quantitatively comparable to that observed on conventional tissue culture polystyrene to rounded morphology and significantly lower adhesion. In addition, pectin-containing matrices improved cell adhesion and proliferation of osteoblast when compared to plain p(LGA) matrices [57], and pectin nanocoating may enhance the biocompatibility of bone and dental implants. In addition, the role of pectin as an inducer of malignant cell apoptosis cannot be ignored [58, 59]. It is hypothesized that the system composed of chitosan, gelatin, and pectin could modulate proliferation and differentiation of MSCs owing to their inherent physiological and biological characteristics. A covalent cross-linking network film of chitosan, gelatin, and pectin can be prepared through the cross-linking reaction with EDC as a cross-linking agent. Different physiochemical, biological, and mechanical microenvironment for cells can be built via adjusting each components.

Polysaccharides compile a big variable family of heterogenic sequences of monomers due to the multiple valencies of monomers and stereospecificity of the glycosidic bonds. The use of natural biopolymers like polysaccharides for various applications has several advantages, such as availability from replenishable agricultural or marine food resources, biocompatibility, and biodegradability. Polysaccharides differ not only in the nature of their component monosaccharides but also in the length of their chains and in the amount of chain branching that occurs. This ability to form branched structures distinguishes polysaccharides from proteins and nucleic acids, which occur only as linear polymers. Polysaccharides have the tendency to be extremely bioactive and can offer unique biochemical functions based on their nanoscale organization and can provide a stable drug and gene delivery platform [60, 61].

Cationic polysaccharides are nontoxic, biodegradable, and biocompatible materials. They are especially suitable for transfection and biological applications because they are water soluble, can be readily transported to cells *in vivo*, and thus act as an effective vehicle for transporting agents complexed with them [62]. Major cationic polysaccharides used in therapeutic purposes are either natural (chitosan and cyclodextrin [CD]) or semisynthetic in origin. Semisynthetic cationic polysaccharides have been synthesized by conjugation of various oligoamines to oxidized polysaccharides and were tested for activity. Polycations of dextran, pullulan, and arabinogalactan (AG) grafted with oligoamines of 2–4 amino groups were also investigated and were found effective in gene delivery [63]. One of the most relevant characteristics of these kind of carriers is that polysaccharide hydroxyl groups can be easily modified and could be possible that sugar-recognition receptor of cell surface may help for internalization [64].

2.3.3 Chitosan

Chitosan is a cationic linear polymer obtained from chitin comprising copolymers of β(1-4)-glucosamine and randomly located *N*-acetyl-D-glucosamine.

Chitosan is a biopolymer composed of units of (1-4) D-glucosamine and (1-4) *N*-acetyl-D-glucosamine. It is analogous to GAG in structure and exhibits numerous interesting physicochemical and biological properties. Due to the unique cationic character, it can interact with polyanions to mimic the ECMs and further modulate the cell behaviors.

It has been proved to be biodegradable, biocompatible, non-antigenic, nontoxic, and biofunctional polymer [65]. The main sources of chitin are two marine crustaceans, shrimp and crab [60]. It can be fully or partially deacetylated form of chitin; the degree of deacetylation of typical commercial chitosan is usually between 70% and 95% and the MW ranges between 10 and 1000 kDa. Chitosan is insoluble in water but soluble in acidic solvents below pH 6. Solubility of chitosan in inorganic acid solvent is quite limited. Chitosan is soluble in 1% hydrochloric acid but insoluble in sulfuric and phosphoric acids. Chitosan solution forms a poly-ion complex with anionic hydrocolloid and provides gel; due to the high positive charge, it can easily form polyelectrolyte complexes with negatively charged nucleotides by electrostatic interactions [60, 66, 67]. The major disadvantage of chitosan is its poor solubility at physiological pH values that results from the deprotonation of the primary amine groups, which have a pKa around 6.5 (25 °C, 100 mM NaCl), slow endosome release, and difficulty in timely unpacking of chitosan/plasmid DNA (pDNA) complexes [68].

The molecular size of chitosan and pH of the medium are two important parameters that dictate its permeabilizing and perturbing effects on the cell membrane. A combined electrostatic–hydrophobic driving force from chitosan might induce the destabilization of cell membranes, and interactions between the carbohydrate backbone of chitosan and cell surface proteins might exert an important role in the chitosan-mediated transfection of cells [60]. Chitosan-based transfection systems are advantageous because of non-immunogenicity, lack of biohazards, and the possibility of introducing larger DNA fragments into targets over viral vectors in gene therapy. However, the high molecular weight of chitosan precluded its usage because of toxicity problems. Nevertheless, the low molecular weights are neither toxic nor hemolytic, and they are shown to form complexes with DNA and protect against nuclease degradation, thereby validating LMWC as components of a synthetic gene delivery system [60].

NOVAFECT oligomeric chitosans are ultrapure chitosan oligomers that were marketed as carriers for nonviral gene therapy [69]. Chitosan has gained increasing interest for siRNA delivery. Nanoparticles prepared from well-defined chitosans of various chemical compositions, degrees of polymerization (DP(*n*)), and chain architectures were reported [70]. Results showed that the most efficient gene silencing was achieved using fully de-*N*-acetylated chitosans with intermediate chain lengths (DP(*n*) 100–300). These chitosans mediated efficient siRNA delivery, and in several cell lines, potent long-term silencing of both exogenous and endogenous target genes with minimal cytotoxicity was obtained. *In vitro* transfection studies demonstrated the ability of oligomeric chitosan–DNA nanoparticles to effectively transfect COS-7 cells [71].

The low physiological solubility of chitosan has limited its gene transfection efficiency. One strategy to improve chitosan properties in gene delivery applications is the tailoring of its structure. Toh et al. [72] synthesized different substitution degrees of succinated chitosans to increase its water solubility. Succinated chitosans with 5% and 10% degrees of substitution showed improved transfection efficiency as compared to nascent with improved safety against HEK 293T cells.

To improve the transfection efficiency of chitosan, PEI (MW 1.8 kDa) was linked to chitosan to form a complex by Gao et al. [73]. Weecharangsan et al. evaluated chitosan lactate (CL) as a DNA complexing agent, for its efficiency in transfecting green monkey fibroblasts and effect on cell viability was compared with PEI [74]. It was found that the transfection efficiency of arginine–chitosan (Arg-Cs)/DNA self-assembled nanoparticles (ACSN) was much higher than that of Cs/DNA self-assembled nanoparticles (CSN) [75]. Similarly, Yu et al. reported copolymers poly(L-lysine)-graft-chitosan (PLL-g-Chi) as an effective gene delivery vectors [76].

2.3.4 Cyclodextrins (CD)

CD is nontoxic and non-immunogenic and is produced from starch by means of enzymatic conversion. It is a cyclic oligomer of glucose units having a unique amphiphilic structure that consists of a hydrophobic cavity and hydrophilic exterior and thus building CD as water soluble. CD contains a number of glucose monomers ranging from six to eight units in a ring, creating a cone shape: (i) α-cyclodextrin: six-membered sugar ring molecule; (ii) β-cyclodextrin: seven sugar ring molecule; and (iii) γ-cyclodextrin: eight sugar ring molecule.

CD forms inclusion complexes with small hydrophobic molecules and has been incorporated into a gene delivery polymer to provide a facile and versatile attachment site for targeting ligand. CD-binding moieties are attached relative to various types of ligands; thus Forrest et al. derivatized PEI with β-CD (CD-PEI), for testing novel cell-specific targeting compounds [77]. The influence of CD in forming grafting complexes showed that the more the grafting percentage of the CD, the less cytotoxic the polymer will be. Transfection efficiency also increased if the grafting percentage was increased [78].

Pun et al. reported linear and branched PEI (LPEI and bPEI), grafted with β-CD. The CD containing polycations were combined with adamantane–PEG (AD–PEG) conjugates to form particles, and the gene expression observed was equal or greater than LPEI [79]. Similarly, series of novel cationic star polymers by conjugating multiple oligoethylenimine arms onto α-CD core were synthesized. Prepared polymers were found to have less toxicity, were biocompatible, and did not elicit immune reactions [80]. Jiang et al. studied target ligand folic acid and cell-penetrating peptide octaarginine coupled with the β-CD and low MW PEI (600 Da) for gene delivery to tumor cells [81].

2.3.5 Dextran: Dextran–Spermine

Dextran is a biodegradable and biocompatible polymer. Its straight chain consists of α-1,6 glycosidic linkages with few α-1,3 glycosidic linkages as the branch linkage. Dextran is biocompatible and used as plasma expander and for various other uses. It is a water-soluble polysaccharide with multiple hydroxyl groups applicable to chemical modification. Furthermore, it has been extensively used for pharmaceutical applications to demonstrate the feasibility in gene expression and in cancer gene therapy [82, 83]. Cationic dextrans were prepared with reagents carrying positive charge, conjugation through primary amines. Cationization of dextran increases its gene delivery efficiency similar to that of PEI [84, 85]. Various degrees of grafting of dextran onto PEI were reported by Tseng et al. in an effort to investigate how the conjugation affected the mechanisms of transgene expression [86].

Semisynthetic cationic polysaccharides were synthesized by conjugation of various oligoamines to oxidized polysaccharides via reductive amination [87]. In few publications, Azzam et al. reported a new class of biodegradable polycations capable of complexing and transfecting various genes to many cell lines in relatively high yields [88, 89]. More than 300 different polycations were prepared starting from various natural polysaccharides and oligoamines having two to four amino groups. These cationic polysaccharides were prepared by reductive amination of the oligoamine and periodate-oxidized polysaccharides. Although most of these conjugates formed stable complexes with DNA, only the dextran–spermine (D-SPM) polycations of defined MW were found to be active in transfecting a wide range of cell lines *in vitro* attributed to spermine residues, which play a crucial role in cell transfection. The complexes formed between D-SPM and pDNA exhibits a hydrophilic surface due to the saccharides building units of the polymer backbone. Such hydrophilic nature limits the cell uptake of the complexes. Higher spermine ratios of grafted oligoamines resulted in high gene expression, whereas low to negligible expressions were obtained with lower spermine contents [62].

Biodegradable water-soluble polysaccharide–SPM polycation conjugates for nucleic acid delivery were synthesized by oxidizing polysaccharides using potassium periodate, followed by SPM conjugation. The polycations differ in their polysaccharide type, AG or dextran (D), and/or in the IO(4)-/saccharide mole ratio used for polysaccharide oxidation (1 : 1, 1 : 3, or 1 : 5), resulting in either D(1 : 1)-SPM, AG(1 : 1)-SPM, D(1 : 3)-SPM, AG(1 : 3)-SPM, or AG(1 : 5)-SPM. Chemical structure of the conjugates was characterized for total nitrogen and primary amino groups. The binding and the electrostatic neutralization of the polycations by plasmid DNA, as well as the relationship between chemical structure, physical parameters, and transfection of NIH3T3 cells, were also studied [90]. D(1 : 1)-SPM, the only polycation that showed efficient cell transfection in culture, was shown to have (i) high SPM content (2000 nmol mg^{-1}); (ii) high levels of cross-linked SPM (39–51%); (iii) at DNA P-/NH3+ ratio of 2.0, a plateau in neutralization of cationic groups (+48 mV, as determined by HC-labeled D(1 : 1)-SPM titration with DNA) and a drop in zeta-potential from +42 mV for the polymer alone to 0 mV for the polyplex, suggesting that some of the charges are hidden from the DNA; (iv) pH(surface) value of 9.2, suggesting that at physiological bulk pH, the polymer is only partially ionized and therefore can act as a "proton sponge" in the endosome; and (v) high sensitivity to serum-rich growth medium. An oleyl derivative, *N*-oleyl-D-SPM (ODS), was synthesized and demonstrated improved transfection efficiency in serum-rich medium.

Hydrophobization of D-SPM was achieved by treating the polymer with *N*-hydroxysuccinimide derivatives of cholesterol and fatty acids. The oleate-modified D-SPM remarkably enhanced the gene expression in contrast to unmodified D-SPM, which resulted with a drastic decrease in the transfection yields. Modified D-SPM derivatives of other fatty acids and cholesterol showed improved transfection yields in comparison to unmodified D-SPM. This improvement in cell transfection was attributed to oleate residues, which probably play a role in increasing stability and uptake of polycation–DNA complexes. Similarly, Hosseinkhani et al. reported modification of the D-SPM polycation with PEG resulted in high transfection yield in serum-rich medium. This offers a new biodegradable polycation based on natural components, which is capable of transfecting cells and tissues *in vitro* and *in vivo* [91]. Figure 2.2 shows synthesis of D-SPM-based conjugate.

2.3.6 Miscellaneous Polysaccharide Polymers

2.3.6.1 Pullulan

Pullulan is a natural water-soluble polysaccharide with a repeated unit of maltotriose condensed through the α-1,6 linkage [92]. It is nontoxic, non-immunogenic, non-mutagenic, and noncarcinogenic in nature. DNA/cationic-modified pullulan stable complexes were produced resulting in stable temperature-sensitive gene carriers [93]. On the same basis, from DNA complexed aminated pullulan and starch microspheres *in vitro* release of DNA lasted for more than 14 days, being the slowest from starch particles without evident DNA degradation [94]. Cationic pullulan, dextran, and mannan complexed with pDNA were also tested in cellular models and *in vivo* mice models [64, 95]. Also, cationic pullulan/pDNA complex was used in transfection of rat brain endothelial and human brain microvascular endothelial cells [96]. Similar kind of complexes was studied for liver targeting gene expression [92]. Thakor et al. studied *in vitro* transfection using conjugated spermine-pullulan/pDNA anionic complexes in rat sensory neurons [97].

2.3.6.2 Schizophyllan

Schizophyllan is a natural β-(1-3)-D-glucan polysaccharide produced by fungus Schizophyllum. After being modified stable complexes were formed with antisense oligonucleotides and studied in different types of cells including melanoma leukemia. The cytotoxicity of modified schizophyllans was negligibly as small as the natural (unmodified). Macrophage-migration inhibitory factor in the inflammatory bowel disease, cytotoxicity, or other adverse effects were reported [98, 99].

2.3.6.3 Alginic Acid

Alginic acid, also called algin or alginate, is an anionic polysaccharide distributed widely in the cell walls of brown algae. Preosteoblastic cells and calcium

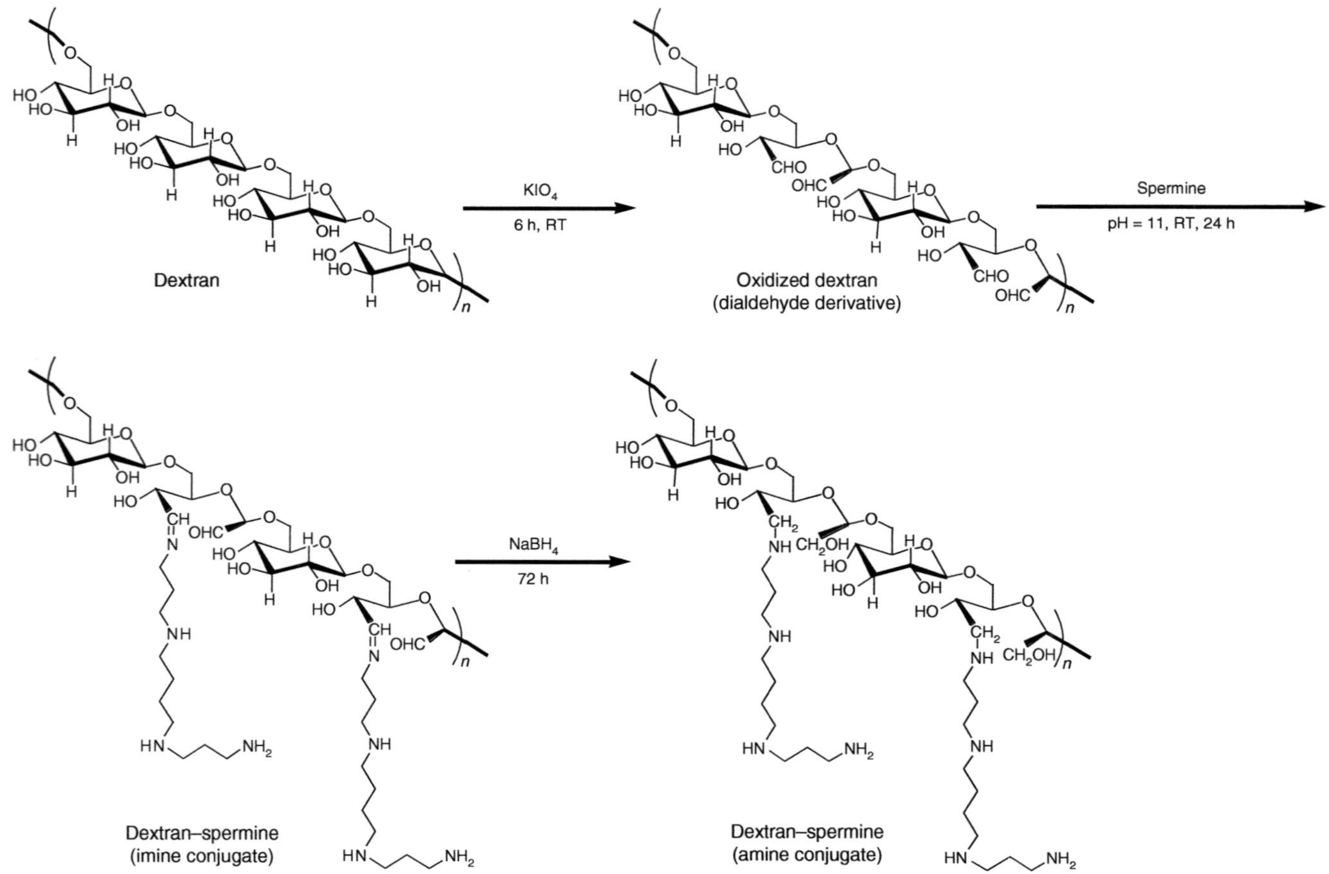

Figure 2.2 Synthesis of dextran–spermine-based conjugate.

phosphate–DNA embodied into alginate hydrogels nanoparticles, had a sustained release of DNA, and was used in mice as an efficient gene delivery system by increasing bone regeneration [100]. Zhao et al. reported the use of alginate/calcium carbonate/DNA nanoparticles for gene transfection in 293T cells and HeLa cells [101]. Alginate spheres were used to encapsulate transfected chondrocytes carrying the *Photinus pyralis* luciferase gene to be used to fix damaged articular cartilage. The chondrocyte viability was not affected by the encapsulation as tested *in vitro*. Also the *in vivo* results evidenced that even after 32 days following transfection still some luciferase activity could be determined [102].

2.3.6.4 Hydroxypropyl Cellulose

Hydroxypropyl cellulose is a derivative of cellulose with both water and organic solubility. Xu et al. reported hydroxypropyl cellulose, with cationic poly((2-dimethyl amino)ethyl methacrylate) as gene vectors. Chains complexed with pDNA were used as an efficient gene delivery system with low cytotoxicity in HEK293 cells [103].

2.3.6.5 Glycosaminoglycans

These disaccharide repeating units charged polyanions at physiological conditions, containing a hexosamine and hexuronic acid or hexose like in keratan sulfate. Examples include chondroitin sulfate, dermatan sulfate, keratan sulfate, heparan sulfate, heparin, and hyaluronan [61].

Hyaluronic Acid HA is water soluble, naturally occurring polysaccharide approved for injections by the Food and Drug Administration (FDA). It is low toxic, non-immunogenic biodegradable, and biocompatible polymer with multiple functional groups for possible conjugations. This polymer can induce receptor-mediated intracellular signaling. Its polyanionic structure allows HA neutralizing the positive charge of polycations [104, 105]. HA has been used as a drug carrier and a ligand on liposomes or nanoparticles to CD44 overexpressing cells. HA PEI copolymer was synthesized by imine reaction. The copolymer/DNA complex displayed higher transfection than the PEI used alone with lower cytotoxicity [104]. Hyaluronan microspheres prepared using adipic dihydrazide-mediated cross-linking resulted in successful transfection when incorporated with plasmid DNA and selectively bind cells when conjugated with monoclonal antibodies to E- and P-selectin via protein A for selective cellular targeting [105].

Hyaluronate-grafted poly-L-lysine copolymer was used to target complexed DNA in liver sinusoidal endothelial cells [106]. In another study, HA chitosan nanoparticles were tested *in vitro* in human corneal epithelial and normal human conjunctival cell lines resulting in efficient transfection possibly mediated by endocytosis internalization [107]. The use of alginate and hyaluronan/PEI nanoparticles complexed with pDNA used for gene transfection was also reported by Pathak et al. [108]. Encapsulated pDNA by hyaluronan nanoparticles release was studied for one month, and expression following transfection was increased when PEI was used [109].

Chondroitin Sulfate Chondroitin sulfate was effective to form ternary stable anionic complexes when added to encapsulate cationic polyplexes and lipoplexes. This final complexes efficiently mediated gene delivery (in B16–F10 cells) with minimal toxicity [110]. This glycosaminoglycan also recognizes a CD44 receptor, which may be important in tumor treatment as it induces receptor-mediated endocytosis, improving tumor targeting gene delivery [111]. Chondroitin sulfate-polyethylenimine/pDNA nanocomplexes showed low toxicity when transfection was investigated in HepG2 and HeLa cell lines. In mice encouraging results were obtained as treatment for Ehrlich ascites tumor [112]. Tomiyama et al. introduced chondroitin sulfate into a polyacrylamide–peptide conjugate/pDNA complex, which resulted in low stability and gene regulation [111].

Heparin and Heparan Sulfate Heparin, a highly sulfated glycosaminoglycan, is widely used as an injectable anticoagulant and has the highest negative charge density of any known biological molecule. Heparin–PEI nanogels delivered efficiently a plasmid expressing HSulf-1 into HSulf-1-deficient SKOV3 human ovarian cancer cells, resulting in successful transfection as inhibition of cell proliferation. The complex increased cell apoptosis with minimal toxicity [113].

Other Polysaccharides Hydroxyethyl starch grafted to polyethylenimine was used to control shielding/deshielding of pDNA complexes transfection [114]. Hexadecyl group-bearing-cationic–cycloamylose nanogel/phospholipase A2 nanoparticles complexed with pDNA were delivered successfully into cells increasing the pDNA expression [115]. A natural polysaccharide with extract from mulberry leaves and conjugated with ethylenediamine with a final MW of around 50 KDa and complexed with the growth factor β1 allowed efficiently transfection on bone marrow MSCs [116].

Modified high mannose *N*-glycan-polyacridine peptide glycopeptides/pDNA proved to be efficient in receptor gene delivery [117]. A novel nonviral gene delivery vector composed of a high-mannose *N*-glycan conjugated to a polyacridine peptide was reported. These studies suggest that a high-mannose *N*-glycan conjugated to a polyacridine peptide may also facilitate receptor-mediated gene delivery in dendritic cells and thereby find utility in the delivery of DNA vaccines [117]. An anionic sugar polymer gellan gum (GG) has been introduced into PEI system to increase transfection efficiency with minimal toxicity. Synthesized GG–PEI nanocomposites (NCs) exhibited negligible toxicity in *in vitro* and *in vivo* as compared to PEI or lipofectamine. The study demonstrated that GG-blended PEI NCs hold promise for future applications in gene delivery both *in vitro* and *in vivo* [118]. Branched PEI, 25 kDa, was blended with GG, an anionic heteropolysaccharide, for partial neutralization of its excess positive charge to form gellan gum–polyethylenimine (GP) NCs. Among all the NCs, one member, named GP3, showed the best transfection efficiency in tested cell lines in comparison with the rest of the series, PEI, lipofectamine, and other commercial transfection agents also exhibited minimum cytotoxicity. The study suggests that GP3 may act as an efficient nonviral gene carrier with diverse biomedical applications [119].

Yao et al. prepared amphoteric HA derivative with PEI chains (HAP) for gene delivery, which was synthesized by an imine reaction of the oxidized HA and the PEI. HAP/DNA complexes were evaluated *in vitro* for gene transfection in HepG2 cells and *in vivo* in mice as a ligand targeted gene delivery system that showed a higher cell transfection efficiency as well as *in vivo* expression in tumor when compared to PEI/DNA complexes [104].

The gene transfection with naked plasmid DNA always shows low efficiency *in vivo* although it is simple and safe [120]. On the other hand, when a plasmid DNA is mixed with a vector, the molecular size and surface charge of the plasmid DNA nanoparticle affect the body fate of the plasmid DNA [121]. For example, the complex of plasmid DNA with cationic liposomes mainly accumulates in the lung, liver, and spleen [122]. In an attempt to alter the organ distribution of lipid complexes, their surface was modified by PEG and other hydrophilic polymers [123]. This modification enabled the plasmid DNA to prolong the half-life in the blood circulation by suppressing their interaction with plasma components and erythrocytes, which strongly modifies the *in vivo* characteristics of DNA. The particle size, charge, and the surface characteristics of the complex have a strong influence on the plasmid DNA body distribution and transfection efficiency [124, 125]. We have shown that dextran-engrafted spermine is capable of complexing and transfecting various genes to different cell lines *in vitro* and *in vivo* [88, 91, 126].

D-SPM-based conjugate (Figure 2.2) was prepared by reductive amination between oxidized dextran and spermine. In our collaboration work [91], cationized dextran was modified with PEG (Figure 2.3). Dextran was initially oxidized with potassium periodate and the obtained dialdehyde derivative was allowed to react under basic conditions with spermine. The corresponding D-SPM polycation was further modified with increasing amount of PEG to obtain a partially shielded polycation. Polyethylene glycol monomethyl ether (mPEG, MW = 2000 Da) was initially treated with *p*-nitrophenyl chloroformate to obtain the active *p*-nitrophenyl carbonate derivative (mPEG–OPNC) in relatively good yields.

Substitution of the PEG moieties to D-SPM was simply achieved by mixing an aqueous solution of the mPEG–OPNC with a concentrate aqueous solution of the polycation at pH 9. The amount of PEG was fixed at 1%, 3%, and 5% (mol/mol) of PEG moieties to primary amines (ε-NH_2), respectively. The mixture was allowed to stir at room temperature overnight, purified by column chromatography, and lyophilized to dryness. Preliminary intravenous (i.v.) injection to mice of free plasmid DNA and its nanoparticle formulation with non-PEGylated D-SPM polycation did not induce gene expression in all tested organs (data not shown). Modification of the D-SPM polycations with the nonionic hydrophilic polymer, PEG, remarkably improved the transfection efficiency of the polycations in serum-rich mediums.

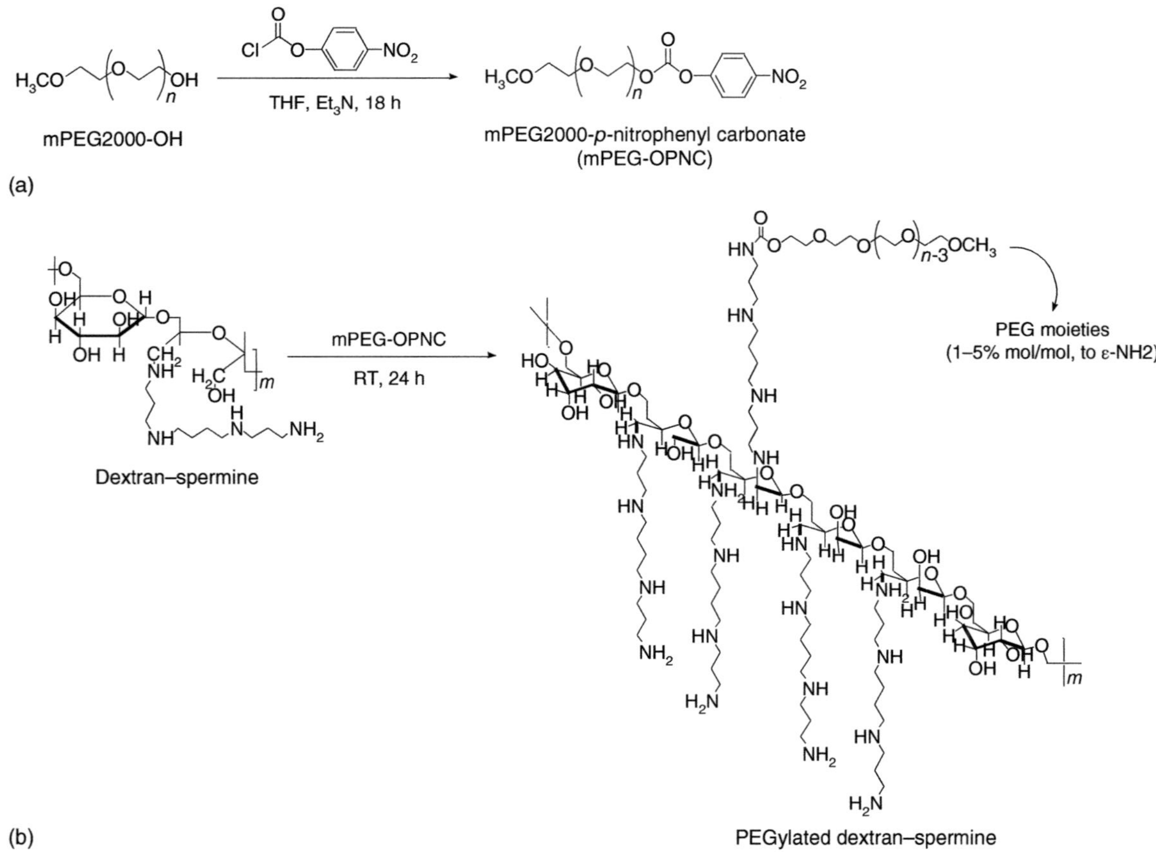

Figure 2.3 (a) Synthesis of mPEG2000-*p*-nitrophenyl carbonate (m-PEG–OPNC) and (b) modification of dextran–spermine-based conjugate with PEG.

References

1 Langer, R. and Peppas, N.A. (2003). *AlChE J.* 49: 2990–3006.
2 Langer, R. (2000). *Acc. Chem. Res.* 33: 94–101.
3 Yoshimoto, H., Shin, Y.M., Terai, H., and Vacanti, J.P. (2003). *Biomaterials* 24: 2077–2082.
4 Zhang, Y.Z., Venugopal, J., Huang, Z.M. et al. (2005). *Biomacromolecules* 6: 2583–2589.
5 Venugopal, J., Zhang, Y.Z., and Ramakrishna, S. (2005). *Nanotechnology* 16: 2138–2142.
6 Barzin, J. and Sadatnia, B. (2007). *Polymer* 48: 1620–1631.
7 Sluma, H.D., Passlack, J., Buttner, B., Scherf, M. (1993), Synthetic hydrophobic membranes and method for their manufacture. US Patent 5246582.
8 Zhang, X., Goosen, F.A., Wyss, U.P., and Pichor, D. (1993). *Rev. Macromol. Chem. Phys. C* 33: 81–102.
9 Herrmann, J.B., Kelly, R.J., and Higgins, G.A. (1970). *Arch. Surg.* 100: 486–490.
10 Chu, C.C. (2002). Biodegradable polymeric biomaterials: an updated overview. In: *Biomaterials Principles and Applications* (ed. J.B. Park and J.D. Bronzino). Taylor & Francis.
11 Choueka, J., Charvet, J.L., Alexander, H. et al. (1995). *J. Biomed. Mater. Res.* 29: 1309–1315.
12 Zhao, N., Xiong, Z.C., Yang, D.J. et al. (2008). *Polym. Compos.* 30: 1771–1778.
13 Hosseinkhani, H., Hosseinkhani, M., Vasheghani, E., and Nekoomanesh, M. (2009). *Adv. Sci. Lett.* 2: 70–77.
14 Me, L.J., Koh, W.G., and Pishko, M.V. (2005). *Biotechnol. Progr.* 21: 926–932.
15 Lahooti, S. and Sefton, M.V. (2000). *Cell Transplant.* 9: 785–796.
16 Sefton, M.V., May, M.H., Lahooti, S., and Babensee, J.E. (2000). *J. Controlled Release* 65: 173–186.
17 Lim, F. and Sun, A.M. (1980). *Science* 210: 908–910.
18 Sawhney, A.S., Pathak, C.P., van Rensburg, J.J. et al. (1994). *J. Biomed. Mater. Res.* 28: 831–838.
19 Burdick, J.A., Frankel, D., Dernell, W.S., and Anseth, K.S. (2003). *Biomaterials* 24: 1613–1620.
20 Burdick, J.A., Mason, M.N., Hinman, A.D. et al. (2002). *J. Controlled Release* 83: 53–63.
21 Burdick, J.A. and Anseth, K.S. (2002). *Biomaterials* 23: 4315–4323.
22 Mann, B., Gobin, A., Tsai, A. et al. (2001). *Biomaterials* 22: 3045–3051.
23 Lee, K.Y. and Mooney, D.J. (2001). *Chem. Rev.* 101: 1869–1879.
24 Rowley, J.A., Madlambayan, G., and Mooney, D.J. (1999). *Biomaterials* 20: 45–53.
25 Merrill, E., Salzman, E., Wan, S. et al. (1982). *Trans. Am. Soc. Artif. Internal Organs* 28: 482–487.
26 Ratner, B.D., Hoffman, A., Schoen, F., and Lemons, J. (2004). *An Introduction to Materials in Medicine*. Academic Press.
27 West, J. and Hubbell, J. (1995). *React. Polym.* 25: 139–147.

28 West, J. and Hubbell, J. (1999). *Macromolecules* 32: 241–244.
29 Hern, D.L. and Hubbell, J.A. (1998). *J. Biomed. Mater. Res.* 39: 266–276.
30 Huh, K. and Bae, Y. (1999). *Polymer* 40: 6147–6155.
31 Leach, J.B. and Schmidt, C.E. (2005). *Biomaterials* 26: 125–135.
32 Koh, W.G., Me, L.J., and Pishko, M.V. (2003). *Anal. Chem.* 75: 5783–5789.
33 Khademhosseini, A., Yeh, J., Eng, G. et al. (2005). *Lab Chip* 5: 1380–1386.
34 Zhan, W., Seong, G.H., and Crooks, R.M. (2002). *Anal. Chem.* 74: 4647–4652.
35 Heo, J., Thomas, K.J., Seong, G.H., and Crooks, R.M. (2003). *Anal. Chem.* 75: 22–26.
36 Khademhosseini, A., Yeh, J., Jon, S. et al. (2004). *Lab Chip* 4: 425–430.
37 Beebe, D.J., Moore, J.S., Bauer, J.M. et al. (2000). *Nature* 404: 588–590.
38 Mann, B., Schmedlen, R., and West, J. (2001). *Biomaterials* 22: 439–444.
39 Hoffman, A.S., Stayton, P.S., Bulmus, V. et al. (2000). *J. Biomed. Mater. Res.* 52: 577–586.
40 Bayer, E.A. and Wilchek, M. (1980). *Methods Biochem. Anal.* 26: 1–45.
41 Bulmus, V., Ding, Z., Long, C. et al. (2000). *Bioconjugate Chem.* 1: 78–83.
42 Ding, Z., Long, C., Hayashi, Y. et al. (1999). *Bioconjugate Chem.* 10: 395–400.
43 Stayton, P.S., Shimoboji, T., Long, C. et al. (1995). *Nature* 378: 472–474.
44 Guo, L.K., Kawazoe, N., Hoshiba, T. et al. (2008). *J. Biomed. Mater. Res. A* 87A: 903–912.
45 Curran, J.M., Chen, R., and Hunt, J.A. (2006). *Biomaterials* 27: 4783–4793.
46 Curran, J.M., Chen, R., and Hunt, J.A. (2005). *Biomaterials* 26: 7057–7067.
47 Georges, P.C. and Janmey, P.A. (2005). *J. Appl. Phys.* 98: 1547–1553.
48 Medrado, G.C.B., Machado, C.B., Valerio, P. et al. (2005). *J. Biomed. Mater.* 1: 155–161.
49 Xia, W.Y., Liu, W., Cui, L. et al. (2004). *J. Biomed. Mater. Res. Part B* 71: 373–380.
50 Shen, F., Cui, Y.L., Yang, L.F. et al. (2004). *Polym. Int.* 49: 1596–1599.
51 Mao, J.S., Zhao, L.G., Yao, K.D. et al. (2003). *J. Biomed. Mater. Res. A* 64A: 301–308.
52 Mao, J.S., Zhao, L.G., Yin, Y.J., and Yao, K.D. (2003). *Biomaterials* 24: 1067–1074.
53 Semde, R., Amighi, A., Devleeschouwer, M.J., and Moes, A.J. (2000). *Int. J. Pharm.* 197: 181–192.
54 Kim, J.H. and Fassihi, R. (1997). *J. Pharm. Sci.* 86: 316–328.
55 Bodmeier, R. and Paeratakul, O. (1994). *Pharm. Res.* 11: 882–888.
56 Morra, M., Cassinelli, C., and Cascardo, G. (2004). *Biomacromolecules* 5: 2094–2104.
57 Liu, L.S., Won, Y.J., Cooke, P.H. et al. (2004). *Biomaterials* 25: 3201–3210.
58 Jackson, C.L., Dreaden, T.M., Theobald, L.K. et al. (2007). *Glycobiology* 17: 805–819.
59 Olano-Martin, E., Rimbach, G.H., Gibson, G.R., and Rastall, R.A. (2003). *Anticancer Res.* 23: 341–346.
60 Sahoo, D., Sahoo, S., Mohanty, P. et al. (2009). *Design. Monomers Polym.* 12: 377–404.

61 Boddohi, S. and Kipper, M.J. (2010). *Adv. Mater.* 22: 2998–3016.
62 Azzam, T., Eliyahu, H., Makovitzki, A. et al. (2004). *J. Controlled Release* 96: 309–323.
63 Yudovin-Farber, I., Azzam, T., Metzer, E. et al. (2005). *J. Med. Chem.* 48: 1414–1420.
64 Jo, J., Okazaki, A., Nagane, K. et al. (2010). *J. Biomater. Sci. Polym.* 21: 185–204.
65 Khor, E. and Lim, L.Y. (2003). *Biomaterials* 24: 2339–2349.
66 Chandy, T. and Sharma, C.P. (1990). *Biomater. Artif. Cells Artif. Organs* 18: 1–24.
67 Douglas, K.L., Piccirillo, C.A., and Tabrizian, M. (2008). *Eur. J. Pharm. Biopharm.* 68: 676–687.
68 Malmo, J., Varum, K.M., and Strand, S.P. (2011). *Biomacromolecules* 12: 721–729.
69 Klausner, E.A., Zhang, Z., Chapman, R.L. et al. (2010). *Biomaterials* 31: 1814–1820.
70 Malmo, J., Sorgard, H., Varum, K.M., and Strand, S.P. (2012). *J. Controlled Release* 158: 261–268.
71 Howard, K.A., Rahbek, U.L., Liu, X. et al. (2006). *Mol. Ther.* 14: 476–484.
72 Toh, E.K., Chen, H.Y., Lo, Y.L. et al. (2011). *Nanomedicine* 7: 174–183.
73 Gao, J.Q., Zhao, Q.Q., Lv, T.F. et al. (2010). *Int. J. Pharm.* 387: 286–294.
74 Weecharangsan, W., Opanasopit, P., Ngawhirunpat, T. et al. (2006). *AAPS Pharm. Sci. Tech.* 7: E74–E79.
75 Saranya, N., Moorthi, A., Saravanan, S. et al. (2011). *Int. J. Biol. Macromol.* 48: 234–238.
76 Yu, H., Chen, X., Lu, T. et al. (2007). *Biomacromolecules* 8: 1425–1435.
77 Forrest, M.L., Gabrielson, N., and Pack, D.W. (2005). *Biotechnol. Bioeng.* 89: 416–423.
78 Park, T.G., Jeong, J.H., and Kim, S.W. (2006). *Adv. Drug Delivery Rev.* 58: 467–486.
79 Pun, S.H., Bellocq, N.C., Liu, A. et al. (2004). *Bioconjugate Chem.* 15: 831–840.
80 Yang, C., Li, H., Goh, S.H., and Li, J. (2007). *Biomaterials* 28: 3245–3254.
81 Jiang, Q.Y., Lai, L.H., Shen, J. et al. (2011). *Biomaterials* 32: 7253–7262.
82 Hosseinkhani, H. and Tabata, Y. (2006). *J. Nanosci. Nanotechnol.* 6: 2320–2328.
83 Hosseinkhani, H., Kushibiki, T., Matsumoto, K. et al. (2006). *Cancer Gene Ther.* 13: 479–489.
84 Thomas, J.J., Rekha, M.R., and Sharma, C.P. (2011). *Int. J. Pharm.* 389: 195–206.
85 Thomas, J.J., Rekha, M.R., and Sharma, C.P. (2010). *Colloids Surf., B* 81: 195–205.
86 Tseng, W.C., Tang, C.H., and Fang, T.Y. (2004). *J. Gene Med.* 6: 895–905.
87 Yudovin-Farber, I., Eliyahu, H., and Domb, A.J. (2011). *Cold Spring Harbor Protoc.* 1: 5553.
88 Eliyahu, H., Makovitzki, A., Azzam, T. et al. (2005). *Gene Ther.* 12: 494–503.
89 Azzam, T., Eliyahu, H., Shapira, L. et al. (2002). *J. Med. Chem.* 45: 1817–1824.
90 Eliyahu, H., Siani, S., Azzam, T. et al. (2006). *Biomaterials* 27: 1646–1655.

91 Hosseinkhani, H., Azzam, T., Tabata, Y., and Domb, A.J. (2004). *Gene Ther.* 11: 194–203.

92 Hosseinkhani, H., Aoyama, T., Ogawa, O., and Tabata, Y. (2002). *J. Controlled Release* 83: 287–302.

93 Constantin, M., Oanea, I., Harabagiu, V. et al. (2011). *Digest J. Nanomater. Biostruct.* 6: 849–861.

94 Constantin, M., Fundueanu, G., Cortesi, R. et al. (2003). *Drug Delivery* 10: 139–149.

95 Rekha, M.R. and Sharma, C.P. (2009). *Biomaterials* 30: 6655–6664.

96 Thomsen, L.B., Lichota, J., Kim, K.S., and Moos, T. (2011). *J. Controlled Release* 151: 45–50.

97 Thakor, D.K., Teng, Y.D., and Tabata, Y. (2009). *Biomaterials* 30: 1815–1826.

98 Matsumoto, T., Numata, M., Anada, T. et al. (2004). *Biochim. Biophys. Acta* 1670: 91–104.

99 Takedatsu, H., Mitsuyama, K., Mochizuki, S. et al. (2012). *Mol. Ther.* 10: 1234–1241.

100 Krebs, M.D., Salter, E., Chen, E. et al. (2010). *J. Biomed. Mater. Res. A* 92: 1131–1138.

101 Zhao, D., Zhuo, R.X., and Cheng, S.X. (2012). *Mol. Biosyst.* 8: 753–759.

102 Madry, H., Cucchiarini, M., Stein, U. et al. (2003). *J. Gene Med.* 5: 502–509.

103 Xu, F.J., Ping, Y., Ma, J. et al. (2009). *Bioconjugate Chem.* 20: 1449–1458.

104 Yao, J., Fan, Y., Du, R. et al. (2010). *Biomaterials* 31: 9357–9365.

105 Yun, Y.H., Goetz, D.J., Yellen, P., and Chen, W. (2004). *Biomaterials* 25: 147–157.

106 Takei, Y., Maruyama, A., Ferdous, A. et al. (2004). *FASEB J.* 18: 699–701.

107 de la Fuente, M., Seijo, B., and Alonso, M.J. (2008). *Invest. Ophthalmol. Visual Sci.* 49: 2016–2024.

108 Pathak, A., Patnaik, S., and Gupta, K.C. (2009). *Nucleic Acids Symp. Ser.* 53: 57–58.

109 Mahor, S., Collin, E., Dash, B.C., and Pandit, A. (2011). *Curr. Drug Delivery* 8: 354–362.

110 Kurosaki, T., Kitahara, T., Fumoto, S. et al. (2010). *J. Pharm. Pharm. Sci.* 13: 351–361.

111 Tomiyama, T., Toita, R., Kang, J.H. et al. (2011). *Nanoscale Res. Lett.* 6: 532.

112 Pathak, A., Kumar, P., Chuttani, K. et al. (2009). *ACS Nano* 3: 1493–1505.

113 Liu, P., Gou, M., Yi, T. et al. (2011). *Oncol. Rep.* 27: 363–370.

114 Noga, M., Edinger, D., Rodl, W. et al. (2012). *J. Controlled Release* 159: 92–103.

115 Toita, S., Sawada, S., and Akiyoshi, K. (2011). *J. Controlled Release* 155: 54–59.

116 Deng, W.W., Cao, X., Wang, M. et al. (2012). *Small* 8: 441–451.

117 Anderson, K., Fernandez, C., and Rice, K.G. (2010). *Bioconjugate Chem.* 21: 1479–1485.

118 Goyal, R., Tripathi, S.K., Tyagi, S. et al. (2011). *J. Biomed. Nanotechnol.* 7: 38–39.

119 Goyal, R., Tripathi, S.K., Tyagi, S. et al. (2011). *Eur. J. Pharm. Biopharm.* 79: 3–14.

120 Yu, L., Suh, H., Koh, J.J., and Kim, S.W. (2001). *Pharm. Res.* 18: 1277–1283.

121 Kircheis, R., Schuller, S., Brunner, S. et al. (1999). *J. Gene Med.* 1: 111–120.

122 Templeton, N.S., Lasic, D.D., Frederik, P.M. et al. (1997). *Nat. Biotechnol.* 15: 647–652.

123 Woodle, M.C., Engbers, C.M., and Zalipsky, S. (1994). *Bioconjugate Chem.* 5: 493–496.

124 Liu, Y., Mounkes, L.C., Liggitt, H.D. et al. (1997). *Nat. Biotechnol.* 15: 167–173.

125 Goula, D., Benoist, C., Mantero, S. et al. (1998). *Gene Ther.* 5: 1291–1295.

126 Azzam, T., Eliyahu, H., Makovitzki, A., and Domb, A.J. (2003). *Macromol. Symp.* 195: 247–262.

3

Recent Advances and Applications of Hydrogels in Medicine

3.1 Introduction

Hydrogels are cross-linked hydrophilic polymeric materials possessing three-dimensional (3D) structures. Despite the uncross-linked polymers or other materials, they swell up to thousands of times their dry weight without being dissolved in the presence of aqueous solutions due to the formation of physical or chemical bonds between polymeric chains [1]. Due to their notable properties such as versatility in fabrication, tunable physical, chemical, and biological characteristics, high biocompatibility, and similarity to living tissues, hydrogels have been widely used in biomedical/pharmaceutical applications [1–6].

Over the past decade, stimuli-responsive hydrogels have been extensively studied due to their excellent applications in various fields ranging from drug delivery [7], tissue engineering [8], 3D printing [9], heart repair [10], and molecular switches [11]. Such materials indicate reversible or irreversible changes in physical properties or chemical structures in response to environmental stimulus such as temperature [9], pH [11], electric field [12], and ionic strength [13]. Therefore, such unique properties allow us to design diverse types of self-regulated and stimuli-responsive systems. The type of stimuli-responsive hydrogels is also regarded as smart and intelligent materials. Among a variety of stimuli that have been utilized, biomolecule-responsive hydrogels have received particular attention recently, since they can undergo structural transition in response to a specific target biomolecule, such as glucose, proteins, nucleic acids, and polypeptides. For example, glucose-responsive hydrogels that mediated structural changes in accordance with the glucose concentration of the surroundings can be applied as suitable materials in diabetes treatment. Diabetes is a disorder of glucose regulation in blood and one of the most significant health concerns in this century that threatens human health seriously [14]. The main therapy for this illness has been insulin injection, which may cause several drawbacks including the diet limitation, the injection difficulty, and the long period of treatment. Therefore, an alternative therapy with the ability of direct responses to blood glucose levels that prepares a safe and continuous administration of insulin from a closed-loop and feedback-controlled insulin-release system seems essential [14]. Consequently,

Biomedical Engineering: Materials, Technology, and Applications, First Edition. Hossein Hosseinkhani.

stimuli-responsive hydrogels have recently emerged to be promising candidates for insulin delivery [15–17].

Enzyme-responsive hydrogels are another important class of responsive materials that are presented in all metabolic and biological processes. The high selectivity and enzyme specificity of these systems distinguish them with typical catalyzed or non-catalyzed chemical reactions. These superb characteristics enable enzyme-responsive hydrogels to be employed as highly valuable systems in site-specific triggered release and tumor tissue-targeted delivery, while conventional stimuli polymers that are sensitive to temperature, pH, or light cannot easily achieve these goals [18–20]. An antigen-responsive hydrogel involves the highly selective and specific interaction of an antibody to its antigen. Many materials are considered antigens by the body, including disease-causing microorganisms such as viruses, fungi, bacteria, and parasites as well as toxic substances such as poisons. Antigen–antibody binding is related to human immune responses used to detect diseases in the body. These interactions are formed via non-covalent reactions including van der Waals forces, hydrophobic bonding, hydrogen bonds, electrostatic forces, or a mixture of all of these interactions [21]. These responsive materials can be effectively applied in the field of drug delivery, sensing, clinical diagnostics, and remote-controlled vaccine depot.

Deoxyribonucleic acid (DNA) and ribonucleic acid (RNA) are the two principal nucleic acids. These biopolymers are formed from nucleotides that are regarded as genetic materials. They can be modified with various functional groups involving carboxyl, amino, and thiol, which may lead to better accessibility and easy fabrication of a DNA synthesizer. Due to their negative charges and large sizes, nucleic acids have poor inherent transfection efficiency and are fragile [22]. These drawbacks can be addressed by encapsulation of nucleic acids within hydrogel networks. Besides the precise interaction of RNA and DNA strands, the highly selective binding of DNA sequences to aptamers can also be considered for preparation of stimuli-responsive hydrogels. Aptamers are single-stranded molecules that can bind to specific targets exhibiting similar properties like antibodies, thus can be promising in recognition of cancer cells. Apart from previously mentioned nucleic acids, there is a chemically synthesized single-stranded nucleic acid with modified backbone called peptide nucleic acid (PNA). Despite RNA and DNA, PNA is stable against temperature and enzymatic cleavage with the ability to form achiral and nonionic PNA–RNA, PNA–DNA, and PNA–PNA bindings.

This chapter discusses current developments concerning recent advances and applications of biomolecule-responsive hydrogels in medicine via emphasizing this research area with novel literature studies. The discussion is categorized into three main types, namely, glucose-responsive hydrogels, protein-responsive hydrogel, and nucleic acid-responsive hydrogels. Their potential applications in drug delivery, cell and cancer research, regenerative medicine, and biosensor are discussed, and some problems as well as the new directions on future developments are also presented.

To use hydrogels in various biomedical applications, it is desirable to greatly control their mechanical properties. The material properties of the hydrogel are responsible for cell attachment, differentiation, viability, and proliferation. To control the mechanical properties of hydrogels, a number of parameters such as the density and chemistry of the cross-links, as well as the concentration, chemistry, and molecular weight of the precursors can be modified. In many cases, the current approaches to fabricate hydrogels do not result in the synthesis of constructs with desired mechanical and chemical properties. Limitations with generating robust hydrogels that can withstand the *in vivo* environment include the need for low overall concentration of material, the requirement for degradation, and the need for cytocompatibility. Interpenetrating networks (IPNs) are a potentially powerful method of modifying hydrogel properties for tissue engineering. An IPN is a mixture of two or more cross-linked networks that are mixed together at the molecular level. When only one polymer of the IPN is cross-linked and the other is left in its linear form, the system is referred to as a semi-IPN. Conversely, when both or all types of polymer are cross-linked, the system is called a full-IPN. IPNs help to improve the mechanical strength and resiliency of the overall polymer and provide it with special physical and chemical characteristics. These characteristics can include temperature-responsive behavior and interfacial compatibility By using 3D technology of biomaterials and hydrogels IPN-based hydrogels can be fabricated with improved biological, chemical, and mechanical properties for tissue engineering. IPNs demonstrated intermediate mechanical properties, which were dependent on the concentrations of the component polymers. Many research works have investigated that the mechanical properties of IPNs interconnected networks have been significantly enhanced.

Engineering hydrogels aim to develop a technique that draws from microscale approaches, novel biomaterials, and biological principles to overcome some of the limitations of the current tissue engineering approaches. These include the inability to generate vascularized tissues, to uniformly seed cells throughout the scaffold, and to mimic the biological complexity and architecture of tissues. To accomplish this task several strategies are required. This strategy aims to address the increasing levels of complexity associated with the process. In strategy 1, novel hydrogels comprising natural and biodegradable materials with improved mechanical properties in which cells can be encapsulated need to be developed. The ability to control the mechanical properties is important for enabling the use of cell-laden hydrogel constructs. In this step, also engineering the hydrogels to deliver desired signals to cells is important. In strategy 2, engineering approaches of adding microvasculature to the scaffolds by developing methods to make microchannel networks or interconnected macropores and seed them with cells must be considered. In strategy 3, additional complexity to cell-laden hydrogel scaffolds by encapsulating cells surrounding the microvessels is necessary to complete the ability of hydrogel for medical application. Tissue-engineered constructs fabricated using the aforementioned approach promise versatility, biocompatibility, and exceptional compliance with many cell types. The functionality, suitability, and success of hydrogels as tissue implants or biomedical devices are strongly dependent on their bulk properties. Therefore, to fabricate microscale vascularized 3D tissue-engineered constructs, it is important to synthesize hydrogel scaffolds and 3D architectures exhibiting desirable bulk properties that can be tailored to the needs of the host. Consequently, the first specific strategy is to synthesize hydrogels with controllable chemical, biological, and mechanical properties. Specifically, composite hydrogels made from IPNs are the most important part to achieve this strategy. In addition, it is important to control the delivery of molecules from these hydrogels for *in vivo* vascularization as well as for maintaining cell function. Thus, controlled release of specific molecules from these IPN networks are important to be considered.

The ability to directly seed cells within macroporous hydrogels will be important for overcoming challenges associated with uniform cell seeding density and vascularization. However, most hydrogels lack the desired mechanical and biological properties that are associated with tissues in the body. Engineering IPNs composed of different types of natural hydrogels to enable uniform cell distribution and to control the mechanical and biological properties of the hydrogel scaffolds would be very attractive to achieve the aforementioned goal. The mechanical properties of the hydrogels need to be analyzed and optimized to effectively mimic *in vivo* tissues and the viability of cells within the hydrogels. Examples of such engineered IPN hydrogels would be the use of photocross-linkable HA, collagen, and self-assembled peptide amphipile (PA). These molecules form hydrogels by various mechanisms and thus can be used to form two (or three) completely independent polymeric networks that will enable their use in making IPN networks. Photocross-linking HA creates cross-links between acrylated functional groups that react in the presence of photoinitiator and UV light. Correspondingly, collagen fibers form hydrogels at physiological temperatures, while self-assembled amphiphilic peptides form hydrogels in presence of appropriate ionic conditions. Therefore, the independent gelation of these hydrogels is essential in building novel composite hydrogels having improved mechanical properties.

IPN hydrogel can be fabricated by use of three different hydrogels such as collagen, photocross-linkable HA, and self-assembled PA. Composite hydrogels comprising IPN gels are synthesized by using various combinations of collagen, HA, and self-assembling precursors and inducing gelation at specific conditions. IPN hydrogels are synthesized (Figure 3.1) comprising various combinations of the three different hydrogel precursors to generate a library of IPN hydrogels. Initially, one can synthesize hydrogels of each precursor solution individually and

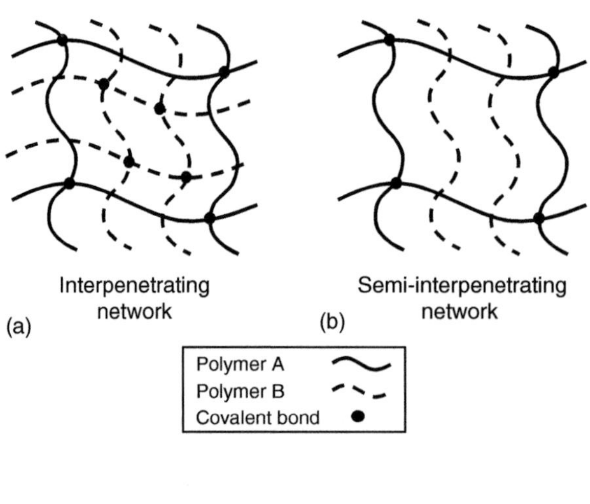

Figure 3.1 Schematic diagram of interpenetrating networks.

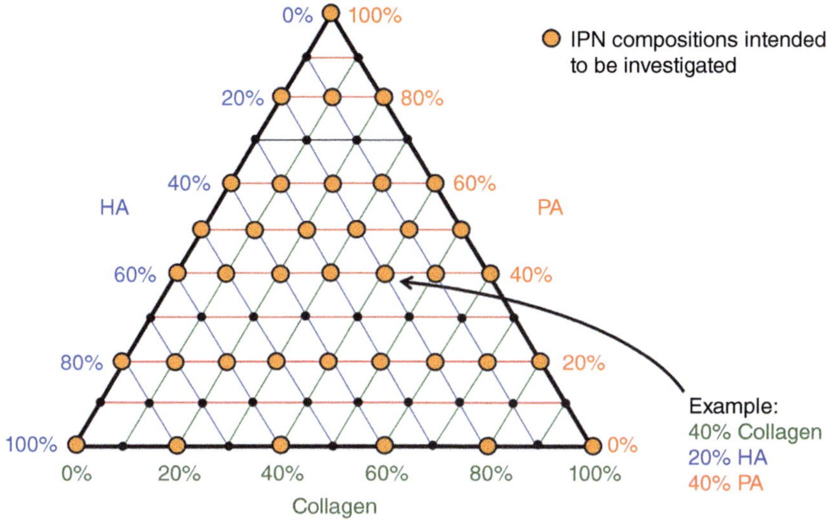

Figure 3.2 Ternary diagrams of the various IPN compositions that will be investigated in this study. Proposed samples include the 3 individual hydrogel precursors, 14 binary IPN combinations, and 20 ternary IPN mixtures.

test them for their mechanical properties. Next, one can construct binary IPNs that consist of two of the three aforementioned hydrogels in various concentrations. An example would be the fabrication of five different IPNs of collagen and HA at the following mass ratios: (i) 20% collagen, 80% HA; (ii) 40% collagen, 60% HA; (iii) 50% collagen, 50% HA; (iv) 60% collagen, 40% HA; and (v) 80% collagen, 20% HA. These concentrations will allow us to measure the mechanical properties of the entire spectrum of each binary IPN. Finally, one can develop IPNs that consist of all three precursor solutions at 20 different concentrations that, again, will allow us to study the properties of the entire spectrum of IPNs. Based on testing at least five different concentrations for each of the three binary IPN conditions (i.e. IPNs of HA–PA, HA–collagen, and collagen–PA) as well as the 20 different concentrations of triple IPNs (i.e. HA–PA–collagen (Figure 3.2), we anticipate that the total number of experiments can be easily performed using standard laboratory techniques. In these approaches it is possible to synthesize and test biological and mechanical properties of the library of hundreds of different polymers by using liquid handling robots. During the aforementioned experiments, one can also vary the polymer composition of the individual hydrogels between 2 and 10 mass/vol%. This will allow us to develop a library of IPN data that will be used in the isolation of various IPNs that offer the desired mechanical and biological properties that most closely mimic those of the natural tissue. Also, it is necessary to compare the various IPN mechanical strengths to the values we obtain for each individual hydrogel. It is hypothesized that the higher degree of cross-linking within the IPNs will significantly increase the mechanical strengths of the hydrogels.

To cross-link the hydrogel precursors for each IPN combination, we can also take advantage of the different cross-linking mechanisms of each individual precursor.

Going back to the example used previously, the synthesis of a collagen/HA IPN must begin by first gelling the cross-linked HA network by exposure to UV light and then increasing the temperature from 4 °C (where collagen is in solution form) to 37 °C (where it will form a gel). To minimize the degree of phase separation and to increase the mobility of one phase in the other, we can also examine partial curing of individual hydrogel precursors. An example would be to "loosely" gel the HA network to ease the diffusion of the collagen component, followed by increasing the temperature. Thus, the sequential manipulation of temperature, interspaced with short UV radiation, can be used to produce IPN networks consisting of collagen and HA (comprising three different precursors). One approach to fabricating such hydrogels will be to induce PA to form supramolecular nanofibers within the collagen–HA IPN. Fortunately, the size of the pores generated using either of the three components is large, thus enabling the diffusion of large molecules. Alternatively, the three components could be added simultaneously and induced to undergo gelation. Permutations in the order and length of the temperature changes and UV exposure and quantity of self-assembling peptide within the matrix will produce IPNs with varying bulk properties. The structures of the IPNs and the presence of the various polymers in the hydrogels are characterized with FTIR spectroscopy, ^{13}C-NMR and DSC. FTIR spectrum is recorded using the potassium bromide pellet technique. The sample is scanned over the range 400–4000 cm^{-1} in a FTIR spectrometer. ^{13}C solid-state CP/MAS NMR can also be used for the characterization of the water-insoluble SPH-IPN with a 300 MHz spectrometer at room temperature. Also, DSC measurements of dried samples can be carried out on a TA instruments DSC. The thermal analysis is performed from 20 to 340 under dry nitrogen. Also, scanning electron microscope (SEM) can be used to analyze the hydrogels. To prevent the morphology of the porous structure from changing, the dried hydrogel is frozen in liquid nitrogen and then cut to expose its inner structure, which will be coated with a thin layer of palladium gold alloy and imaged in a SEM.

To analyze the distribution of cross-linked polymers in the IPN, fluorescein isothiocyanate (FITC)-labeled hydrogel precursors can be used. Briefly, the hydrogel precursor and FITC is dissolved in sodium bicarbonate buffer (pH 8.7) and mixed and stirred at ambient temperature away from light. After being precipitated with absolute ethanol, the product is purified against double distilled water (DDW). Purified FITC-labeled hydrogel precursors are freeze-dried and used for synthesis of the IPNs. After IPNs had been put in liquid nitrogen and cut to expose their inner structure, the inner surfaces of the polymer can be imaged under a confocal laser-scanning microscope. It is also important to measure the porosity of IPN hydrogels by using the solvent replacement method. The dried hydrogel is immersed in absolute ethanol overnight and weighed after excess ethanol on the surface is removed. The porosity is calculated according to the following equation:

$$\text{Porosity} = (M2 - M1)/\rho V \qquad (3.1)$$

where $M1$ and $M2$ represent the mass of the hydrogel before and after being immersed in absolute ethanol, respectively, ρ is the density of absolute ethanol, and V is the volume of the swollen hydrogel. Also, the swelling characteristics of the IPNs can be analyzed by immersing dried test samples to swell in excess DDW (pH 7.0) at room temperature. At specific time intervals, the samples is removed from DDW and then blotted with a piece of paper towel to absorb excess water on the surfaces. The swelling ratios (Qs) of IPN hydrogels are calculated from the following expression:

$$\mathrm{Qs} = (W_s - W_d)/W_d \tag{3.2}$$

where W_d is the weight of the dry hydrogel, while W_s is the weight of the swollen hydrogel at certain time intervals. To examine the mechanical properties of IPN hydrogels, atomic force microscope (AFM) and Instron devices can be used. The bulk material properties of the hydrogel materials are characterized by measuring either the tensile or compressive stress as a function of the strain using an Instron 5542 apparatus. This will characterize the material by its Young's modulus, compressive modulus, and yield point. The hydrogels is prepared in dimensions of $0.4\,\mathrm{mm} \times 5\,\mathrm{mm} \times 11\,\mathrm{mm}$, and tested at a strain rate of $0.05\,\mathrm{mm\,s^{-1}}$ until failure. The mechanical properties is tested without cells because cells may degrade the hydrogels. In addition, nanoscale hydrogel stiffness, relevant to actual cell sensing, is also determined by AFM. The elastic modulus, E, is obtained by specific measurement. Briefly, E is measured by using a Hertz cone model to fit the first 5–50 nm of indentation profiles obtained from fully hydrated gel samples with unsharpened, pyramid-tipped cantilevers of spring constant $60\,\mathrm{pN\,nm^{-1}}$. Measurements are performed on multiple regions of a gel. Also, bulk elastic measurements are made on the same samples by a simple tension method, with E calculated from the slope of the tensile force per instantaneous cross section versus large deformation uniaxial strain.

To engineer and optimize mechanical properties, a number of candidate hydrogels for subsequent studies need to be formulated. These hydrogels will be chosen to maximize cell viability and optimized mechanical properties. For mechanical properties, IPN hydrogels that most closely mimic the tissue microenvironment such as compliance and contractility of cardiac muscle need to be designed. For example, the average moduli of muscle is $E \sim 5$ to 15 kPa at strains up to 10–30% or nearly $E_{\mathrm{Muscle}} = 12\,\mathrm{kPa}$ based on AFM indentation studies. Muscle cells cultured on substrates with such tissue-like stiffnesses (8–11 kPa) striate to show similar phenotype as cells in the native muscle. Thus, it is necessary to use IPN networks that mimic the native tissue mechanical properties and maintain cell viability for subsequent studies. There are many benefits to using HA as a hydrogel; however, it can be cytotoxic at high concentrations. Although we envision that this will not be a problem, we need to consider decreasing the HA and photoinitiator concentrations or using other materials such as calcium alginate if excessive cytotoxicity appears. Also, we can make hybrid structures that incorporate a fibrous scaffold into the IPN gels. Fibrous scaffolds could be made of collagen, PLA, poly(lactic-*co*-glycolic acid) (PLGA), poly(glycerol sebacate) (PGS), or silk. However, this method does not result in a homogeneous distribution of cells.

Figure 3.3 shows a schematic of an ideal IPN and a nonideal IPN. Nonideal IPNs form when the thermodynamic immiscibilities of the individual hydrogel precursor

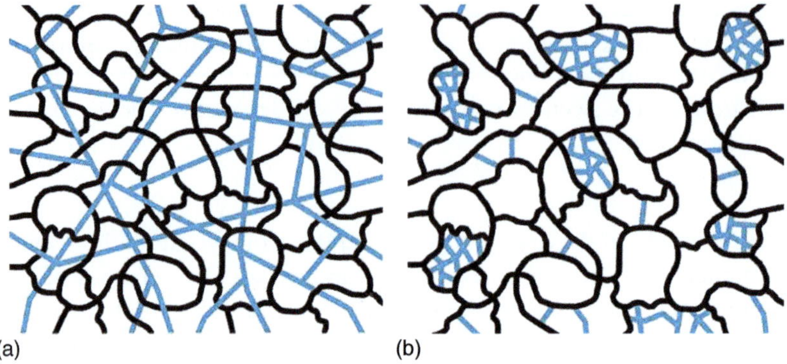

Figure 3.3 Model nanostructures of IPNs. (a) Ideal IPN structure where the two networks are fully developed and interpenetrated. (b) More realistic structure of a hybrid IPN, in which the one network domain is not necessarily connected but rather dispersed in the softer polymer matrix.

solutions. We do not envision this affecting IPN hydrogels as all of proposed precursors' solutions exhibit hydrophilic properties (even the PA nanofibers). However, in the case that we do encounter nonideal hydrogels, we still can propose two methods to ensure that engineered hydrogels are fully developed. First, it is necessary to reduce the molecular weight of the individual hydrogels to ensure that they fully interpenetrate each other. Second, each gelation process must be optimized. By analyzing the structures of the IPNs and their gelation process, we will be able to determine the exact source of nonideality and make adjustments accordingly.

3.2 Glucose-Responsive Hydrogels

3.2.1 Glucose Oxidase Hydrogels

Glucose oxidase (Gox) hydrogels are formed by the reaction of glucose with oxygen. Briefly, the response of polymeric hydrogels to the bioproducts obtained from the enzymatic oxidation of glucose is caused by glucose sensitivity. Gox catalyzes the oxidation of glucose to bioproducts of gluconic acid and H_2O_2 (Eq. (3.1)). Consequently, the glucose-responsive hydrogels can be indirectly obtained through the addition of polymeric hydrogels, which are sensitive to these biomolecules. The pH-responsive hydrogels are considered as suitable candidates for glucose applications. As they are conjugated or loaded with Gox, glucose is transformed to gluconic acid by Gox, which lowers the pH in the hydrogels and changes the swelling of pH-responsive hydrogels. This property allows pH-responsive hydrogels to release entrapped insulin, making them significantly valuable for modulated (self-regulated) insulin delivery. More specifically, the self-modulated insulin release in the pH-responsive hydrogels is controlled by two different mechanisms: (i) swelling (high pH) and (ii) shrinking or collapsing (low pH).

$$\text{Glucose} + H_2O + O_2\ \text{Gox gluconic acid} + H_2O_2 \tag{3.3}$$

Ishihara et al. for the first time introduced a Gox hydrogel based on N,N-diethylaminoethyl methacrylate (DEA) and 2-hydroxypropyl methacrylate (HPMA) copolymer immobilized with Gox for the regulation of insulin permeation [23]. It was observed that the permeation of insulin was enhanced in the presence of glucose. Yang and Kim entrapped hydrophobically modified Gox (HmGox) inside a β-cyclodextrin grafted polyethyleneimine (PEI-βCD), to construct a glucose-responsive hydrogel [24]. The HmGox was formed via a reaction of Gox with palmitic acid-*N*-hydroxysuccinimide ester (PA-NHS) at a ratio of 1 : 40, to obtain strong encapsulation of Gox within the hydrogel network. As the concentration of glucose increased in the enzymatic reaction, gluconic acid was formed, resulting in lower pH, thus improving the swelling and promoting the release of FITC dextran from an HmGox-loaded hydrogel. In a fascinating study, Peppas and coworkers synthesized a poly(methacrylic acid (PMAA)-graft-ethylene glycol) hydrogel in the presence of Gox [25]. At low pH, carboxyl and etheric groups formed a complex causing the hydrogel to collapse. At high pH values, the gels swelled to 20 times the size of its dry state. The rate of expansion for hydrogels containing Gox was higher than for hydrogels without Gox. Similar hydrogels have been fabricated by copolymerization of sulfadimethoxine monomer with *N*,*N*-dimethylacrylamide (DMA) [26] and *N*-isopropylacrylamide (NIPAM) with methacrylic acid [27].

In recent years, a great deal of research has been dedicated to the preparation of biocompatible Gox-based hydrogels using poly(ethylene glycol) (PEG) grafts [28–31] as well as chitosan [32, 33] as a biodegradable, biocompatible, nontoxic, and non-immunogenic natural polymer. The PEG grafts were expected to minimize subsequent immunoreaction and rejection by the body, and to postpone the degradation of proteins and enzymes within the hydrogels [34]. These hydrogels, with diameters between 30 and 300 μm, and microparticle sizes are conducive to fast swelling/deswelling dynamics in the response to pH changes, with a faster response to the smallest particles. In less than five minutes, reversible swelling was obtained in response to pulsatile variations of glucose [31]. Podual et al. fabricated glucose-responsive hydrogels by encapsulation of Gox microparticles within a poly diethylaminoethyl-graft-ethylene glycol (PDEAE-PEG)-graft-ethylene glycol (PEG) by using tetra(ethylene glycol) dimethacrylate as a cross-linker in the presence of silicon oil. The pH-sensitivity of this hydrogel indicated rapid swelling/deswelling behavior between a pH of 3 and 7.4 with smaller particles showing a faster response. In addition, it was concluded that the cross-linking concentration had a strong effect on the extent and speed of the swelling; PEG grafts with longer lengths increased the hydrogel swelling relaxation. Another important biocompatible polymer for preparation of Gox hydrogels is chitosan. Chitosan, a natural polysaccharide polymer, which can be found in animals, has the capability to form *in situ* hydrogels when incorporated with a polyol counter-ionic dibase salt (such as phosphate disodium and β-glycerol). As the solution reaches body temperature, a sol–gel transition is achieved [32]. This biocompatible hydrogel, with a glucose concentration of $3\,mg\,ml^{-1}$, indicated maximum swelling. Although insulin was released from hydrogels in a pulsatile manner over two hours, the swelling behavior did not exhibit a linear relationship with glucose concentration. This formulation, with a dose of 3 IU kg^{-1}, was used for *in vivo* experiments on diabetic rats, where the insulin release occurred in response to glucose concentration.

Liu et al. synthesized alginate–chitosan microsphere (ACMS) hydrogels via an emulsification-internal gelation–adsorption–polyelectrolyte coating technique for encapsulation of Gox [35]. The obtained ACMS was spherical in a wet state with a diameter of 138 μm. This pH-dependent Gox showed greater loading and entrapment efficiency when the pH of the medium was lower than the isoelectric point (PI) of Gox. The Gox encapsulation efficiency and loading were reduced with increasing chitosan concentrations in the coating solution and improved with increasing Gox concentrations in the loading solution. The activity of Gox–ACMS was maintained and indicated prolonged production of H_2O_2 as compared to free Gox.

3.2.2 Con A Hydrogels

Glucose-responsive hydrogels can also be developed from the combination of glycopolymer–lectin systems and glucose. Lectins are special types of proteins, which bind carbohydrates and serve as natural receptors in the fabrication of glucose-responsive hydrogels. They interact with glycolipids and glycoproteins on the cell surface and have a natural role in hormone regulation and cell adhesion. Among lectins, concanavalin A (Con A), obtained from the jack bean plant *Canavalia ensiformis*, is considered the most important protein, with four binding sites [36]. This property makes the Con A complex a suitable candidate for a glucose-monitoring and glucose-responsive drug delivery system. Brownlee and Cerami were the first researchers to develop glucose-responsive systems using Con A for insulin release [37]. The authors prepared a complex of Con A with a stable, biologically active glycosylated insulin derivative. In the presence of free glucose, the glycosylated insulin derivative competitively binds to Con A, facilitating the release of glucose from the complex.

Several considerations should be taken into account for the use of Con A-based hydrogel systems, including the leakage and biocompatibility of Con A, as well as the solubility and stability of the Con A complex in water [14]. Therefore, efforts were made to improve the glucose-responsive complexes for drug delivery, to ensure suitable release of insulin in different conditions and, more importantly, to minimize the leakage of Con A.

Kim and Park established a method to enhance the low aqueous stability and solubility of Con A in the sol–gel phase-reversible hydrogels [38]. Con A was modified by mixing with PEG molecules. The gels were built up by a combination of glycosylated polymer chains and PEGylated-Con A. The glucose-binding affinity reached a maximum by increasing the number of grafted PEG chains to five. In comparison to Con A, modified PEG–Con A not only showed better long-term stability and aqueous solubility but also improved the glucose sensitivity.

A covalent link with Con A was obtained through the copolymerization of 2-glucosyloxyethyl methacrylate (GEMA) with Con A, which created higher stability compared to mixing with Con A [39]. Tanna et al. chemically cross-linked the amine groups of Con A with carboxylic moieties on various carbomer resins such as Carbopol 934P NF, 941P NF [40], and 974P NF [41] by using carbodiimide chemistry before mixing with dex, to synthesize glucose-responsive gels. These chemical couplings decreased the hydrogel leakage and improved gel stability.

Tanna et al. used dex for the fabrication of hydrogels [42–45]. The acrylic derivatives of dex were chemically cross-linked with Con A through photopolymerization. Based on the results, the leakage of Con A was efficiently reduced by the cross-linking, and the gel showed sol–gel reversible transition in the presence of glucose.

Hubble and colleagues reported cross-linked carboxymethyl dextran (CM-dex)–Con A hydrogel membranes by applying carbodiimide chemistry [46]. Through this facile method, hydrogels with a controllable degree of carboxylic group replacement, and different protein replacements amount, could be obtained. In addition, the CM-dex/Con A hydrogels displayed pH and ionic sensitivity, with no potential toxic bioproducts. The mechanical properties and gelation time of the resulting hydrogels were demonstrated by the content of coupled Con A as well as by the initial degree of carboxylic group replacement; results indicated that the CM-dex/Con A hydrogel gelation time is shorter than that of the CM-dex hydrogel. They further studied the effect of molar mass and degree of branching on the interaction of commercially available dex with Con A [47]. Microcalorimetry analysis was used to study the bio-specific interactions between dex and Con A. Based on these results, the interaction of these materials was suitably explained by a two-site model. This model was able to determine the relevance between molecular mass, association constant, and binding energy.

Nie and colleagues established systems using chitosan, Con A, and dex for the preparation of glucose-responsive composite microparticles [48]. Chitosan microparticles were created, grafting Con A by Schiff's bases and dex layers through special affinity. It was reported that the use of dex provided strong affinity to Con A, and chitosan enhanced the loading capacity and encapsulation efficiency of insulin.

Nie and coworkers further synthesized photo-cross-linked glucose-responsive hydrogels [49]. Glycidyl methacrylate-modified dextran (dex-G)-ethylene glycol and acrylate methacrylate-modified Con A (Con A-E) were fabricated using poly(ethylene glycol) dimethacrylate (PEGDMA) as a cross-linking agent. The Con A-E and dex-G were created by Michael addition and a ring-opening reaction, respectively. The medium containing different glucose concentrations had a direct effect on the size of the hydrogels. In addition, concentrations of both the components, specifically PEGDMA, played a key role in glucose sensitivity. In this system, methacrylated Con A increased the biocompatibility and glucose sensitivity by reducing the lectin leakage and reversible binding to dex-G.

Similar systems have been prepared via reversed-phased emulsion cross-linking for microhydrogel insulin delivery [50]. These sphere-like microhydrogels showed diameters less than 5 μm. Based on the release test, insulin was reversibly released from these gels in response to different concentrations of glucose. In addition, the released insulin remained active without annihilation of the tertiary network. The degree of substitution (Ds) of dex-G showed significant effects on both the surface burst release and release rate of the microhydrogels and limited the glucose sensitivity of the microhydrogels at high Ds, especially Ds 32. The biocompatibility of these systems for insulin delivery was confirmed using an MTT assay in the L929 cell line. More recently, Yin et al. exploited a similar concept to construct glucosyloxyethyl acrylated chitosan (CS-GEA)–Con A microgels in the presence of genipin as a cross-linker for insulin release [51]. The GEA–chitosan, with cross-linking of functional groups and a strong affinity to Con A, prevented the loss of Con A during the modification process.

Yao and Hu reported triply responsive films, including the pH-responsive Con A–dex and the temperature, salt-responsive poly(*N*,*N*-diethylacrylamide) (PDEA)-encapsulated horseradish peroxidase (HRP) hydrogels [52]. Con A–dex layer-by-layer (LbL) assembly films were built up, via Con A, on the surface of pyrolytic graphite (PG) electrode. Then, by polymerization of PDEA and HPR on the surface of the LbL inner layers, the [(Con A–dex–PDEA–HRP)] films were formed. The electroactive probe $Fe(CN)_6{}^{3-}$ represented a triply responsive on–off cyclic voltammetric response at the film electrodes. The mechanism of the pH-responsive Con A–dex inner layer was related to the electrostatic interaction between the probe and the layers at different pH, while the structural change of PDEA at various sulfate concentrations and temperatures was responsible for the temperature and salt-responsive properties of PDEA–HRP outermost layers.

In addition to these approaches, a novel method was proposed by Miyata and coworkers [53]. They synthesized bioconjugated gel particles composed of Con A–GEMA in the presence of hydrophobic *N*,*N*-diethylaminoethyl methacrylate (DEAEMA) and hydrophilic PEGDMA as cross-linkers, via surfactant-free emulsion copolymerization. The GEMA–Con A gel particles with diameters of about 750 nm dispersed stably in a phosphate buffer solution (PBS). It was reported that the complex had a core–shell structure consisting of a hydrophobic core (DEAEMA) and hydrophilic shell (GEMA–PEGDMA) with Con A. The complex underwent changes in size in the presence of glucose in PBS. Although the GEMA–Con A swelling remained constant with galactose in PBS, it was slowly improved by increasing the glucose content. These smart gel particles were suggested as suitable candidates for diabetes therapy.

3.2.3 Phenylboronic Acid Hydrogels

In Sections 2.1 and 2.2, we demonstrated that Gox and Con A are the two most common protein-based glucose-responsive systems. However, they are subject to environmental changes and usually suffer from antigenicity, instability, and expensive cost. Hence, they are not suitable to be stored or used for relatively long periods. The third class of glucose-responsive hydrogels is known as phenylboronic acid (PBA), using synthetic and nonnatural materials. Kuivila et al. were pioneers in the development of PBA-saccharide systems [54]. The mechanism of PBA became clear through the following research [55–59]. Particularly, PBA and its derivatives do not contain any protein and can reversibly form a complex with *cis*-diol, such as glucose, in aqueous solutions [60–62]. PBA complexes exhibit two different forms, charged (hydrophilic) and uncharged (relatively hydrophobic) forms. In an alkali medium, due to the dissociation constant (pK_a) of the 3-acrylamidophenylboronic acid (AAPBA) moiety (8.2) in the NIPAM–AAPBA copolymer, the PBA structure changed to a boronic anion tetrahedral form, via mixing of a hydroxyl residue with boron. However, in a neutral or an acidic medium, boronic acid moieties of PBA could not easily form a complex with *cis*-diol due to its plane triangle structure [63, 64]. A stable network was formed between glucose and the tetrahedral structure [65].

3.2.3.1 Hydrogel Swelling

The first research of glucose-responsive hydrogels using a lower critical solution temperature (LCST) was conducted by Kataoka et al. by using poly[*N*,*N*-dimethylacrylamideco-3-(acrylamido)phenylboronic acid] (DMA-*co*-APBA) [66] and APBA, *N*,*N*-dimethylaminopropyl acrylamide (DMAPAA) [67]. The LCST of this polymer increased through the addition of glucose, whereas it decreased in the presence of a PBA moiety. This shows that the PBA derivatives swell in presence of glucose and shrink without glucose. They further built up a glucose-responsive network based on an APBA hydrogel, which showed an on–off regulation of insulin release in response to glucose content [68]. Thereafter, the modified polymer with APBA was regarded as the main network for drug delivery systems through this swelling. Kataoka and coworkers reported a discontinuous change in swelling behavior of phenylboronic-derivative-based hydrogel networks for different temperatures at each critical glucose concentration at a pH of 9 [69]. The equilibrium swelling of hydrogels was obtained with high glucose content after 500 minutes.

Comb-type hydrogel, as a novel system, was also introduced into glucose-responsive complexes that exhibited a fast response. Okano and coworkers synthesized comb-type-grafted PNIPAM hydrogels, including grafted PNIPAM side chains with mobile ends, indicating a fast swelling–deswelling response to temperature changes [70–72]. A similar concept was exploited for fabrication of poly(NIPAM-*co*-AAPBA) networks grafted poly(NIPAM-*co*-AAPBA) side chains [65]. Owing to the structure, this system responded to glucose at physiological temperatures.

The APBA system, with pK_a of 8.2, cannot be used for *in vivo* application due to its higher pH value than the human body. A novel approach was proposed by Hoare and Pelton, using a charge-switching polymer [73]. The combination of the cationic monomer of *N*,*N*-dimethylamino ethylacrylate (DMAEA) with the anionic monomer of acrylic acid (AA), functionalized the APBA response to glucose concentrations in the physiochemical environment, due to its amphoteric network. In addition, high insulin release was obtained by altering the ionic strength and pH of the system.

Another interesting method was created through modifying the chemical structure of PBA in response to the physiological environment. Kataoka and coworkers fabricated synthetic gels that involved the use of a phenylborate derivative [4-(1,6-dioxo-2,5-diaza-7-oxamyl) phenylboronic acid] (DDOPBA), with a low pK_a (7.8), as a glucose sensor moiety [63, 64]. The DDOPBA, with a near-psychochemical pH, was a suitable replacement for APBA in the same system. Beads of this hydrogel, with diameters of approximately hundreds of micrometers, were synthesized through an inverse emulsion as a template for polymerization [69]. Applying alternative groups may open up a novel area in the development of self-regulated glucose-responsive systems for *in vivo* application.

A colorimetric sensor, using a periodic array inside the gel, was another system used to make the PBA-derived hydrogel system. This system can be prepared using three different methods, including the formation of an array via hydrogels in the case of a hologram [74], a periodic array of voids [75–77], and the inclusion of a colloidal crystal within the gel [78, 79]. In comparison to inverse opal hydrogels, which possess a high porous network, the crystalline colloidal array (CCA) is more accurately correlated with the diffraction wavelength of the incorporated photonic crystal [80, 81]. In CCA sensors, the changes in the hydrogel network are attributed to the chemical interactions of an analyte and functionalized hydrogel network. The array diffracts light based of Bragg's law and the material can be observed as having color when the distance between the particles and the light wavelength is the same. The interparticle distance, which affects the swelling behavior of the hydrogel, has a significant effect on the crystal color. Hence, by mixing a periodic array with a molecule-responsive hydrogel, colorimetric sensors could be obtained. Hamachi and coworkers fabricated a supramolecular hydrogel-based fluorocolorimetric sensor for polyols [82]. The boronic acid-appended as an acceptor, formed boronate ester complexes with polyols, and became anionic and hydrophilic, thus moving from the hydrophobic part. The fluorescence color was changed by increasing the distance between the acceptor and donor. Braun and colleagues studied the interaction between polymerized crystalline colloidal arrays (PCCAs) and PBA networks under physiological conditions to explain the influence of PBA chemistry on the volumetric changes in hydrogels [83]. PCCA-modified PBAs, which formed heterocyclic rings, had a linear and slow response, while similar systems without formation of heterocyclic rings showed a nonlinear and fast response. The substitutions of PBAs could be either improved or decreased. The response rate increased through the use of electron-withdrawing substituents. However, the response magnitude was highly reduced by introducing a methylene linkage between the hydrogel backbone and PBA.

Several glucose-sensing systems were prepared via holographic grating [84–86], the inclusion of latex particles [87, 88], and an inverse opal with the hydrogel [89, 90]. Holographic sensors are popular due to their mass preparation and faster diffusion through large voids, using an inverse opal, which speeds up the response time.

3.2.3.2 Hydrogel Shrinking

PBA-containing synthetic hydrogels usually swell through addition of glucose; however, in some cases, these complexes shrink in these conditions. Phenylboronates bind all molecules with a cis-diol group as a mono-bidentate (1 : 1) system. In contrast, bis-bidentate (2 : 1) systems are able to form one sugar with two boronates and two cis-diols. The 1 : 1 binding model would cause swelling, whereas the 2 : 1 model would lead to hydrogel shrinking. Glucose is considered as the only saccharide that involves a bis-bidentate network containing two diols. The bis-bidentate network creates linkages with the hydrogel and, therefore, shrinks with addition of glucose [33].

Asher and coworkers first introduced a PBA-responsive hydrogel, including PEG or crown ether functional groups [87]. Glucose increased the cross-linking density of the hydrogel through simultaneous binding with boronates, resulting in a reduction in swelling. By increasing the glucose concentration from 0 to ~5 mM, the binding model was 2 : 1, resulting in deswelling of the gel. A further increase showed hydrogel swelling with a 1 : 1 binding model. They further enhanced the photonic glucose-sensing properties of the sensors to allow sensing at physiological pH and salinity, using 4-amino-3-fluorophenylboronic acid and 4-carboxy-3-fluorophenylboronic acid [91]. This sensor was used in an artificial tear fluid, with no interposition from the main materials of the tear fluid, and thus was offered as suitable candidate for contact lenses.

Asher and coworkers applied a bis-bidentate glucose–boronate complex for preparation of a glucose-sensing component, which acted as a hydrogel cross-link [92]. These sensors were applied without using sodium-chelating agents (such as PEG or crown ethers). The fast-response kinetics were achieved through controlling the hydrophilic–hydrophobic balance and elasticity of the hydrogel network. The results indicated that the shrinkage rate increased by incorporation of *n*-hexyl acrylate onto the hydrogel network. Lowe and coworkers synthesized the tetrahedral 2-acrylamidophenylboronic acid (2-APB) glucose-selective sensor within hydrogel films [93]. The 2-APB-based hydrogel was copolymerized with PEG, [2-(acryloyloxy)ethyl]-trimethylammonium chloride (AETA), and (3-acrylamidopropyl) trimethylammonium chloride (ATMA). These three comonomers were applied to enhance the kinetic property of 2-APB and resulted in a faster response to glucose, compared to the 2-APB sensor. By addition of glucose, the contraction was observed for 2-APB-ATMA sensors due to the formation of a 2 : 1 complex between glucose and 2-APB, which resulted in shrinkage. This shrinking was observed as a blue shift for these sensors. In comparison to 3-APB, the 2-APB-ATMA sensors showed a rapid response to glucose and much less dependency on physiological pH values. Similar hydrogel shrinking behavior, wherein glucose generated a 2 : 1 complex with the 2-APB sensor, has also been reported [94, 95].

In an interesting study, Lowe and coworkers investigated the cross-linking mechanism of a 2-APB sensor to demonstrate glucose-induced hydrogel contraction [96]. Three different sugars, including mannose, galactose, and fructose, with relatively similar structures to glucose were used. In the sensor diffraction wavelength, most of the sugars led to a blue shift. The 1 : 1 binding model for three sugars was much higher than glucose, showing that the formation of a 1 : 2 glucofuranose: boronate bis-bidentate complex controlled the cross-linking mechanism in response to glucose. This higher glucose response, compared to mannose and galactose, was attributed to its greater hydrogel shrinking (1 : 2 complex), compared to hydrogel swelling (1 : 1 complex).

Strongin and coworkers developed a synthetic chemomechanical hydrogel sensor with high glucose selectivity over other blood sugars (e.g. galactose and fructose) [97]. This hydrogel was obtained via modification of PMMA with

3-aminophenylboronic acid, diethylenetriamine, dodecylamine, and butylamine and indicated a fast, selective, and reversible size change in human blood plasma in the presence of glucose. The bis-bidentate complexes were formed between glucose and PBA receptors and acted as a cross-link, thus resulting in hydrogel shrinkage.

Novel research, establishing spectroscopic evidence for glucose-bis (boronate) cross-links formed in hydrogels, was proposed by Magda and coworkers [98]. They fabricated a zwitterionic hydrogel via copolymerization of glucose-responsive hydrogels, including boronic acid, with tertiary amines (Lewis bases). As the glucose concentration increased, the hydrogel shrank, and the glucose selectivity was increased. Previously, it was reported that the formation of reversible cross-links by glucose resulted in the shrinkage of zwitterionic hydrogels, while fructose cross-links did not have this effect [87, 93, 94, 99–102]. However, Magda and coworkers reported that another important factor is a change in the free energy of mixing (Π_{mix}), described as the thermodynamic mixing interactions between the solvent and the system. The zwitterionic hydrogel swelled in response to an increase in fructose concentration, due to a large and positive change in Π_{mix}, and shrank upon the addition of glucose, due to a large and negative change in Π_{mix}. By increasing the temperature in fructose in PBS solutions, the hydrogel shrank, whereas it swelled in a glucose solution. These responsive behaviors could be attributed to the degree of hydrogen bonding, which was high in fructose and PBS and relatively low in a glucose solution.

On the way to practical application of zwitterionic hydrogels, plenty of factors need to be addressed, such as the mole fraction of tertiary amines and the geometry and size of the samples within the system. Smaller networks could exhibit faster responses. In addition, the properties of a polymer such as the number of charged groups, density of cross-linking, monomer distribution, and chemical quality play key roles in the dynamic characteristics. For example, using a low concentration of tertiary amines in a boronic acid hydrogel was found to highly increase glucose selectivity and cause hydrogel shrinkage in the response to glucose addition [98].

Apart from applying glucose-shrinking hydrogels to biosensing, they have also been studied for insulin release. Siegel et al. prepared glucose-shrinking hydrogels composed of AAm hydrogels derived with PBA [103]. Based on the pH of system, the hydrogel exhibited both swelling and shrinking behavior upon the addition of glucose. The hydrogel swelled at pH 7.4 and shrank at pH 9.0 through the increase in glucose. In a later report, they fabricated a microvalve prototype inserted in microfluidic device for insulin delivery [104]. Poly(acrylamide) (PAAm)-PBA hydrogels swelled and shrank in the presence of both pH and glucose. The insulin was released in response to glucose concentration, showing an on–off regulation. The response time was controlled by hydrogel thickness. Hydrogels with the thickness of 30 μm had the fastest response time of 7 minutes for a pH concentration cycle. At relatively long response times, the swollen hydrogel stopped the insulin flow. The system was further enhanced via introducing a micromachined silicon membrane valve including an array of micro-orifices to anchor the hydrogel [105]. The system-loaded hydrogel controlled the flow across the membrane in response to glucose, pH, and temperature. In the shrunken state, the collapsed hydrogel allowed the fluid to flow through an opened annular gap. However, in the swelled state, the orifice void spaces were completely covered by the hydrogel, restricting fluid flow. The response times for glucose, pH, and temperature were 10 minutes, 4 minute, and 10 seconds, respectively.

3.2.3.3 Microgels and Nanogels

Microgels and nanogels are defined as 3D miniature hydrogel particles with a size of ten nanometers to a few microns [106–110]. They have been extensively used as scaffolds [111–113] for drug delivery [114–118] and for biosensing applications [119–121]. Microgels indicate better response to external stimuli, higher stability, and larger surface areas over other drug delivery systems. While PBA hydrogels are used as sensing materials, microgels containing PBA can undergo a continuous volume phase transitions and disperse uniformly in a medium with a faster response time [122]. In addition, minimally invasive methods are used for administration of microgels, whereas their bulky counterparts must be implanted through invasive means. Therefore, microgels are considered more effective as candidates for the treatment of diabetes.

One of the most common methods for the preparation of PBA microgels is direct copolymerization of a PBA monomer with a thermoresponsive monomer, such as NIPAM, which are usually monodisperse [123–127]. The PNIPAM causes microgel swelling at room temperature due to its hydrophilicity, whereas above a critical temperature it becomes hydrophobic, thus inducing microgel shrinkage [108]. Typically, the addition of a hydrophobic monomer, such as 3-AAPBA, decreases the VPTT of PNIPAM-based microgels. Therefore, it was expected that hydrophobic PBA microgels would have a lower VPTT compared with PNIPAM microgels. However, only a slight difference in microgel VPTT was reported using this method [123–125]. The glucose-sensitivity of these microgels was highly reliant on their initial state, which was related to the functionalization degree and initial temperature of the particle. They also showed weak glucose selectivity [124]. This behavior was possibly due to an "island/ocean"-like heterogeneous microstructure of the microgels, having PBA-rich domains (islands) and PNIPAM parts (oceans), which cause sophisticated thermo- and glucose-responsive behaviors [125].

The modification of P(NIPAM-*co*-AAc) microgels with PBA groups is another method for fabrication of PBA-based microgels [128–133]. The incorporation of hydrophobic PBA groups significantly decreased the VPTT of obtained microgels. At room temperature, the PNIPAM-PBA microgels, with a PBA concentration of 10.0 mol%, were in a collapsed state. However, the addition of 0.01 M glucose improved the microgel hydrodynamic radius by ~100%, and caused dramatic swelling [128]. Therefore, compared to the direct copolymerization method, the modified microgels not only have better glucose-selectivity but also show higher performance, which might correspond to equal distribution of PBA groups within the P(NIPAM-*co*-AAc) microgel networks [134]. Siddiq and coworkers synthesized phenylboronic acid microgels, using different concentration of 4-vinylpyridine (4-VP) [135]. The p(NIPAM-4-PV-PBA) microgels were prepared via modification of p(NIPAM-4-VP-AAc) with 3-AAPBA. The addition of hydrophilic 4-VP significantly improved the VPTT and swelling behavior of microgels due to its good solubility in water and appropriate pK_a (~5.4), which reduced the pK_a of the PBA moieties. Surprisingly, the 4-VP moiety is not glucose-responsive; however, its addition to the microgels stabilized the PBA–glucose network by the Lewis acid/base interactions, resulting in high glucose-sensitivity at a physiological temperature and pH. The obtained microgels were proposed for self-regulated insulin delivery.

Most of the PBA-containing microgels swell upon glucose addition [62, 124, 128]. Typically, they are formed via a reversible reaction of PBA with 1,2- or 1,3-diols, creating five or six-membered cyclic boronate esters [136, 137]. PBA is in equilibrium with dissociated (tetrahedral) and undissociated (trigonal) forms in aqueous medium. The 1 : 1 binding model became dominant and shifted the equilibrium from an uncharged to a negatively charged form, causing an increase in the ionization degree and generated a Donnan potential for microgel swelling [122]. Zhang and coworkers investigated the kinetics of swelling behavior of P(NIPAM–AAPBA) microgels through a turbidity process [131]. This process occurred in a time of 10^2 seconds, slower than temperature-induced (de)swelling of PNIPAM microgels. They reported that the slow reaction between functionalized PBA and glucose, which was determined to be the rate-determining step, caused the slow swelling of glucose-responsive microgels. However, the (NIPAM–AAPBA) microgels were thought to be better choices for biosensor and drug delivery applications, in comparison with macroscopic hydrogels with the same networks, because swelling occurred much faster than hydrogels (~several hours) [69]. This kind of microgel was also applied to self-regulated insulin release [132, 138].

Contraction-type microgels are another type of glucose-responsive system, which shrinks in the presence of glucose and has drawn significant interest in self-regulated insulin release. The microgel particles were used as chemical valves for the membrane pores [139, 140]. As the glucose concentration increased, the microgels shrunk and opened the pores, thus improving the drug diffusion rate. Although, the contraction-type glucose-responsive microgels are promising, they are not extensively investigated. The only reports to date are by Hoare and Pelton [73] and Ravaine and coworkers [141]. The ability of these microgels to either swell or shrink at different glucose concentrations has limited their applications in sensing and self-regulated insulin release [73, 141]. To improve this, Zhang and colleagues synthesized P(NIPAM-2-AAPBA) contraction-type microgels via modification of the P(NIPAM-AAc) microgel with 2-AAPBA [122]. The size of the microgel shrank uniformly upon glucose addition through formation of a 1 : 2 binding model. The glucose sensitivity of the microgels was not influenced by a change to physiological pH, whereas it was affected by PBA concentration, ionic strength, and temperature. In addition, P(NIPAM-2-AAPBA) showed good glucose sensitivity at ionic strength and physiological pH. Moreover, interference of sugars, including fructose, galactose, and lactate, was insignificant.

PBA-based microgels can also be designed to incorporate fluorophores within their networks. In the presence of both PBA and glucose, the microgel fluorescence is changed. Zenkl et al. prepared PBA-PNIPAM microgels that included fluorescence resonance energy transfer (FRET) donor–acceptors [126]. The fluorescent acceptors and donors were in close proximity in the absence of glucose and fluorescence of the donor was quenched. The incorporation of glucose increased both swelling and the distance between the fluorescent acceptor and donor, thereby decreasing the FRET effect. However, this microgel suffered from several disadvantages, such as very slow swelling, poorly resolved emission peaks, and very high content of carbohydrates required to operating swelling. To address this problem, a novel fluorescent microgel was developed by Liu and coworkers [127]. P(NIPAM–AAPBA) microgels were synthesized with 4-(2-acryloyloxyethylamino)-7-nitro-2,1,3 benzoxadiazole as a FRET donor (NBDAE) and rhodamine B as the FRET acceptor (RhBEA) through emulsion polymerization. The resultant microgel indicated two emission peaks observed for both NBDAE and RhBEA moieties. Glucose and temperature-induced swelling was monitored by FRET efficiencies between the donor and acceptor. Wu et al. synthesized an optical glucose sensor through the copolymerization of nanostructure fluorescent CdSe quantum dots (QDs) with a p(NIPAM-AAm-PBA) microgel [142]. The microgels exhibited swelling and shrinking in response to varying glucose concentrations, which led to the fluorescence in the system being reversibly quenched. The QD-based microgels had excellent characteristics, including high quantum yield, size-controlled fluorescence properties, easy functionalization, potential biocompatibility, tunable dimension, uniform size distribution, fast response time, and stability against photobleaching from QDs. This system was further enhanced by the same research group [143]. In microgels, the use of a 2-acrylamidomethyl-5-fluorophenylboronic acid (FPBA), with diminished interference and a lower pK_a, enabled the system to monitor glucose at physiological pH. Zhou and coworkers synthesized multifunctional hybrid nanogels via *in situ* polymerization of temperature- and pH-responsive hydroxypropyl cellulose–poly(acrylic acid) (HPC–PAA) polymers with CdSe QD nanoparticles [144]. The HPC chains stabilized the CdSe QDs, whereas the pH-responsive PAA chains improved the surface physiochemical features of CdSe QDs, due to their ability to control the swelling/shrinking behavior of the gel networks. HPC–PAA–CdS nanogels indicated a pH-responsive emission at 741 nm and a visible excitonic emission at 592 nm, which make them valuable for sensing and bioimaging applications. In addition, high stability of the nanogels for drug delivery resulted in high drug loading as well as pH-responsive prolonged release.

The core–shell nanostructure was also an effective way to synthesize microgels with excellent properties. They were fabricated via two-stage free radical polymerization [145]. PBA microgels with core–shell structures were prepared by direct incorporation of a P(NIPAM-*co*-3-AAPBA) shell to a PNIPAM core [138] or by modification of a core–shell microgel [129, 130, 146]. The glucose-induced swelling degree was reduced due to the presence of PBA groups only in the shell or core. By contrast, the shell could control the rate of drug release and the core may act as drug reservoir, thus making them favorable candidates for controlled release systems [130]. Sun et al. developed thermo- and glucose-responsive microgels with core–shell structures for controlled release of insulin [147]. The silica (SiO_2) nanoparticle core was covered by a colloidally stable p(NIPAM-*co*-AAPBA-*co*-Dex-Ma) shell through radical polymerization.

As the temperature increased, the shell underwent a distinct transition from a swollen to a shrunken state. Addition of Dex-Ma as a cross-linker improved the glucose-sensitivity, biocompatibility, and hydrophilicity under physiological conditions. The insulin release was controlled by glucose concentration, whereas glucose sensitivity was dependent on the pH. Wu and coworkers synthesized a polymer-inorganic hybrid microgel with core–shell structure, with silver nanoparticle (Ag) cores coated with poly(3-acrylamidophenylboronic acid-*co*-2-(dimethylamino)ethyl acrylate) [p(3-VAPBA-*co*-DMAEA)] inthepresence of *N*,*N*′-methylenebisacrylamide (MBA) as a cross-linker [148]. The volume phase transition behavior of the microgels was designed for uniform swelling and shrinking through addition of glucose at physiological pH. This property was obtained by changing the feeding ratio $r_{\text{DMAEA/3-VAPBA}}$ of DMAEA (resulting in positive charges through ionization) to 3-VAPBA (resulting in negative charges via the formation of glucose–boronate complex). It was demonstrated that the balance of positive and negative charges, at certain concentrations of glucose, played a significant role for this characteristic. Upon addition of glucose, the swelling- and shrinking-responsive microgels were turned off and turned on, respectively. Results revealed that the shrinking-responsive microgels improved the glucose-sensing efficiency in both urine and tear fluid.

Most of the aforementioned microgels are not degradable and thus cannot be easily removed from blood circulation in the body. To solve this, Zhao et al. recently fabricated core–shell microgels by copolymerization of PEG diacrylate (PEGDA), methoxyl PEG acrylate (mPEGA), pentaerythritol tetra(3-mercaptopropionate) (QT), and AAPBA through one-pot thiol–ene click chemistry [149]. Through the addition of glucose, the fluorescence intensity decreased, while the hydrodynamic radius (R_h) of the microgel increased at pH of 7.4. Moreover, the alizarin red S (ARS) release from microgel was highly affected by glucose concentration, showing a higher and faster release rate upon glucose addition to the PBS. In addition, the microgels exhibited high glucose-selectivity due to presence of PBA and good biocompatibility, which makes this system a potential candidate for self-regulated drug delivery. Liu and coworkers synthesized multifunctional microgels of NIPAM, AAPBA, and 2-dimethylamino ethyl methacrylate (DMAEMA) via precipitation emulsion polymerization, using *N*,*N*′-bis(acryloyl)cystamine (BAC) as a degradable reductive cross-linker [150]. This microgel, with an average size of ~250 nm, possessed not only the advantages of controlled insulin release and high stability at physiological pH, glucose, and temperature (pH 7.4, 0.15 M NaCl, 37 °C) but also slow degradation with time, which resulted in better biocompatibility. Therefore, this multifunctional microgel may be a promising candidate for self-regulated drug delivery and nanobiosensor applications.

Wu et al. prepared an injectable glucose-responsive nanogel for controlled insulin release [151]. PNIPAM-PBA monodisperse nanogels were fabricated in the presence of maleic acid-Dex as a cross-linker. The nanogel swelled with an increase in glucose concentration. The insulin release was controlled by the glucose concentration in the release medium and dex content in the nanogels. Based on the *in vivo* results, the nanogels containing insulin exhibited good biocompatibility, caused a reduction in blood glucose levels in diabetic rats, and led to constant blood glucose levels without blood sugar fluctuations.

3.2.3.4 Multilayer Hydrogel Films

LbL assembly is a simple method for the fabrication of multilayer networks via the intermittent dipping of a substrate within the polymer solution, which contains supplementary functional groups [152, 153]. The driving force for preparation of LbL films is electrostatic interactions [154, 155], hydrogen bonding [156–159], covalent bonding [160, 161], hydrophobic interactions [162, 163], charge transfer interactions [164, 165], and biological recognition [166]. Compared with other methods, the LbL technique has excellent advantages such as versatility, ease of fabrication, and high control of the film structures (e.g. composition, thickness, size, and shape). Therefore, they have been applied in various fields of studies [167]. In addition, as drug delivery systems, these films acted as diffusion barriers [168] and were a building block for the drugs [169].

PBA-based LbL films were first synthesized by De Geest et al. via electrostatic interactions between the polycation of PBA and a polyanion [170]. The 3D hollow capsules and planer LbL hydrogel films were obtained by LbL assembly of a polyanion with poly(3 acrylamidophenylboronic acid-*co*-dimethyl aminoethyl acrylate) as a polycation. The capsules were completely disassembled in less than five minutes, upon contact with a certain amount of glucose. The mechanism was attributed to the addition of glucose, which changed the electrostatic interactions between the polyanion and polycation. However, the pK_a of this thin film was higher than physiological pH, which limits its *in vivo* applications.

The modification of a thin film with PBA functionalities was another method for fabrication of PBA-based LbL films. Zhang and coworkers synthesized self-assembly LbL films of poly(vinyl pyrrolidone) (PVP) and PAA via a hydrogen bonding driving force [169]. The films were cross-linked with ethylenediamine and EDC and modified with isopropylamine and 3-aminophenylboronic acid, simultaneously. The films indicated temperature and pH-responsive swelling behavior, and swelling was increased with glucose or fructose. The ARS, as a model drug, bound covalently to functional PBA in the films and displayed faster release in the presence of glucose. The mechanism of drugs containing diol structures was thought to be through binding and release from films, akin to ARS release.

In situ formation of covalent boronate ester bonds between a diol-including polymer and a PBA-including polymer was also used for fabrication of glucose-responsive LbL assembly hydrogel films [171–175]. Applying this method, Ishihara and coworkers prepared multilayered hydrogels on the surfaces of titanium alloys (Ti), using alternating layers of a hydrophilic PVA and phospholipid polymer (PMBV) [172]. PMBV showed a polymeric aggregated network with hydrophobic domains in an aqueous media, which was able to sufficiently dissolve paclitaxel (PTX), a hydrophobic anticancer drug. In the multilayer hydrogels, the PMBV-based PTX functioned as a reservoir. The PTX was loaded in the bottom layer (bottom-type) or the top layer (top-type) of hydrogels to create various release behaviors. Other PMBV–PVA layers acted as diffusion barriers in the absence of

PTX. The cell culture results revealed that the bottom-type hydrogels could maintain stable cell cytotoxicity (through sustained release of PTX), whereas the top-type hydrogels, with an initial burst release of PTX, indicated inordinate inhibition of human cell proliferation after more than five days. The PTX release in the cell culture medium was controlled by both the dissociation and diffusion barrier of the hydrogel via an exchange reaction between D-glucose and PBA functional groups. It was concluded that the concentration of PTX had a significant effect on cell proliferation. They suggested that this LbL hydrogel had the potential to be applied in chronic biomedical implants.

Borax-mediated LbL self-assembly was also introduced into the glucose-responsive hydrogel films. In a novel example, Manna and Patil utilized a borax complex with PVA and chitosan through an LbL assembly technique both on a colloidal substrate and on planar substrate [176]. Although PVA is an uncharged polymer, it was negatively charged after forming a complex with borax. The negatively charged PVA (polyanion) and chitosan (polycation) were successfully assembled with the borax complex through electrostatic interactions. They also investigated the release behavior of doxorubicin (Dox), an anticancer drug, into both an isotonic solution (pH 7) and PBS buffer (pH 7.4). The Dox release was pH-dependent and showed higher release in pH 7.4 due to the higher solubility of Dox at this pH. They further studied the influence of glucose concentration on a similar network and explored the decomposition of multilayer hydrogels in the presence of glucose [177].

3.3 Protein-Responsive Hydrogels

3.3.1 Enzyme-Responsive Hydrogels

Enzyme-responsive hydrogels are chemically cross-linked polymeric networks, which form by attaching enzyme-responsive molecules to the hydrogel, followed by subsequent exposure to sensitive enzymes. They usually form under mild conditions, and most of the enzymes used in preparation of these types of hydrogels are naturally occurring [178–183]. Enzymatic reactions are catalyzed at moderate temperature, with neutral pH, and in an aqueous medium in the presence of most enzymes. These features provide an excellent strategy for cross-linking natural polymers that cannot tolerate unfavorable chemical conditions. Furthermore, by regulating the enzyme activity, the direct control of polymerization can be achieved. Additionally, these reactions can be applied for the formation of *in situ* polymerizing hydrogels without causing toxicity or unwanted side reactions, due to the high selectivity and specificity of enzyme [184]. *In situ*-formed hydrogels are regarded as minimally invasive and have been extensively applied to tissue engineering, drug delivery, and cell encapsulation [185–188]. Enzyme-responsive hydrogels are also used in the fabrication of degradable biomaterials and native extracellular matrices (ECMs) [189]. ECM hydrogels have several favorable features, including remodeling response, site specificity, and ability to fill irregular-shaped spaces, and are thus used for a variety of biomedical applications, such as in the peripheral nervous system [190], skeletal muscle [191–193], and cardiac tissue [194].

3.3.1.1 Formation of Enzyme-Responsive Hydrogels

The chemically cross-linked enzyme-responsive hydrogels can be formed in the presence of transglutaminases (TGs), as a mild and naturally cross-linked protein *in vivo*. These blood-clotting enzymes, which are highly stable against proteolytic degradation, catalyze the formation of chemical bonds between the side chains of lysine (Ly) residues and glutamine (Gl) residues within or across peptide chains [195].

Sperinde and Griffith exploited this system to prepare cross-linked hydrogels of lysine-containing polypeptides and functionalized PEG [196, 197]. Hubbell and coworkers synthesized fibrin hydrogels, using TG factor XIIIa to cross-link bi-domain peptides with fibrin during polymerization [198, 199]. Controlled release of beta-nerve growth factor (β-NGF) was observed under mild conditions, suggesting the hydrogel system is a potential candidate for nerve regeneration [199]. In novel research, Messersmith and colleagues reported thermal activation of factor XIII, which caused an enzymatic *in situ* cross-linking of four-arm star PEG–peptide hydrogels [200]. Schlosshauer and coworkers developed an injectable two-component hydrogel consisting of gelatin-loaded AraC (cytosine arabinoside) and TG, for the treatment of incontinence and erectile dysfunction in prostate cancer patients [201]. The *in situ* formation of chemically cross-linked hydrogels, which adhered to abdominal tissue and improved nerve regeneration, typically occurred within two minutes. It was observed that the gelatin–TG-loaded AraC hydrogel effectively suppressed the growth of prostate carcinoma and sciatic fibroblast cells. However, this chemotherapy drug had no effect on axonal outgrowth. The hydrogel not only showed suitable drug release and coagulation kinetics but also exhibited great biocompatibility and viscoelastic and adhesive properties. Recently, Sun and colleagues introduced a novel method for the preparation of *in situ*-forming hydrogel scaffolds [202]. The large, macroporous gelatin hydrogels containing Huh7, NIH 3T3, and ADSC cells were obtained using microbial transglutaminase (mTG), which prevented cell dispersion after delivery with no toxicity. The hydrogel filaments served as preliminary microvasculature systems after the consistent assembly of human umbilical vein endothelial cells on their surfaces. This cell-laden hydrogel was suggested to be a suitable candidate for cell seeding, mass and oxygen delivery in 3D tissue engineering, and cell culture applications.

Another important enzyme, exclusively found in animal tissues, plants, and fungi, is tyrosinase. Tyrosinase is a copper-containing monooxygenase enzyme that catalyzes the hydroxylation of phenol substrates (e.g. tyramine or tyrosine) to catechol derivatives such as dopamine and 3,4-dihydroxyphenylalanine (L-DOPA), which is then oxidized to orthoquinone materials [203]. Tyrosinase enzymatic reactions are highly important in the synthesis of the pigment melanin, which removes reactive oxygen species (ROS), gives color to hair and skin, and protects DNA in skin cells exposed to ultraviolet (UV) [203]. A high dose of tyrosinase results in an excessive aggregation of melanin and leads to acute skin diseases such as freckles, age spots, and melasma and malignant melanoma in humans [204–206]. On the other hand, a lack of tyrosinase results in oculocutaneous albinism type I skin disorder [207]. Tyrosinase may participate in dopamine neurodegeneration and neurotoxicity related to Parkinson's disease [208].

Payne and coworkers developed pH-responsive hydrogels via enzymatically cross-linking chitosan to a green fluorescent protein (GFP) using tyrosinase [209]. By applying voltage, the hydrogel was selectivity deposited onto a micropatterned surfaces. They also investigated the role of tyrosinase and mTG enzymes in the formation of chitosan and gelatin hydrogels [210]. In the presence of mTG, the chitosan–gelatin hydrogels underwent the reverse process and were stronger. By contrast, the chitosan was shown to be necessary for the formation of tyrosinase-catalyzed hydrogels, which contained a collagen-like coil-to-helix transition, with faster hydrogel formation, compared to mTG enzyme. It was concluded that the concentrations of chitosan and gelatin had a significant effect on the strength of mTG and tyrosinase-catalyzed hydrogels. A similar concept was employed by Jin et al. for the *in situ* formation of hydrophilic glycolic acid (GA)–chitosan–tyrosinase hydrogels, under mild conditions in the presence of either HRP or tyrosinase [211]. The hydrogel gelation time was highly dependent on the Ds of tyrosinase and polymer and enzyme concentrations. HRP-cross-linked hydrogels showed faster gelation time and a greater storage modulus. However, better biocompatibility, biodegradability, higher elasticity, and gel content were observed for tyrosinase hydrogels. Tyrosinase was also combined with other cross-linkers for the *in situ* formation of hydrogels [212]. Maiti and coworkers utilized a mixture of p-cresol and tyrosinase cross-linkers for *in situ* formation of gelatin–carboxymethyl chitosan (CMC)-loaded nano-hydroxyapatite (nHA) hydrogels under physiological temperature. Based on the results, the stability of the hydrogel was controlled by both CMC concentration and cross-linking density. The tyrosinase hydrogels that can be formed under physiological conditions with excellent properties such as low toxicity are regarded as desirable candidates for biomedical applications such as bone, cartilage, and tissue engineering.

HRP-catalyzed hydrogels was also an effective way to make the system available for the *in situ* formation of enzymatically cross-linked hydrogels. These were able to catalyze the coupling of phenol or aniline derivatives via hydrogen peroxide (H_2O_2). HRP is regarded as the most significant enzyme in the preparation of hydrogel due to its rapid hydrogelation, excellent biocompatibility, and great stability [213]. In addition, by manipulating the HRP activity, the porous structure and mechanical property can be controlled easily [214]. Ren et al. developed novel injectable hydrogels composed of poly(L-glutamic acid), tyramine, PEG, HRP, and H_2O_2 under mild conditions [214]. They also investigated the gelation process, swelling, and mechanical and biocompatibility properties of hydrogels. The resultant enzymatically cross-linked hydrogels possessed good biocompatibility. A similar strategy was employed by Park and coworkers for fabrication of *in situ*-forming tissue adhesive hydrogels [215]. Two different gelating derivatives with various phenolic concentrations were obtained in the presence of HRP/H_2O_2. These hydrogels not only exhibited 2–3 times higher tissue adhesiveness than commercially available fibrin glues but also showed good biocompatibility and biodegradability properties. In addition, they could be applied as both sprayable and injectable hydrogels.

HRP/H_2O_2-catalyzed hydrogels have been widely recognized as suitable materials for wound dressing applications due to their mild preparation conditions (e.g. physiological temperature and pH) [216–218]. However, the hydrogelation process for these hydrogels occurs just before application to the wounds. To solve this, Sakai et al. reported the on-wound and *in situ* formation of PVA-containing phenolic hydroxyl moieties (PVA–Ph) through the co-enzymatic hydrogelation process of HRP and Gox in the presence of glucose in the exudate [219]. Results revealed that the PVA–Ph hydrogel could rapidly form within 20 seconds, and possessed a high durability to stretching and compression. More importantly, this provided a faster wound healing process than recent conventional hydrogel wound dressings. Similar concepts have been developed for the *in situ* formation of enzyme-catalyzed cross-linkable hydrogels based on different polymeric networks, including alginate [220, 221], Dex [180, 222], chitosan [223, 224], gelatin [182, 225], and HA [226, 227]. Additionally, although these examples are hydrogels that rely on the use of TG, tyrosinase, and HRP enzymes, it is also possible to use other enzymes to catalyze hydrogel formation such as phosphopantetheinyl transferase [228], esterase [229], lysyl oxidase [230], or acid phosphatase [231].

3.3.1.2 Hydrogel Degradation

Degradable hydrogels are of particular interest for a wide range of biomedical applications, as they are removed from the body in a relative timely fashion, without the requirement for a surgical procedure. They are able to control the release of therapeutic agents [2], enable cell migration and proliferation, facilitate synthetic matrix remodeling [232], and permit infiltration of blood vessels [233, 234], during gene and drug delivery, cell culture, and tissue engineering applications. Their degradation can be varied, from rapid to prolonged rates, depending on the polymer nature, density of cross-linking, and added degradable agents [235–237]. It is worth mentioning that the *in vivo* residential time and biocompatibility of hydrogels are associated with their degradation rates. Hydrogels with rapid degradation rates are removed from the body in a short period compared to prolonged ones. Typically, a hydrogel with a short residence reduces the risk of undesired side effects, due to its rapid clearance from the body, whereas prolonged hydrogel degradation increases cytotoxicity. Hydrogel degradation can be accomplished by either bulk degradation, surface erosion, or a combination of the two. Low cross-linked polymeric networks, such as physically cross-linked hydrogels possessing relatively high permeability and high water content, exhibit bulk degradation. Surface erosion, typically occurs in highly cross-linked hydrogels for which the rate of water and enzyme diffusion is relatively low.

Enzymatically degradable hydrogels can be obtained by the cleavage of biodegradable polymer backbones. Hogvaard and Brondsted were the first to identify the degradation of dextran hydrogels in the presence of dextranase [238]. Since this pioneering research, naturally occurring biodegradable polysaccharides including hyaluronic acid [227], alginate [239], dextrane [240], gelatin [241], and guar galactomannan [242] have been extensively utilized in the preparation of enzymatically degradable hydrogels. For instance, Yu and Pishko reported the fabrication of pH-responsive biodegradable hydrogels composed of dextran maleic acid (Dex-MA) and glycidyl methacrylated dextran-*co*-AA (Dex-GMA-*co*-AA) for targeted drug delivery in the colon [243]. An interesting feature of this system was that the pH and dextranase enzyme could significantly modulate the swelling, release mechanism, and degradation of Dex-based hydrogels. Typically, both hydrogels were completely degraded in simulated intestinal fluid (SIF), within 8–10 hours, in the absence or presence of dextranase, while they remained stable in simulated gastric fluid (SGF). In addition, through the introduction of dextranase, Dex-MA hydrogels were completely degraded and showed degradation-controlled release, while Dex-GMA-*co*-AA with a higher swelling ratio exhibited diffusion-controlled release. Ren and coworkers developed site-specific cancer drug delivery systems based on enzymatically degradable HA hydrogels [244]. In the presence of hyaluronidase, the HA hydrogels were degraded, and cancer cell growth was significantly inhibited. Considering the fact that hyaluronidase expression is markedly higher in many carcinomas, enzymatically degradable hydrogels may be applied for the synthesis of various targeted drug delivery systems for cancer therapy. Jin et al. synthesized biodegradable hydrogels in situ via the enzymatic cross-linking of tyrosinase with chondroitin sulfate–tyramine (CS-TA) [245]. The degradation of CS-TA, via chondroitinase ABC, over a specific time period could be adjusted by changing the concentration of tyrosinase. This novel type of *in situ*-forming hydrogel showed good biocompatibility, biodegradability, elasticity, a porous network, and sustained-release behavior. These properties indicate that this system will perform well in drug delivery and tissue engineering applications.

The enzymatic cleavage of cross-linkers is also an effective strategy to obtain degradable hydrogels. Anseth and coworkers designed cell-degradable PEG hydrogels through thiol–ene photopolymerization for differentiation and culture of human mesenchymal stem cells (hMSCs) [246]. These novel degradable hydrogels formed in less than two minutes, at neutral pH, and were modified by the enzymatic breakdown of matrix metalloproteinase (MMP) as a cross-linker. The hMSCs were not affected by the rate of MMP degradation and survived well in the hydrogels, showing cell spreading and proliferation ability. Encapsulated hMSCs in degradable hydrogels, which were cultured in adipogenic, chondrogenic, or osteogenic medium, indicated higher cell differentiation in comparison to nondegradable hydrogels. Landfester and colleagues reported the preparation of enzymatically degradable PAAm nanogels containing dextran methacrylate cross-linkers via an inverse mini-emulsion technique [247]. The resultant nanogels could be degraded by cleavage of methacrylate groups. Based on the results, both the DS of the dextran-methacrylate and methacrylate units played a significant role in the degradation properties of nanogels. Fascinating research by Burdick and coworkers described the first example of MMP degradable and injectable hydrogels for recombinant tissue inhibitor of MMP (rTIMP-3) delivery in the presence of MMP activity [248]. It was observed that the cleavage of peptide cross-links by MMP caused rTIMP-3 release within the tissue. The results highlighted that the release of rTIMP-3 from hydrogels could effectively minimize the activity of MMP in the myocardial infarction (MI) area without increasing the concentration of rTIMP-3.

3.3.1.3 Hydrogel Swelling

Enzyme-induced hydrogel swelling can be achieved by manipulating the sol–gel transitions or pore sizes. These morphological transformations can lead to changes in shape, size, and properties of hydrogels. This concept was extensively studied by the Ulijn research group by using poly(ethylene glycol acrylamide) (PEGA)-based hydrogels [249–252]. They synthesized hydrogel beads through modification of PEGA polymers with peptides composed of a charged section and a cleavable linker part [249]. The cationic section of the peptides resulted in electrostatic repulsion of polymer chains, causing hydrogel swelling, which improved molecular accessibility. However, upon subjection to proteases (enzymes with ability to break down peptide bonds), these charged sections were removed, causing hydrogel shrinkage. Ulijn and coworkers further reported the fabrication of PEGA-functionalized peptide hydrogels for the release of payload (fluorescently labeled Dex and avidin) through enzyme-triggered charge-induced swelling [250]. The obtained hydrogels exhibited actuation and sensing characteristics due to the zwitterionic (neutral) peptides comprising two oppositely charged amino acids and an enzyme cleavable linker, respectively. It was concluded that the enzymatic cleavage of the sequence, which released doubly positive charged amine groups within the polymers, resulted in hydrogel swelling. In a recent example, a novel type of degradable enzyme-responsive hydrogel-induced swelling consisting of a PAAm-functionalized dextran methacrylate cross-linker was synthesized [247]. The swelling of the nanogel was observed by cleavage of a Dex cross-linker in the presence of dextranase.

3.3.1.4 Supramolecular Hydrogels

Supramolecular hydrogels are formed by self-assembly of small molecules (hydrogelators) through non-covalent interactions such as metal–ligand, π–π stacking, electrostatic interactions, and hydrogen bonding [253, 254]. Thanks to self-assembly and non-covalent interactions of hydrogelators, supramolecular hydrogels are biocompatible and biodegradable and have the capability to respond to external stimuli such as anions [255, 256], cations (ionic strength) [255, 257–259], and pH [257, 260, 261]. Moreover, due to the small molecular weights of hydrogelators (usually below 2000 Da), their clearance from the body is more effective than that of polymeric hydrogels, which possess higher molecular weights [262]. In addition, the cellular toxicity and impurities of cross-linking reactions in polymeric hydrogels could be highly reduced through the formation of supramolecular hydrogels with ordered structures [263]. These promising characteristics have resulted in the extensive use of supramolecular hydrogels in a wide range of biomedical applications including regenerative medicine [257, 258, 264], cell encapsulation [265, 266], drug delivery [267–270], and wound healing [271, 272].

Enzyme Self-Assembly The main common type of enzyme-responsive supramolecular hydrogels is based on the enzymatic formation of self-assembled networks. Peptides or peptide amphiphilic materials are mostly used for the construction of these enzyme-responsive hydrogels due to their unique capability to participate in intermolecular interactions [273]. The first enzyme-responsive supramolecular hydrogel was introduced via the dephosphorylation of the *N*-[fluorenylmethyl(oxy) carbonyl] tyrosine phosphate (Fmoc-*p*Y) by alkaline phosphatase [273]. In this system, the breakdown of the enzymatic gelation bond was found to change the hydrophilicity of the precursor (gelators). Afterward, the polymeric networks containing phosphatase-responsive peptide-based hydrogelator materials became the main stream of the enzyme-responsive hydrogels such as Fmoc-F*p*Y [274], Nap-FFGE*p*Y [275], acyl-YYY*p*Y-OMe [276], and Nap-FFpY [231] (F: phenylalanine; Y: tyrosine; Nap: naphthyl; G: glycine; Y-OME: amino acid ester).

Changing the availability or solubility of the hydrogelator was an effective way to adjust the self-assembly of hydrogels. In a notable example, Ulijn and coworkers reported the synthesis of supramolecular hydrogels using a biocatalyst (subtilisin enzyme) [277]. As the enzyme concentration increased, the self-assembly formation, as well as the supramolecular order of hydrogels, was significantly improved, and surprisingly, the hydrogelator concentrations remained constant. These findings revealed that the enzyme remarkably affected the network and self-assembly process of the hydrogels. Zhao and colleagues also reported a similar strategy to obtain enzyme-triggered supramolecular hydrogels via the addition of phosphate groups within thermoresponsive block copolymers [278]. They prepared an ABA triblock copolymer, P(DEGEA-*co*-OPBA)-*b*-PEG-*b*-P(DEGEA-*co*-OPBA), composed of two random blocks of ethoxydi(ethylene glycol) acrylate (DEGEA) and ((dihydroxyphosphoryl) oxy)butyl acrylate (OPBA) with PEG as a middle block polymer, through atom transfer radical polymerization (ATRP), using a PEO microinitiator. The dephosphorylation study of P(DEGEA-*co*-OPBA) block at 37 °C showed a decrease in the critical micellization temperature (CMT) or LCST of the ABA copolymers due to the cleavage of phosphate groups, which reduced the hydrophilicity of hydrogels. The breakdown of phosphate groups in the presence of 7.9 wt% aqueous solution of polymer at 37 °C caused an enzyme-induced thermoresponsive sol–gel transition. The authors believed that the gelation process of ABA triblock copolymer (which took more than five hours) can be enhanced by manipulating several factors such as enzyme content and phosphate concentration and through selecting a hydrophilic thermoresponsive polymer with a lower LCST, among others.

A β-sheet forming peptide was another strategy to create enzyme-responsive supramolecular hydrogels. This feature, with sequence-defined manner capability, can be employed for the fabrication of well-tunable self-assembled hydrogel networks. Ulijn and colleagues reported the synthesis of supramolecular hydrogels by self-assembly of Fmoc-diphenylalanine using π–π interlocked β-sheets [279]. This novel system was proposed to improve the biological availability of bioactive ligands by placing them on the surface of the structures; hence, it could be used for several applications such as 3D cell culture. Zhou et al. developed a minimalist approach for the synthesis of peptide-based bioactive scaffolds via the self-assembly of a combination of Fmoc-FF and Fmoc-RGD (arginine–glycine–aspartate) as building blocks into β-sheets interlocked via π–π interactions [280]. The presence of RGD sequences as a biological ligand on the surface of fiber mimicked certain vital properties of the ECM. In addition, the RGD hydrogels were found to enhance the adhesion of dermal fibroblast cells; therefore, these might be suitable candidates of cost-effective scaffolds for anchorage-dependent cells.

In addition to β-sheets forming peptides, the self-assembly process of enzyme-responsive hydrogels may be driven by either a thermodynamically or a kinetically controlled protease-catalyzed reaction. For example, Ulijn and coworkers demonstrated the self-assembly of aromatic dipeptide hydrogels under thermodynamic control [281]. To investigate the sequence/structure relationships of peptides, four relatively similar dipeptide amphiphiles, including Fmoc–TL–OMe (TL), Fmoc–TF–OMe (TF), Fmoc–SL–OMe (SL), and Fmoc–SF–OMe (SF), were utilized. All dipeptides, TL, TF, SL, and SF were self-assembled into β-sheets, which were interlocked by the π–π stacking of Fmoc. The results revealed no kinetically trapped aggregates with delicate structural differences. A main advantage related to the thermodynamically controlled peptide synthesis is the formation of stable structures. However, the relatively slow self-assembly process, high concentration of enzyme and substrate, and side reactions restrict their applications [282, 283]. By contrast, kinetically controlled protease-catalyzed reactions can take place in a rapid time at a relatively low enzyme concentration, without considerable reshuffling of the hydrogelator sequence. In a recent example, Gross and colleagues designed a novel method for the rapid oligomerization of the hydrophilic ethyl ester (OEt) dipeptide lego, KL-OEt, in which K is a hydrophilic lysine and L is a hydrophic leucine, triggered via protease catalysis [284]. The hydrogel was formed in less than 10 seconds at alkaline pH upon enzyme addition. It was concluded that the use of both α-chymotrypsin, with high specificity, and kinetically controlled peptide reactions played a significant role in the generation of alternating peptide sequences and fast reaction rates.

Enzyme-Mediated Disassembly The supramolecular disassembly of enzyme-responsive hydrogels can be achieved either via destruction of building blocks or by specific molecules or peptide sequences that lead to the cleavage of building blocks. Hartgerink and colleagues developed an example based on the enzymatic disassembly of peptide amphiphile supramolecular hydrogels, in the presence of MMP-2, for tissue engineering [285]. This novel self-assembly hydrogel was slowly broken down into shorter fibers upon the cleavage of peptide sequences through the addition of MMP-2. Langer and coworkers exploited a similar strategy for fabrication of self-assembled MMP-cleavable arginine–alanine–aspartate–alanine (RADA) peptide templates into nanofibrous hydrogels [286]. The hydrogel was found to mimic the natural ECM and was degraded upon exposure to MMP-2. In another example, enzyme-mediated disassembly was developed via the introduction of an MMP-2-sensitive cleavable motif within the β-sheet forming multi-domain peptide hydrogels [287]. These self-assembled amphiphilic peptides possessed a modular ABA triblock motif with the B block flanking the A block, creating self-assembly. The gel indicated both proteolytic degradation in the presence of collagenase IV and specific degradation by MMP-2. It was concluded that using a mixture of RGD (a cell adhesion motif) and MMP-2 in the hydrogel enhanced cell spreading, viability, and migration. Schneider and colleagues reported the synthesis of MMP-13-sensitive self-assembling β-hairpin peptide hydrogels [288]. In their work, the self-supporting hydrogels were obtained by triggered folding and self-assembly of degrading peptides (which were defined as a mixture of five peptides with different degradation properties upon addition of MMP-13) under mild condition (150 mM, pH 7.6). The MMP-13 proteolysis of degrading peptide hydrogels not only exhibited a time (14 days), sequence, and network morphology-dependent behavior but also showed well-defined sequence location. Degrading peptide hydrogels indicated higher sensitivity to MMP-13 in comparison to MMP-3.

The examples described earlier mainly deal with enzyme-triggered disassembly of supramolecular hydrogels in which MMP could break down the peptide sequences and consequently alter the structure of polymeric networks. On the other hand, two significant issues need to be addressed for the application of these types of supramolecular hydrogels. First, as MMP sensitive peptides were introduced into the building blocks of these systems, their exact effect on the self-assembly process might not be clear, resulting in difficulties in anticipating the properties of assembled hydrogels such as network morphologies and structural characteristics. Hence, an absolute characterization and design of the assembly hydrogels would be required for the use of these peptide-based supramolecular systems with various types of MMPs. Second, the encapsulation of a peptide substrate within supramolecular hydrogels, and its accessibility to MMPs, would affect the enzyme–substrate formation. For example, an embedded peptide with high sensitivity to MMP facilitates the formation of enzyme–substrate mixtures, which further simplify the enzymatic reaction [289]. Consequently, the peptide sequence, which is deeply incorporated into the assembled hydrogels with low accessibility to the targeted MMP, would probably have a poor degradation rate. Thus, it would be highly favorable to develop a peptide-based hydrogel with the ability to separate MMP degradability from the assembly process within supramolecular networks [289].

To support this, a novel type of MMP-2-sensitive supramolecular filament was designed by Cui and coworkers using amyloid-derived amphiphilic peptides [289]. The addition of an oligoproline sequence caused the immediate destruction of the assembled β-sheet-forming nanostructures, through the cleavage of the peptide cross-linkers. Moreover, the peptide substrates on the filament surfaces were found to be accessible for targeted MMP degradation. Surprisingly, the assembled morphology of the filaments remained unchanged in response to different types of cross-linkers at pH 7.5. However, these filaments showed a slow degradation due to the breakdown of peptide cross-linkers upon incorporation of MMP-2. Here, the molecular assembly process was nicely separated from the enzymatic cleavage reaction; therefore, the authors stated that this filament might be avidly explored as an avenue for enzymatically degradable networks and could be potentially utilized with other types of enzymes and MMPs.

In a departure from the aforementioned examples, which are mainly concerning disassembly of peptide-based supramolecular hydrogels in the presence of MMPs, other types of proteases can also be applied in the preparation of enzymatically degradable hydrogels. For example, Tung and coworkers fabricated a protease degradable gelator consisting of two β-sheet forming peptide sequences and a protease degradable sequence [290]. The disassembly of the gel occurred through the breakdown of the peptide linker in the presence of urokinase plasminogen activator (uPA). Gel degradation was controlled by the concentration of uPA. Liu and colleagues reported the synthesis of a novel type of DNA-based self-assembly hydrogel via hybridization of the sticky ends of building blocks (Y-scaffold and linker) [291]. These hydrogels indicated reversible thermoresponsive behavior and were degraded in the presence of restriction endonucleases (restriction enzymes). Notably, the use of BamHI and EcoRI enzymes resulted in cleavage of the restriction sites of the linker, leading to a gel-to-sol transition of DNA hydrogels.

In addition to enzyme-induced disassembly of supramolecular hydrogels, Gajanayake and coworkers used proteolytic enzymes, including lipase (esterase), MMP-2, and MMP-9 for the release of the immunosuppressive drug tacrolimus (FK506) [292]. The hydrogel degradation occurred upon cleavage of the ester linkage of the triglycerol monostearate (TGMS) amphiphile by the lipase and MMPs. An interesting feature of this system was that the enzymatically degradable tacrolimus-laden hydrogel could significantly enhance long-term graft survival (up to 100 days), making them desirable alternative candidates for reconstructive surgery.

Reversible Self-Assembly Reversible/dynamic supramolecular hydrogels, with the enzymatic ability to both form and dissociate self-assembled hydrogels, are of particular interest for enzyme-responsive materials. A significant advantage of the reverse reaction is that no by-products except water are formed for peptide bonds. More importantly, they can provide spatiotemporal control of self-assembly and other properties of the hydrogels. The first example of a reversible supramolecular hydrogel was reported by Xu and coworkers, who utilized a kinase/phosphatase to adjust the hydrogel formation and the phosphorylation/dephosphorylation of the hydrogelator [275]. In the presence of adenosine triphosphates (ATPs), the kinase led to a gel-to-sol transition while the phosphatase caused a gel-to-sol transition in a Nap-FFGEY hydrogelator. Further research was reported by Stupp and colleagues by using a similar approach [293]. Here, it was observed that the specificity of employed enzymes in this nanostructure helped to create more accurate biomimetic hydrogels. This nanostructure exhibited gel-to-sol (phosphorylation) and sol-to-gel (dephosphorylation) transitions, respectively, upon treatment with kinase A and alkaline phosphatase. In addition, the disassembly of peptide nanostructure via kinase A caused the enzyme-triggered release of DOX as an anticancer drug.

Reversible supramolecular hydrogels can also be obtained using thermolysin and α-chymotrypsin. For example, Williams et al. fabricated reversible hydrogels through the self-assembly of Fmoc peptides by π–π interlocked β-sheets under thermodynamic control [294]. They utilized the thermolysin enzyme to create reversible building blocks. This system not only indicated fully reversible behavior but also exhibited stable self-assembled nanostructure and high control of the molecular self-assembly. Very recently, Ulijn and colleagues used α-chymotrypsin in the preparation of nonequilibrium biocatalytic self-assembled peptides, which showed reversible instability [295]. To obtain this, Nap-dipeptides were self-assembled into unidirectional nanofibers via a mixture of hydrogen bonding and π-stacking interactions. The use of α-chymotrypsin catalyzed transacylation for the creation and degradation processes during kinetic- and thermodynamically controlled reactions, respectively. They also employed the thermolysin enzyme to catalyze the dynamic dissociation of nonequilibrium hydrogelation and assembly. It was concluded that the critical gelation concentration (CGC) of the peptide played a key role on the final state of nanofibers (solution or hydrogel), and in either case, the self-assembly of Nap-peptide-NH_2 dictated this state.

3.3.2 Antigen-Responsive Hydrogels

An antigen-responsive hydrogel involves the highly selective and specific interaction of an antibody to its antigen. Antibodies (immunoglobulins) are spherical proteins that use special binding sites to identify and capture foreign molecules called antigens. Many materials are considered antigens by the body, including disease-causing microorganisms such as viruses, fungi, bacteria, and parasites as well as toxic substances such as poisons. Antibody structures consist of four polypeptide chains, containing two identical heavy chains with a molecular weight of 55 kDa and two identical light chains with a molecular weight of 25 kDa, which form a Y-shaped structure through disulfide bonding [296]. Antigen–antibody binding is related to human immune responses used to detect diseases in the body. These interactions are formed via non-covalent reactions including van der Waals forces, hydrophobic bonding, hydrogen bonds, electrostatic forces, or a mixture of all of these interactions [21]. Antibodies have been utilized for the detection of biological and nonbiological substances in a variety of immunological assays [297].

In a pioneer work, Miyata et al. synthesized antigen-responsive hydrogels by coupling *N*-succinimidylacrylate (NSA) to rabbit immunoglobulin G (IgG) [298]. In the presence of acrylamide, MBA, and goat IgG (as an antibody), the modified monomer was polymerized and resulted in the formation of a hydrogel cross-linked both by antigen–antibody interactions and chemically. The addition of free antigen (rabbit IgG) caused hydrogel swelling and loss of the antigen–antibody cross-linkers. Miyata et al. further enhanced this system by designing the reversible responsive hydrogel [299]. Both antibody (goat IgG) and antigen (rabbit IgG) were functionalized with the vinyl groups. To fabricate a semi-interpenetrating (semi-IPN) hydrogel, the modified antibody was polymerized with acrylamide, and the modified antigen was polymerized with MBA and acrylamide in the presence of the polymerized antibody. Addition of free antigen (rabbit IgG) disassociated the linkage between hydrogel and antigen–antibody pairs, hence decreasing the cross-link density of the hydrogel and induced swelling. The authors demonstrated that this hydrogel has reversible swelling behavior due to the chemically cross-linking of both antibody and antigen within the semi-IPN network.

Recently, the same group ameliorated the design concept of antigen-responsive hydrogels exploiting a molecular imprinting approach [300], an attractive technique for the preparation of biomaterials. Applying this technique, Miyata et al.

developed a ternary protein complex from α-fetoprotein (AFP, a tumor-specific marker glycoprotein), Con A (a lectin that binds with a saccharide chain of AFP), and AFP-specific antibody. The Con A and antibody were acrylated by NSA and hydrogels were formed through the addition of a low molecular weight (MBA) and high molecular weight (PEGDA and PEGDMA) cross-linkers via radical polymerization. In response to AFP, the imprinted hydrogels cross-linked with PEGDA, and PEGDMA shrank more dramatically than MBA cross-linked hydrogels. However, the swelling ratio of hydrogels fabricated with PEGDMA did not change in the presence of AFP, suggesting that the cross-linker chain length played a key role in responsive behavior and dynamic recognition of imprinted hydrogels.

A new single-step microfluidic immunoassay principle, using antigen-responsive hydrogels, was established by Hisamoto and coworkers [301]. This immunoassay was constructed via the mixture of a soluble coating containing enzyme-labeled antibody and a fluorescent enzyme substrate-copolymerized hydrogel. Based on the variation in molecular size, the hydrogel distinguished the free enzyme-conjugated antibody from its complexed form. This unique microfluidic immunoaasay exhibited a selective and sensitive response in the range of 10–1000 ng ml^{-1}. The design concept mentioned in this research might be employed in the fabrication of multiplexed-bioassay microdevices as effective tools for clinical diagnostics.

In a fascinating research, Jung and Yi reported a facile post-fabrication protein conjugation method for sensing platforms and to enhance protein conjugation [302]. To achieve this, they prepared hydrogel microparticle platforms and tobacco mosaic virus (TMV) nanotemplates by rapid bioorthogonal reactions. Protein conjugation examination via a red fluorescent protein R-Phycoerythrin (R-PE) indicated notably improved protein conjugation capacity of TMV particles in comparison with hydrogel microparticles or planar substrates. Further investigation of protein conjugation kinetics through strain-promoted alkyne-azide cycloaddition (SPAAC) reaction and tetrazine (Tz)-trans-cyclooctene (TCO) cycloaddition illustrated that TMV-particles caused a less restricted environment for protein conjugation. Moreover, the anti-R-PE antibody (R-Ab)-R-PE pair study demonstrated significantly enhanced capture capacity of R-Ab conjugated TMV particles than hydrogel microparticles. Furthermore, TMV assembly density was found to be controllable via altering the concentration of TMV which exhibited minor effect on capture capacity and protein conjugation.

Apart from the previously mentioned antigen-based hydrogels that were only applied *in vitro*, a breakthrough study, achieving remote-controlled vaccine depot *in vivo*, was developed by Gübeli and colleagues [303]. They fabricated a hydrogel depot based on a vinylsulfone-functionalized eight-arm PEG, a humanized single-chain variable fragment (scFv), and a fluorescein for the remote-controlled delivery of a high-risk oncogenic human papilloma virus type 16 (HPV16) vaccine in mice. The authors showed that this hydrogel overcome the main disadvantages of stimuli-sensing materials including adverse (immune) reactions and biocompatibility. This system also indicated a comparable therapeutic efficacy with the common injection-based vaccines and thus confirms the efficiency of remote-controlled hydrogel depots to be further replaced with the current multi-injection systems.

3.4 Nucleic Acid-Responsive Hydrogels

Nucleic acids are large biomolecules made from monomers (smaller molecules) called nucleotides. Along with other biological macromolecules such as proteins, amino acids, carbohydrates, and lipids, they are essential for all living beings. DNA and RNA are the two principal nucleic acids. Due to their negative charges and large sizes, nucleic acids have poor inherent transfection efficiency and are fragile [22]. These drawbacks can be addressed by encapsulation of nucleic acids within hydrogel networks.

3.4.1 RNA-Responsive Hydrogels

RNA is a linear and a single-stranded biomolecule, which is used for short-term genetic information storage. Furthermore, it is the major molecule for encoding the genetic information of viruses. However, RNA is less stable than DNA because it is more susceptible to degradation due to having ribose sugars rather than deoxyribose sugars within its structure. The three well-known types of RNA are messenger RNA (mRNA), transfer RNA (tRNA), and ribosomal RNA (rRNA). There are also other types, regarded as small/short RNA, including Piwi-interacting RNA (piRNA), small interfering RNA (siRNA), and microRNA (miRNA). They have been considerably utilized in cancer treatments [304–306]. However, the successful delivery of these RNA to specific cells *in vivo* is yet to be addressed. For example, the *in vivo* siRNA delivery to target cells using current methods containing direct injection of siRNA in saline, complexation with positively charged polymers or peptides, incorporation into liposomes, and encapsulation with micro- or nano-polymeric spheres remained a real challenge [307]. This is because the siRNA is very susceptible to degradation, the silencing effect in cells only remains for few days [308], and targeting at a specific site is difficult [309]. To solve these problems, a new approach for controlled delivery of silencing RNA at the specific site was established by Alsberg and coworkers [310]. In this study, three injectable, biodegradable hydrogels including collagen, photocross-linked alginate, and calcium cross-linked alginate were utilized for sustained, localized delivery of siRNA. Each hydrogels showed a different release behavior. It was observed that the addition of positively charged polymers (PEI and chitosan) to the calcium cross-linked alginate hydrogels (negatively charged) prolonged the siRNA release because of the electrostatic interactions between positively charged polymers and nucleotides. Cells treated with siRNA indicated higher fluorescence release from hydrogels compared with untreated cells. The use of injectable hydrogels (as biopolymer scaffolds) caused the minimally invasive delivery to the specific site. Additionally, cells incorporated within these scaffolds providing cell transplantation and siRNA delivery. The design concept described in this research, achieving localized and controlled gene delivery to both transplanted and host cells might be further applied in regenerative medicine and tissue engineering.

3.4.2 DNA-Responsive Hydrogels

Of all stimuli-responsive materials, DNA-responsive hydrogels are of particular interest due to their highly specific DNA interactions and efficient stability. In a pioneering research, a DNA-responsive hydrogel was developed by Lin et al. [311]. The addition of single-stranded DNA resulted in adjusting hydrogel stiffness. Two single-stranded DNA molecules were modified and copolymerized with an acryl group and acrylamide, respectively. In the presence of a linker DNA-strand possessing complementary sequences, polymer-modified strands detached and formed the hydrogel. The author showed that the incorporation of a fuel strand hybridizing to the linker strand improved the hydrogel stiffness and led to the more rigid linker. However, the addition of a DNA strand to the fuel strand decreased the stiffness. Using this principle, Murakami and Maeda fabricated DNA-responsive hydrogels that underwent swelling and shrinking through the addition of particular motifs [312]. The hydrogel was formed by cross-linking of single-stranded DNA with polyacrylamide and shrunk in response to a complementary sequence. This resulting shrinking was due to the higher chain length of single-stranded DNA than that of double-stranded one. To obtain hydrogel swelling, the single-stranded DNA formed a dense secondary structure, and then the addition of a complementary DNA strand resolved this dense structure, causing hydrogel swelling.

3.4.3 Aptamer-Responsive Hydrogels

Besides the precise interaction of RNA and DNA strands, the highly selective binding of DNA sequences to aptamers can also be considered for preparation of stimuli-responsive hydrogels. Aptamers are single-stranded molecules that can bind to specific targets. They are produced from systematic evolution of ligands by exponential enrichment (SELEX) process. Aptamers not only exhibit similar target-bindings with antibodies but also show simple preparation, efficient thermal stability, and easy labeling [313].

In a notable study, Tan and coworkers utilized aptamer responsive-hydrogels by copolymerization of two single-stranded DNA to PAAm and cross-linking to a hydrogel using a single-stranded linker DNA [314]. The linker aptamer with complementary sequences to both PAAm-coupled DNA strands was specific to adenosine. In the presence of adenosine, the aptamer–target binding was weakened, therefore decreasing the cross-link density and leading to the dissolution of hydrogel. This dissolution was shown to control the release of a model drug (gold nanoparticles [AuNPs]). A similar concept was applied for the fabrication of hydrogels that detect non-glucose targets via the mixture of the personal glucose meter (PGM) with a glucoamylase-trapped aptamer-bound hydrogel [315]. In the presence of target molecules, aptamers coupled with targets and formed aptamer–target network, resulting in the dissociation of hydrogel and glucoamylase release, which accelerates amylose hydrolysis to generate a high level of glucose for PGM readout. By using different aptamer sequences, this simple and rapid method can be further utilized for quantitative and portable detection of a broad range of non-glucose targets because only DNA hybridization is required to synthesize the hydrogel. Using this concept, Wang and Wang developed aptamer-functionalized hydrogel diffraction gratings through the combination of soft lithography and photofabrication process for detection of human thrombin [316]. The hydrogels were prepared by microcontact printing via employing the precursor solution as ink and poly(dimethylsiloxane) (PDMS) stamps. Both chemical and physical (aptamer and its complementary sequence) cross-linkers were applied to functionalized thrombin-responsive hydrogel. When subjected to the thrombin solution, the aptamer–bound thrombin interaction was broken, thus reducing the cross-link density and caused hydrogel swelling. The gratings not only indicated high sensitivity to the human thrombin in the wide range (0.01–0.20 mg ml^{-1}) but also exhibited fast responsive rate due to the micrometer-scale dimension of the gratings. Interestingly, no extra signal improvement treatment or fluorophore-labeling is needed for this system. Recently, Liu et al. developed an aptamer-responsive hydrogel using two complementary polymer DNA strands (PS-A and PS-B) and the OTA aptamer for visual detection of ochratoxin A (OTA) [317]. Through the incorporation of OTA, the aptamers coupled with OTA result in the breakdown of the hydrogel, thus releasing the preloaded AuNPs. Next, to obtain responsive quantitative and visual detection, Au@Pt core–shell nanoparticles (Au@PtNPs) were encapsulated within hydrogel network to produce quantitative readout in a volumetric bar-chart chip (V-Chip). In the V-Chip, the movement of the ink bar was pushed by O_2 through the oxidation of H_2O_2, and it was commensurate with target concentration. Additionally, OTA was enhanced from real sample by applying an immunoaffinity column (IAC) prior to the incorporation with the hydrogel. After the enhancement, the V-Chip was able to detect 1.27 nM (0.51 ppb) OTA, which met the test standard (2.0 ppb); therefore it can be used to certify human health and food safety.

3.4.4 PNA-Responsive Hydrogels

PNA is a synthetic oligonucleotide with modified structure in which 2-aminoethylglycine linkage is replaced the phosphodiester backbone [318]. In comparison to RNA and DNA, PNA is not sensitive to enzymatic cleavage, hence it is stable in biological systems and cells. PNA is capable of binding with both RNA and DNA targets in a sequence-specific pattern to generate a PNA–RNA double helical network and a PNA–DNA Watson–Crick type. It can also form other specific and complex helical structures with DNA sequences including triplex or quadruplex [319–321]. Due to lack of the electrostatic repulsion, PNA is a stronger binder than RNA and DNA.

Recently, Kopecek and colleagues creatively designed novel approaches to prepare oligonucleotide-hybrid hydrogels based on PNA–DNA complexation [322]. The HPMA copolymer–PNA conjugates were synthesized via reversible addition–fragmentation chain transfer (RAFT) polymerization and then cross-linked by the introduction of the linker DNA. The self-assembly of PNA–DNA complexation resulted in the formation of hydrogels. It was demonstrated that the hydrogels can be synthesized by two methods of complexations. Type I hydrogel was produced through the PNA/DNA double-helix hybridization, applying Watson–Crick interaction between two complementary sequences. Type II hydrogel employed P-form oligonucleotide triplex consisted of one DNA and two PNA sequences. Compared to type II, type I hydrogel indicated a higher CGC as well as greater pore sizes. As a consequence, type II hydrogel possessed higher elasticity and denser structure. Although the proposed methods displayed low mechanical properties due to the physical cross-linking, these new self-assembled hydrogels may be further applied for pharmaceutical and biomedical applications.

3.5 Future Prospects

Biomolecules such as glucose, proteins, nucleic acids, and polypeptides are involved in all living organisms, serving as the most fundamental and vital parts of alive systems that vary based on the conditions of surrounding environment. In this chapter, recent advances and applications of biomolecule-responsive hydrogels according to three categories, namely, glucose-responsive hydrogel, protein-responsive hydrogels, and nucleic acid-responsive hydrogels were highlighted. It is clearly obvious that biomolecule-responsive hydrogels will further design in terms of fundamental research, and their practical applications in various areas involving controlled drug/gene delivery, tissue engineering, sensing, regenerative medicine, and cell and cancer research will be extensively investigated. However, achieving the ultimate purpose of using biomolecule-responsive hydrogels in preclinical and clinical areas is still at the very early stage.

Glucose-responsive hydrogels are typically based on Gox, Con A, and PBA, which are able to respond to glucose. Therefore, they have been applied in the fabrication of self-regulated insulin release and glucose sensing systems. In spite of numerous promising properties, there are still many challenges remaining in this field that should be addressed before their real applications. While Gox and Con A are natural glucose-based systems, possessing good biodegradability and biocompatibility, they show poor stability due to the changes in environment conditions. These drawbacks can be overcome by using PBA-responsive hydrogels as synthetic glucose-based materials that indicate good stability properties. On the other hand, some barriers such as biodegradability, poor glucose selectivity, insufficient operation at physiological pH, and their slow response rate raise concern about the clinical use of PBA-responsive hydrogels. The modification of PBA as well as the rational design of hydrogels might solve this issue. It is expected that merging the field of glucose-responsive hydrogels with biologically significant molecules such as carbohydrates and their derivatives can expand the field of glucose-responsive hydrogels.

Protein-responsive hydrogels are classified into two major types, including enzyme-responsive hydrogels and antigen-responsive hydrogels. Enzyme-responsive hydrogels with high substrate specificity and high efficiency are of particular interest since they can be carried out under mild conditions. However, it is not easy to obtain enzyme-responsive properties for synthetic hydrogels because of the poor enzyme-liable functionalities in their chemical structures. The current ones are formed via the enzymatic phosphorylation and dephosphorylation reactions. Hence, the mixture of common synthetic hydrogels or development of advanced ones with other highly suitable enzymes are key steps that will further this research area. Another bottleneck to overcome is the synthesis of enzyme-responsive hydrogels that can respond to multi-enzyme reactions that will further enrich this field. The conventional enzyme systems are mostly responsive to one type of specific enzyme. Thus, it is highly favorable to design enzyme-responsive hydrogels with multi-responses to the enzyme concentrations. Similarly, antigen-responsive hydrogel involves the highly selective and specific interaction of an antibody to its antigen. Due to their fascinating characteristics in the human body to detect diseases, they might be widely investigated in various biomedical areas. Finally, the application of nucleic acid-responsive hydrogels based on four main categories, RNA-responsive hydrogels, DNA-responsive hydrogels, aptamer-responsive hydrogels, and PNA-responsive hydrogels, was discussed. Although the RNA, DNA, and the natural nucleic acid-responsive hydrogels provide the advantages of biocompatibility, they are not stable against temperature and enzymatic cleavage. PNA, serving as a synthetic nucleic acid not only solved these issues but also indicated stronger binding properties than RNA and DNA. Therefore, PNA-responsive hydrogels will expect to play important role in the biomedical fields in the future including target drug delivery, cancer research, tissue engineering, regenerative medicine, and food safety. On the way toward the real/clinical application of biomolecule-responsive hydrogels, plenty of factors should be extensively studied and examined under both *in vitro* and *in vivo* conditions. For example, biocompatibility, biointegration, and toxicity of biomolecule-responsive hydrogels should be carefully evaluated. From the living body point of view, biocompatibility is seriously depended on the interactions at the tissue/polymer interface. These interactions are influenced by physical nature, chemical structure, surface properties, and degradation of the materials. In addition, the development of advanced hydrogels with tunable biological and mechanical properties that cause no/low side effects is of great importance.

References

1 Peppas, N.A., Hilt, J.Z., Khademhosseini, A., and Langer, R. (2006). *Adv. Mater.* 18: 1345.

2 Lin, C.C. and Anseth, K.S. (2009). *Pharm. Res.* 26: 631.

3 Slaughter, B.V., Khurshid, S.S., Fisher, O.Z. et al. (2009). *Adv. Mater.* 21: 3307.

4 Drury, J.L. and Mooney, D.J. (2003). *Biomaterials* 24: 4337.

5 Branco, M.C., Pochan, D.J., Wagner, N.J., and Schneider, J.P. (2010). *Biomaterials* 31: 9527.

6 Zhao, F., Ma, M.L., and Xu, B. (2009). *Chem. Soc. Rev.* 38: 883.

7 Sudarsan, S., Franklin, D.S., Sakthivel, M., and Guhanathan, S. (2016). *Carbohydr. Polym.* 148: 206.

8 Lin, Z., Cao, S., Chen, X. et al. (2013). *Biomacromolecules* 14: 2206.

9 Abbadessa, A., Blokzijl, M.M., Mouser, V.H.M. et al. (2016). *Carbohydr. Polym.* 149: 163.

10 Wang, X., Chun, Y.W., Zhong, L. et al. (2016). *Int. J. Cardiol.* 190: 177.

11 Wei, Y., Xie, R., Lin, Y. et al. (2016). *Int. J. Biol. Macromol.* 89: 353.

12 Indermun, S., Choonara, Y.E., Kumar, P. et al. (2014). *Int. J. Pharm.* 462: 52.

13 Zhang, R., Tang, M., Bowyer, A. et al. (2005). *Biomaterials* 26: 4677.

14 Wu, Q., Wang, L., Yu, H. et al. (2011). *Chem. Rev.* 111: 7855.

15 Gu, Z., Dang, T.T., Ma, M. et al. (2013). *ACS Nano* 7: 6758.

16 Gordijo, C.R., Shuhendler, A.J., and Wu, X.Y. (2010). *Adv. Funct. Mater.* 20: 1404.

17 Farahani, B.V., Ghasemzadeh, H., and Afraz, S. (2016). *J. Chin. Chem. Soc.* https://doi.org/10.1002/jccs.201500511.

18 Shi, J.J., Xiao, Z.Y., Kamaly, N., and Farokhzad, O.C. (2011). *Acc. Chem. Res.* 44: 1123.

19 Kamaly, N., Xiao, Z., Valencia, P.M. et al. (2012). *Chem. Soc. Rev.* 41: 2971.

20 Kumar, A., Srivastava, A., Galaev, I.Y., and Mattiasson, B. (2007). *Prog. Polym. Sci.* 32: 1205.

21 Janeway, C.A., Travers, P., Walport, M., and Shlomchik, M.J. (2001). *Immunobiology: The Immune System in Health and Disease*. New York: Garland Science.

22 Danhier, F., Ansorena, E., Silva, J.M. et al. (2012). *J. Controlled Release* 161: 505.

23 Ishihara, K., Kobayashi, M., Ishimaru, N., and Shinohara, I. (1984). *Polym. J.* 16: 625.

24 Yang, X. and Kim, J.C. (2011). *Int. J. Biol. Macromol.* 48: 661.

25 Hassan, C.M., Dolye, F.J. III, and Peppas, N.A. (1997). *Macromolecules* 30: 6166.

26 Kang, S.I. and Bae, Y.H. (2003). *J. Controlled Release* 86: 115.

27 Huang, H.Y., Shaw, J., Yip, C., and Wu, X.Y. (2008). *Pharm. Res.* 25: 1150.

28 Schwarte, L.M. and Peppas, N.A. (1998). *Polymer* 39: 6057.

29 Podual, K., Doyle, F.J. III, and Peppas, N.A. (2000). *Polymer* 41: 3975.

30 Podual, K., Doyle, F.J. III, and Peppas, N.A. (2000). *J. Controlled Release* 67: 9.

31 Podual, K., Doyle, F.J. III, and Peppas, N.A. (2000). *Biomaterials* 21: 1439.

32 Kashyap, N., Viswanad, B., Sharma, G. et al. (2007). *Biomaterials* 28: 2051.

33 Ravaine, V., Ancla, C., and Catargi, B. (2008). *J. Controlled Release* 132: 2.

34 Harris, J.M. (1992). *PolyEthylene Glycol. Chemistry: Biotechnical and Biomedical Applications*. New York: Plenum Press.

35 Liu, Q., Rauth, A.M., and Wu, X.Y. (2007). *Int. J. Pharm.* 339: 148.

36 Sharon, N. and Lis, H. (1972). *Science* 177: 949.

37 Brownlee, M. and Cerami, A. (1979). *Science* 206: 1190.

38 Kim, J.J. and Park, K. (2001). *J. Controlled Release* 77: 39.

39 Miyata, T., Jikihara, A., Nakamae, K., and Hoffman, A.S. (2004). *J. Biomater. Sci., Polym. Ed.* 15: 1085.

40 Tanna, S., Sahota, T.S., Clark, J., and Taylor, M.J. (2002). *J. Pharm. Pharmacol.* 54: 1461.

41 Tanna, S., Sahota, T.S., Clark, J., and Taylor, M.J. (2002). *J. Drug Targeting* 10: 411.

42 Tanna, S., Sahota, T.S., Sawicka, K., and Taylor, M.J. (2006). *Biomaterials* 27: 4498.

43 Tanna, S., Taylor, M.J., Sahota, T.S., and Sawicka, K. (2006). *Biomaterials* 27: 1586.

44 Taylor, M.J., Tanna, S., Sahota, T.S., and Voermans, B. (2006). *Eur. J. Pharm. Biopharm.* 62: 94.

45 Taylor, M.J., Tanna, S., and Sahota, T.S. (2008). *Drug Dev. Ind. Pharm.* 34: 73.

46 Zhang, R., Tang, M., Bowyer, A. et al. (2006). *React. Funct. Polym.* 66: 757.

47 Benzeval, I., Bowyer, A., and Hubble, J. (2012). *Eur. J. Pharm. Biopharm.* 80: 143.

48 Yin, R., Han, J., Zhang, J., and Nie, J. (2010). *Colloids Surf., B* 76: 483.
49 Yin, R., Wang, K., Han, J., and Nie, J. (2010). *Carbohydr. Polym.* 82: 412.
50 Yin, R., Tong, Z., Yang, D., and Nie, J. (2012). *Carbohydr. Polym.* 89: 117.
51 Yin, R., Wang, K., Du, S. et al. (2014). *Carbohydr. Polym.* 103: 369.
52 Yao, H. and Hu, N. (2011). *J. Phys. Chem. B.* 115: 6691.
53 Kawamura, A., Hata, Y., Miyata, T., and Uragami, T. (2012). *Colloids Surf., B* 99: 74.
54 Kuivila, H.G., Keough, A.H., and Soboczenski, E.J. (1954). *J. Org. Chem.* 19: 780.
55 Lorand, J.P. and Edwards, J.O. (1959). *J. Org. Chem.* 24: 769.
56 Springsteen, G. and Wang, B. (2002). *Tetrahedron* 58: 5291.
57 Yan, J., Springsteen, G., Deeter, S., and Wang, B. (2004). *Tetrahedron* 60: 11205.
58 Barker, S.A., Chopra, A.K., Hatt, B.W., and Somers, P.J. (1973). *Carbohydr. Res.* 26: 33.
59 Kimura, T. and Shinkai, S.J. (1999). *Syn. Org. Chem. Jpn.* 57: 1084.
60 Boeseken, J. (1947). *Carbohydr. Chem.* 4: 189.
61 Aronoff, S., Chen, T., and Cheveldayoff, M. (1975). *Carbohydr. Res.* 40: 299.
62 Hoare, T. and Pelton, R. (2007). *Macromolecules* 40: 670.
63 Matsumoto, A., Yoshida, R., and Kataoka, K. (2004). *Biomacromolecules* 5: 1038.
64 Matsumoto, A., Ikeda, S., Harada, A., and Kataoka, K. (2003). *Biomacromolecules* 4: 1410.
65 Zhang, S.B., Chu, L.Y., Xu, D. et al. (2008). *Polym. Adv. Technol.* 19: 937.
66 Kataoka, K., Miyazaki, H., Okano, T., and Sakurai, Y. (1994). *Macromolecules* 27: 1061.
67 Hisamitsu, I., Kataoka, K., Okano, T., and Sakurai, Y. (1997). *Pharm. Res.* 14: 289.
68 Kataoka, K., Miyazaki, H., Bunya, M. et al. (1998). *J. Am. Chem. Soc.* 120: 12694.
69 Matsumoto, A., Kurata, T., Shiino, T., and Kataoka, K. (2004). *Macromolecules* 37: 1502.
70 Yoshida, R., Uchida, K., Kaneko, Y. et al. (1995). *Nature* 374: 240.
71 Kaneko, Y., Sakai, K., Kikuchi, K. et al. (1995). *Macromolecules* 28: 7717.
72 Kaneko, Y., Sakai, K., Kikuchi, K. et al. (1996). *Macromol. Symp.* 109: 41.
73 Hoare, T. and Pelton, R. (2008). *Biomacromolecules* 9: 733.
74 Blyth, J., Millington, R.B., Mayes, A.G. et al. (1996). *Anal. Chem.* 68: 1089.
75 Takeoka, Y. and Watanabe, M. (2002). *Langmuir* 18: 5977.
76 Takeoka, Y. and Watanabe, M. (2003). *Adv. Mater.* 15: 199.
77 Lee, Y.J. and Braun, P.V. (2003). *Adv. Mater.* 15: 563.
78 Holtz, J.H. and Asher, S.A. (1997). *Nature* 389: 829.
79 Holtz, J.H., Holtz, J.S.W., Munro, C.H., and Asher, S.A. (1998). *Anal. Chem.* 70: 780.
80 Goponenko, A.V. and Asher, S.A. (2005). *J. Am. Chem. Soc.* 127: 10753.
81 Aguirre, C.I., Reguera, E., and Stein, A. (2010). *Adv. Funct. Mater.* 20: 2565.
82 Ikeda, M., Fukuda, K., Tanida, T. et al. (2012). *Chem. Commun.* 48: 2716.
83 Zhang, C., Losego, M.D., and Braun, P.V. (2013). *Chem. Mater.* 25: 3239.

84 Kabilan, S., Marshall, A.J., Sartain, F.K. et al. (2005). *Biosens. Bioelectron.* 20: 1602.

85 Lee, M.C., Kabilan, S., Hussain, A. et al. (2004). *Anal. Chem.* 76: 5748.

86 Kabilan, S., Blyth, J., Lee, M.C. et al. (2004). *J. Mol. Recognit.* 17: 162.

87 Alexeev, V.L., Sharma, A.C., Goponenko, A.V. et al. (2003). *Anal. Chem.* 75: 2316.

88 Asher, S.A., Alexeev, V.L., Goponenko, A.V. et al. (2003). *J. Am. Chem. Soc.* 125: 3322.

89 Nakayama, D., Takeoka, Y., Watanabe, M., and Kataoka, K. (2003). *Angew. Chem. Int. Ed.* 42: 4197.

90 Lee, Y.J., Pruzinsky, S.A., and Braun, P.V. (2004). *Langmuir* 20: 3096.

91 Alexeev, V.L., Das, S., Finegold, D.N., and Asher, S.A. (2004). *Clin. Chem.* 50: 2353.

92 Ben-Moshe, M., Alexeev, V.L., and Asher, S.A. (2006). *Anal. Chem.* 78: 5149.

93 Yang, X., Pan, X., Blyth, J., and Lowe, C.R. (2008). *Biosens. Bioelectron.* 23: 899.

94 Horgan, A.M., Marshall, A.J., Kew, S.J. et al. (2006). *Biosens. Bioelectron.* 21: 1838.

95 Yang, X.P., Lee, M.C., Sartain, F. et al. (2006). *Chem. Eur. J.* 12: 8491.

96 Pan, X., Yang, X., and Lowe, C.R. (2008). *J. Mol. Recognit.* 21: 205.

97 Samoei, G.K., Wang, W., Escobedo, J.O. et al. (2006). *Angew. Chem. Int. Ed.* 45: 5319.

98 Horkay, F., Cho, S.H., Tathireddy, P. et al. (2011). *Sens. Actuators B* 160: 1363.

99 Tierney, S., Volden, S., and Stokke, B.T. (2009). *Biosens. Bioelectron.* 24: 2034.

100 Tierney, S., Falch, B.M.H., Hjelme, D.R., and Stokke, B.T. (2009). *Anal. Chem.* 81: 3630.

101 Lin, G., Chang, S., Hao, H. et al. (2010). *Sens. Actuators, B* 144: 332.

102 Dean, K.E.S., Horgan, A.M., Marshall, A.J. et al. (2006). *Chem. Commun.* 350: 3507.

103 Siegel, R.A., Gu, Y., Baldi, A., and Ziaie, B. (2004). *Macromol. Symp.* 207: 249.

104 Baldi, A., Gu, Y., Loftness, P.E. et al. (2003). *J. Microelectromech. Syst.* 12: 613.

105 Baldi, A., Lei, M., Gu, Y. et al. (2006). *Sens. Actuators, B* 114: 9.

106 Pelton, R. (2000). *Adv. Colloid Interface Sci.* 85: 1.

107 Saunders, B.R. and Vincent, B. (1999). *Adv. Colloid Interface Sci.* 80: 1.

108 Guan, Y. and Zhang, Y. (2011). *Soft Matter* 7: 6375.

109 Oh, J.K., Lee, D.I., and Park, J.M. (2009). *Prog. Polym. Sci.* 34: 1261.

110 Reese, C.E., Mikhonin, A.V., Kamenjicki, M. et al. (2004). *J. Am. Chem. Soc.* 126: 1493.

111 Gan, T., Zhang, Y., and Guan, Y. (2009). *Biomacromolecules* 10: 1410.

112 Gan, T.T., Guan, Y., and Zhang, Y.J. (2010). *J. Mater. Chem.* 20: 5937.

113 Wang, D., Cheng, D., Guan, Y., and Zhang, Y. (2011). *Biomacromolecules* 12: 578.

114 Chacko, R.T., Ventura, J., Zhuang, J., and Thayumanavan, S. (2012). *Adv. Drug Delivery Rev.* 64: 836.

115 Zha, L., Banik, B., and Alexis, F. (2011). *Soft Matter* 7: 5908.

116 Raemdonck, K., Demeester, J., and De Smedt, S. (2009). *Soft Matter* 5: 707.

117 Kabanov, A.V. and Vinogradov, S.V. (2009). *Angew. Chem. Int. Ed.* 48: 5418.
118 Hamidi, M., Azadi, A., and Rafiei, P. (2008). *Adv. Drug Delivery Rev.* 60: 1638.
119 Kim, J.S., Singh, N., and Lyon, L.A. (2006). *Angew. Chem. Int. Ed.* 45: 1446.
120 Liu, Y., Zhang, Y.J., and Guan, Y. (2009). *Chem. Commun.* 1867.
121 Wu, W., Zhou, T., Aiello, M., and Zhou, S. (2010). *Biosens. Bioelectron.* 38: 1343.
122 Tang, Z., Guan, Y., and Zhang, Y. (2014). *Polym. Chem.* 5: 1782.
123 Elmas, B., Onur, M.A., Senel, S., and Tuncel, A. (2002). *Colloid. Polym. Sci.* 280: 1137.
124 Lapeyre, V., Gosse, I., Chevreux, S., and Ravaine, V. (2006). *Biomacromolecules* 7: 3356.
125 Ge, H., Ding, Y., Ma, C., and Zhang, G. (2006). *J. Phys. Chem. B* 110: 20635.
126 Zenkl, G., Mayr, T., and Khmant, I. (2008). *Macromol. Biosci.* 8: 146.
127 Wang, D., Liu, T., Yin, J., and Liu, S.Y. (2011). *Macromolecules* 44: 2282.
128 Zhang, Y.J., Guan, Y., and Zhou, S.Q. (2006). *Biomacromolecules* 7: 3196.
129 Zhang, Y.J., Guan, Y., and Zhou, S.Q. (2007). *Biomacromolecules* 8: 3842.
130 Luo, Q.F., Liu, P.X., Guan, Y., and Zhang, Y.J. (2010). *ACS Appl. Mater. Interfaces* 2: 760.
131 Xing, S., Guan, Y., and Zhang, Y. (2011). *Macromolecules* 44: 4479.
132 Liu, P.X., Luo, Q.F., Guan, Y., and Zhang, Y.J. (2010). *Polymer* 51: 2668.
133 Zhang, Y., Liu, K., Guan, Y., and Zhang, Y. (2012). *RSC Adv.* 2: 4768.
134 Hoare, T. and McLean, D. (2006). *J. Phys. Chem. B* 110: 20327.
135 Farooqi, A.H., Wu, W., Zhou, S., and Siddiq, M. (2011). *Macromol. Chem. Phys.* 212: 1510.
136 Guan, Y. and Zhang, Y. (2013). *Chem. Soc. Rev.* 42: 8106.
137 Nishiyabu, R., Kubo, Y., James, T.D., and Fossey, J.S. (2011). *Chem. Commun.* 47: 1106.
138 Lapeyre, V., Ancla, C., Catargi, B., and Ravaine, V. (2008). *J. Colloid Interface Sci.* 327: 316.
139 Ogawa, K., Wang, B., and Kokufuta, E. (2001). *Langmuir* 17: 4704.
140 Zhang, K. and Wu, X.Y. (2002). *J. Controlled Release* 80: 169.
141 Ancla, C., Lapeyre, V., Gosse, I. et al. (2011). *Langmuir* 27: 12693.
142 Wu, W., Zhou, T., Shen, J., and Zhou, S. (2009). *Chem. Commun.* 4390.
143 Wu, W., Zhou, T., Aiello, M., and Zhou, S. (2010). *Biosens. Bioelectron.* 25: 2603.
144 Wu, W., Aiello, M., Zhou, T. et al. (2010). *Biomaterials* 31: 3023.
145 Jones, C.D. and Lyon, L.A. (2000). *Macromolecules* 33: 8301.
146 Wu, W., Mitra, N., Yan, E.C.Y., and Zhou, S. (2010). *ACS Nano* 4: 4831.
147 Sun, L., Zhang, X., Zheng, C. et al. (2012). *RSC Adv.* 2: 9904.
148 Ye, T., Jiang, X., Xu, W. et al. (2014). *Polym. Chem.* 5: 2352.
149 Zhao, L., Xiao, C.S., Ding, J.X. et al. (2013). *Acta Biomater.* 9: 6535.
150 Zhang, X., Lu, S., Gao, C. et al. (2013). *Nanoscale* 5: 6498.
151 Wu, Z., Zhang, X., Guo, H. et al. (2012). *J. Mater. Chem.* 22: 22788.
152 Decher, G. (1997). *Science* 277: 1232.
153 De Geest, B.G., Sanders, N.N., Sukhorukov, G.B. et al. (2007). *Chem. Soc. Rev.* 36: 636.
154 Zhang, Y.J., Guan, Y., and Zhou, S.Q. (2005). *Biomacromolecules* 6: 2365.

155 Wang, C.Y., Ye, S.Q., Dai, L. et al. (2007). *Biomacromolecules* 8: 1739.
156 Zhang, Y.J., Guan, Y., Yang, S.G. et al. (2003). *Adv. Mater.* 15: 832.
157 Lee, H., Mensire, R., Cohen, R.E., and Rubner, M.F. (2012). *Macromolecules* 45: 347.
158 Stockton, W.B. and Rubner, M.F. (1997). *Macromolecules* 30: 2717.
159 Wang, L., Wang, Z., Zhang, X. et al. (1997). *Macromol. Rapid Commun.* 18: 509.
160 Zhang, Y.J., Yang, S.G., Guan, Y. et al. (2003). *Macromolecules* 36: 4238.
161 Kohli, P. and Blanchard, G.J. (2000). *Langmuir* 16: 4655.
162 Kotov, N.A. (1999). *Nanostruct. Mater.* 12: 789.
163 Lojou, E. and Bianco, P. (2004). *Langmuir* 20: 748.
164 Zhang, Y.J. and Cao, W.X. (2001). *Langmuir* 17: 5021.
165 Shimazaki, Y., Mitsuishi, M., Ito, S., and Yamamoto, M. (1997). *Langmuir* 13: 1385.
166 Anzai, J.I., Kobayashi, Y., Nakamura, N. et al. (1999). *Langmuir* 15: 221.
167 Tang, Z., Wang, Y., Podsiadlo, P., and Kotov, N.A. (2006). *Adv. Mater.* 18: 3203.
168 Qi, W., Yan, X., Fei, J. et al. (2009). *Biomaterials* 30: 2799.
169 Ding, Z., Guan, Y., Zhang, Y., and Zhu, X.X. (2009). *Polymer* 50: 4205.
170 De Geest, B.G., Jonas, A.M., Demeester, J., and De Smedt, S.C. (2006). *Langmuir* 22: 5070.
171 Zhang, X., Guan, Y., and Zhang, Y. (2012). *Biomacromolecules* 13: 92.
172 Choi, J., Konno, T., Takai, M., and Ishihara, K. (2012). *Biomaterials* 33: 954.
173 Ding, Z.B., Guan, Y., Zhang, Y., and Zhu, X.X. (2009). *Soft Matter* 5: 2302.
174 Zhang, X., Guan, Y., and Zhang, Y. (2012). *J. Mater. Chem.* 22: 16299.
175 Levy, T., Dejugnat, C., and Sukhorukov, G.B. (2008). *Adv. Funct. Mater.* 18: 1586.
176 Manna, U. and Patil, S. (2009). *J. Phys. Chem. B* 113: 9137.
177 Manna, U. and Patil, S. (2010). *ACS Appl. Mater. Interfaces* 2: 1521.
178 Davis, N.E., Ding, S., Forster, R.E. et al. (2010). *Biomaterials* 31: 7288.
179 Jin, R., Teixeira, L.S.M., Dijkstra, P.J. et al. (2011). *J. Controlled Release* 152: 186.
180 Jin, R., Teixeira, L.S.M., Dijkstra, P.J. et al. (2010). *Tissue Eng. Part A* 16: 2429.
181 Jin, R., Teixeira, L.S.M., Dijkstra, P.J. et al. (2010). *Biomaterials* 31: 3103.
182 Sakai, S., Hirose, K., Taguchi, K. et al. (2009). *Biomaterials* 30: 3371.
183 Sakai, S., Ogushi, Y., and Kawakami, K. (2009). *Acta Biomater.* 5: 554.
184 Sofia, S.J., Singh, A., and Kaplan, D.L. (2002). *Macromol. Sci. Pure A* 39: 1151.
185 Kwon, J.S., Kim, S.W., Kwon, D.W. et al. (2014). *Biomaterials* 35: 5337.
186 Hoban, D.B., Newland, B., Moloney, T.C. et al. (2013). *Biomaterials* 34: 9420.
187 Yu, J., Xu, X., Yao, F. et al. (2014). *Int. J. Pharm.* 470: 151.
188 Weng, L., Rostambeigi, N., Zantek, N.D. et al. (2013). *Acta Biomater.* 9: 8182.
189 Hubbell, J.A. (2003). *Curr. Opin. Biotechnol.* 14: 551.
190 Nagao, R.J., Lundy, S., Khaing, Z.Z., and Schmidt, C.E. (2011). *Neurol. Res.* 33: 600.
191 Valentin, J.E., Turner, N.J., Gilbert, T.W., and Badylak, S.F. (2010). *Biomaterials* 31: 7475.
192 Turner, N.J., Yates, A.J. Jr., Weber, D.J. et al. (2010). *Tissue Eng. Part A* 16: 3309.

193 Wolf, M.T., Daly, K.A., Reing, J.E., and Badylak, S.F. (2012). *Biomaterials* 33: 2916.

194 Quarti, A., Nardone, S., Colaneri, M. et al. (2011). *Interact. Cardiovasc. Thorac. Surg.* 6: 569.

195 Ulijn, R.V. (2006). *J. Mater. Chem.* 16: 2217.

196 Griffith, L.G. and Sperinde, J.J. (1997). *Macromolecules* 30: 5255.

197 Griffith, L.G. and Sperinde, J.J. (2000). *Macromolecules* 33: 5476.

198 Schense, J.C. and Hubbell, J.A. (1998). *Bioconjugate Chem.* 10: 75.

199 Sakiyama-Elbert, S.E. and Hubbell, J.A. (2000). *J. Controlled Release* 69: 149.

200 Sanborn, T.J., Messersmith, P.B., and Barron, A.E. (2002). *Biomaterials* 23: 2703.

201 Franke, K., Baur, M., Daum, L. et al. (2013). *Neurosci. Lett.* 541: 248.

202 Chen, P.Y., Yang, K.C., Wu, C.C. et al. (2014). *Acta Biomater.* 10: 912.

203 Briganti, S., Camera, E., and Picardo, M. (2003). *Pigm. Cell Res.* 16: 101.

204 Mapunya, M.B., Hussein, A.A., Rodriguez, B., and Lall, N. (2011). *Phytomedicine* 18: 1006.

205 Chen, Q.X. and Kubo, I. (2002). *J. Agric. Food Chem.* 50: 4108.

206 Prezioso, J.A., Epperly, M.W., Wang, N., and Bloomer, W.D. (1992). *Cancer Lett.* 63: 73.

207 Oetting, W. and King, R. (1994). *J. Invest. Dermatol.* 103: 131S.

208 Tessari, I., Bisaglia, M., Valle, F. et al. (2008). *J. Biol. Chem.* 283: 16808.

209 Chen, T., Small, D.A., Wu, L.Q. et al. (2003). *Langmuir* 19: 9382.

210 Chen, T., Embree, H.D., Brown, E.M. et al. (2003). *Biomaterials* 24: 2831.

211 Jin, R., Lin, C., and Cao, A. (2014). *Polym. Chem.* 5: 391.

212 Mishra, D., Bhunia, B., Banerjee, I. et al. (2011). *Mater. Sci. Eng., C* 31: 1295.

213 Teixeira, L.S., Feijen, J., van Blitterswijk, C.A. et al. (2012). *Biomaterials* 33: 1281.

214 Ren, K., He, C., Cheng, Y. et al. (2014). *Polym. Chem.* 5: 5069.

215 Lee, Y., Bae, J.W., Oh, D.H. et al. (2013). *J. Mater. Chem. B* 1: 2407.

216 Tran, N.Q., Joung, Y.K., Lih, E., and Park, K.D. (2011). *Biomacromolecules* 12: 2872.

217 Lih, E., Lee, J.S., Park, K.M., and Park, K.D. (2012). *Acta Biomater.* 8: 3261.

218 Kurisawa, M., Chung, J.E., Yang, Y.Y. et al. (2005). *Chem. Commun.* 14: 4312.

219 Sakai, S., Tsumura, M., Inoue, M. et al. (2013). *J. Mater. Chem. B* 1: 5067.

220 Sakai, S. and Kawakami, K. (2008). *J. Biomed. Mater. Res. Part A* 85A: 345.

221 Sakai, S., Hirose, K., Moriyama, K., and Kawakami, K. (2010). *Acta Biomater.* 6: 1446.

222 Jin, R., Hiemstra, C., Zhong, Z.Y., and Feijen, J. (2007). *Biomaterials* 28: 2791.

223 Jin, R., Teixeira, L.S.M., Dijkstra, P.J. et al. (2009). *Biomaterials* 30: 2544.

224 Sakai, S., Yamada, Y., Zenke, T., and Kawakami, K. (2009). *J. Mater. Chem.* 19: 230.

225 Wang, L.S., Chung, J.E., Chan, P.P.Y., and Kurisawa, M. (2010). *Biomaterials* 31: 1148.

226 Lee, F., Chung, J.E., and Kurisawa, M. (2008). *Soft Matter* 4: 880.

227 Lee, F., Chung, J.E., and Kurisawa, M. (2009). *J. Controlled Release* 134: 186.

228 Mosiewicz, K.A., Johnsson, K., and Lutolf, M.P. (2010). *J. Am. Chem. Soc.* 132: 5972.
229 Zhao, F., Gao, Y., Shi, J. et al. (2010). *Langmuir* 27: 1510.
230 Bakota, E.L., Aulisa, L., Galler, K.M., and Hartgerink, J.D. (2010). *Biomacromolecules* 12: 82.
231 Wang, Q., Yang, Z., Gao, Y. et al. (2008). *Soft Matter* 4: 550.
232 Tibbitt, M.W. and Anseth, K.S. (2009). *Biotechnol. Bioeng.* 103: 655.
233 Kretlow, J.D. and Mikos, A.G. (2008). *AlChE J.* 54: 3048.
234 Novosel, E.C., Kleinhans, C., and Kluger, P.J. (2011). *Adv. Drug Delivery Rev.* 63: 300.
235 Franssen, O., Vos, O.P., and Hennink, W.E. (1997). *J. Controlled Release* 44: 237.
236 Hahn, S.K., Park, J.K., Tomimatsu, T., and Shimoboji, T. (2007). *Int. J. Biol. Macromol.* 40: 374.
237 Truong, V.X., Barker, I.A., Tan, M. et al. (2013). *J. Mater. Chem. B* 1: 221.
238 Hovgaard, L. and Brondsted, H. (1995). *J. Controlled Release* 36: 159.
239 Ashton, R.S., Banerjee, A., Punyani, S. et al. (2007). *Biomaterials* 28: 5518.
240 Kumashiro, Y., Ooya, T., and Yui, N. (2004). *Macromol. Rapid Commun.* 25: 867.
241 Yamamoto, N., Kurisawa, M., and Yui, N. (1996). *Macromol. Rapid Commun.* 17: 313.
242 Burke, M.D., Park, J.O., Srinivasarao, M., and Khan, S.A. (2005). *J. Controlled Release* 104: 141.
243 Yu, X. and Pishko, M.V. (2011). *Soft Matter* 7: 8898.
244 Tan, J.X., Wang, X.Y., Li, H.Y. et al. (2011). *Int. J. Cancer* 128: 1303.
245 Jin, R., Lou, B., and Lin, C. (2013). *Polym. Int.* 62: 353.
246 Anderson, S.B., Lin, C.C., Kuntzler, D.V., and Anseth, K.S. (2011). *Biomaterials* 32: 3564.
247 Klinger, D., Aschenbrenner, E.M., Weiss, C.K., and Landfester, K. (2012). *Polym. Chem.* 3: 204.
248 Purcell, B.P., Lobb, D., Charati, M.B. et al. (2014). *Nat. Mater.* 13: 653.
249 Thornton, P.D., McConnell, G., and Ulijn, R.V. (2005). *Chem. Commun.* 5913.
250 Thornton, P.D., Mart, R., and Ulijn, R.V. (2007). *Adv. Mater.* 19: 1252.
251 Thornton, P.D., Mart, R.J., Webb, S.J., and Ulijn, R.V. (2008). *Soft Matter* 4: 821.
252 McDonald, T.O., Qu, H., Saunders, B.R., and Ulijn, R.V. (2009). *Soft Matter* 5: 1728.
253 Estroff, L.A. and Hamilton, A.D. (2004). *Chem. Rev.* 104: 1201.
254 Schneider, H.J. and Strongin, R.M. (2009). *Acc. Chem. Res.* 42: 1489.
255 Piepenbrock, M.O.M., Lloyd, G.O., Clarke, N., and Steed, J.W. (2010). *Chem. Rev.* 110: 1960.
256 Kim, H.J., Lee, J.H., and Lee, M. (2005). *Angew. Chem. Int. Ed.* 44: 5810.
257 Silva, G.A., Czeisler, C., Niece, K.L. et al. (2004). *Science* 303: 1352.
258 Haines-Butterick, L., Rajagopal, K., Branco, M. et al. (2007). *Proc. Natl. Acad. Sci. U.S.A.* 104: 7791.
259 Komatsu, H., Matsumoto, S., Tamaru, S.I. et al. (2009). *J. Am. Chem. Soc.* 131: 5580.

260 Aggeli, A., Bell, M., Boden, N. et al. (1997). *Nature* 386: 259.

261 Xu, X.D., Lin, B.B., Feng, J. et al. (2012). *Macromol. Rapid Commun.* 33: 426.

262 Du, X., Zhou, J., and Xu, B. (2014). *Chem. Asian J.* https://doi.org/10.1002/asia.201301693.

263 Lee, K.Y. and Mooney, D.J. (2012). *Prog. Polym. Sci.* 37: 106.

264 Holmes, T.C., de Lacalle, S., Su, X. et al. (2000). *Proc. Natl. Acad. Sci. U.S.A.* 97: 6728.

265 Wu, D., Wang, T., Lu, B. et al. (2008). *Langmuir* 24: 10306.

266 Wang, T., Jiang, X.J., Tang, Q.Z. et al. (2009). *Acta Biomater.* 5: 2939.

267 Abu Hashim, I.I., Higashi, T., Anno, T. et al. (2010). *Int. J. Pharm.* 392: 83.

268 Chen, Y., Pang, X.H., and Dong, C.M. (2010). *Adv. Funct. Mater.* 20: 579.

269 Zhang, X.L., Huang, J., Chang, P.R. et al. (2010). *Polymer* 51: 4398.

270 Ni, X.P., Cheng, A., and Li, J. (2009). *J. Biomed. Mater. Res. Part A* 88A: 1031.

271 Ulijn, R.V., Bibi, N., Jayawarna, V. et al. (2007). *Mater. Today* 10: 40.

272 Yang, Z. and Xu, B. (2007). *J. Mater. Chem.* 17: 2385.

273 Yang, Z.M., Gu, H.W., Fu, D.G. et al. (2004). *Adv. Mater.* 16: 1440.

274 Sadownik, J.W., Leckie, J., and Ulijn, R.V. (2011). *Chem. Commun.* 47: 728.

275 Yang, Z.M., Liang, G.L., Wang, L., and Xu, B. (2006). *J. Am. Chem. Soc.* 128: 3038.

276 Gao, J., Zheng, W.T., Kong, D.L., and Yang, Z.M. (2011). *Soft Matter* 7: 10443.

277 Hirst, A.R., Roy, S., Arora, M. et al. (2010). *Nat. Chem.* 2: 1089.

278 Woodcock, J.W., Jiang, X.G., Wright, R.A.E., and Zhao, B. (2011). *Macromolecules* 44: 5764.

279 Smith, A.M., Williams, R.J., Tang, C. et al. (2008). *Adv. Mater.* 20: 37.

280 Zhou, M., Smith, A.M., Das, A.K. et al. (2009). *Biomaterials* 30: 2523.

281 Hughes, M., Frederix, P.W.J.M., Raeburn, J. et al. (2012). *Soft Matter* 8: 5595.

282 Schellenberger, V. and Jakubke, H.D. (1991). *Angew. Chem. Int. Ed. Engl.* 30: 1437.

283 Bordusa, F. (2002). *Chem. Rev.* 102: 4817.

284 Qin, X., Xie, W., Butterfoss, G.L. et al. (2013). *Chem. Commun.* 49: 4839.

285 Jun, H.W., Yuwono, V., Paramonov, S.E., and Hartgerink, J.D. (2005). *Adv. Mater.* 17: 2612.

286 Chau, Y., Luo, Y., Cheung, A.C.Y. et al. (2008). *Biomaterials* 29: 1713.

287 Galler, K.M., Aulisa, L., Regan, K.R. et al. (2010). *J. Am. Chem. Soc.* 132: 3217.

288 Giano, M.C., Pochan, D.J., and Schneider, J.P. (2011). *Biomaterials* 32: 6471.

289 Lin, Y.A., Ou, Y.C., Cheetham, A.G., and Cui, H. (2014). *Biomacromolecules* 15: 1419.

290 Law, B., Weissleder, R., and Tung, C.H. (2006). *Biomacromolecules* 7: 1261.

291 Xing, Y.Z., Cheng, E.J., Yang, Y. et al. (2011). *Adv. Mater.* 23: 1117.

292 Gajanayake, T., Olariu, R., Leclère, F.M. et al. (2014). *Science* 6: 249ra110.

293 Webber, M.J., Newcomb, C.J., Bitton, R., and Stupp, S.I. (2011). *Soft Matter* 7: 9665.

294 Williams, R.J., Smith, A.M., Collins, R. et al. (2009). *Nat. Nanotechnol.* 4: 19.

295 Debnath, S., Roy, S., and Ulijn, R.V. (2013). *J. Am. Chem. Soc.* 135: 16789.

296 Glick, B.R. and Pasternak, J.J. (1998). Microbial production of therapeutic agents. In: *Molecular Biotechnology: Principles Applications of Recombinant DNA* (ed. B.R. Glick and C.L. Patten), 235. Washington, DC: ASM Press.

297 Miyata, T. and Uragami, T. (2001). *Polymeric Biomaterials. Biological Stimulus-Responsive Hydrogels*, 959. New York: CRC Press.

298 Miyata, T., Asami, N., and Uragami, T. (1999). *Macromolecules* 32: 2082.

299 Miyata, T., Asami, N., and Uragami, T. (1999). *Nature* 399: 766.

300 Miyata, T., Hayashi, T., Kuriu, Y., and Uragami, T. (2012). *J. Mol. Recognit.* 25: 336.

301 Wakayama, H., Henares, T.G., Jigawa, K. et al. (2013). *Lab Chip* 13: 4304.

302 Jung, S. and Yi, H. (2014). *Langmuir* 30: 7762.

303 Gübeli, R.J., Hövermann, D., Seitz, H. et al. (2013). *Adv. Funct. Mater.* 23: 5355.

304 Tseng, S.J., Zeng, Y.F., Deng, Y.F. et al. (2013). *Chem. Commun.* 49: 2670.

305 Bader, A.G., Brown, D., Stoudemire, J., and Lammers, P. (2011). *Gene Ther.* 18: 1121.

306 Qureshi, I.A. and Mehler, M.F. (2012). *Nat. Rev. Neurosci.* 13: 528.

307 de Fougerolles, A.R. (2008). *Hum. Gene Ther.* 19: 125.

308 Dykxhoorn, D.M., Palliser, D., and Lieberman, J. (2006). *Gene Ther.* 13: 541.

309 Song, E., Zhu, P., Lee, S.K. et al. (2005). *Nat. Biotechnol.* 23: 709.

310 Krebs, M.D., Jeon, O., and Alsberg, E. (2009). *J. Am. Chem. Soc.* 131: 9204.

311 Lin, D.C., Yurke, B., and Langrana, N.A. (2005). *J. Mater. Res.* 20: 1456.

312 Murakami, Y. and Maeda, M. (2005). *Biomacromolecules* 6: 2927.

313 Liu, J., Cao, Z., and Lu, Y. (1948). *Chem. Rev.* 2009: 109.

314 Yang, H., Liu, H., Kang, H., and Tan, W. (2008). *J. Am. Chem. Soc.* 130: 6320.

315 Yang, L., Zhu, Z., Zou, Y. et al. (2013). *J. Am. Chem. Soc.* 135: 3748.

316 Wang, X. and Wang, X. (2013). *Chem. Commun.* 49: 5957.

317 Liu, R., Huang, Y., Ma, Y. et al. (2015). *ACS Appl. Mater. Interfaces* 7: 6982.

318 Egholm, M., Buchardt, O., Nielsen, P.E., and Berg, R.H. (1992). *J. Am. Chem. Soc.* 114: 1895.

319 Betts, L., Josey, J.A., Veal, J.M., and Jordan, S.R. (1995). *Science* 270: 1838.

320 Egholm, M., Nielsen, P.E., Buchardt, O., and Berg, R.H. (1992). *J. Am. Chem. Soc.* 114: 9677.

321 Ishizuka, T., Yang, J., Komiyama, M., and Xu, Y. (2012). *Angew. Chem. Int. Ed.* 51: 7198.

322 Chu, T.W., Feng, J., Yang, J., and Kopecek, J. (2015). *J. Controlled Release* 220: 608.

4

Self-Assembled Nanomaterials

4.1 Introduction

Many diseases including cardiovascular, diabetes, osteoporosis, and cancers cannot be treated effectively with current clinical therapy because of significant reduction in the level of tissue regeneration. Every day, thousands of people of all ages are admitted to hospitals because of the malfunction of some vital organ. Because of the dearth of transplantable organs, many of these people will die. The ever-growing demand for donor organs to meet the needs of individuals on waiting lists will likely never be met. In short, the need for organs cannot be met by traditional methods of transplantation. Regenerative medicine may change that. Regenerative medicine is a new field of science that uses stem cells to regenerate biological tissues and improve tissue functions. Replacing these lost cells using stem cell-based therapies offers the possibility of exciting alternative treatments of current clinical medicine. Regenerative medicine will only be a topic of the future rather than the present and is based on combinational technology of materials science, stem cells technology, and reconstructive surgery that aim to regenerate natural tissues as well as to create biological substitutes for defective or lost organs and tissues.

The design of materials that can regulate cell behavior such as proliferation and differentiation is a key component in regenerative medicine therapy. From the viewpoint of immune system response of the body, the implanted biomaterials should mimic the structure and biological function of native extracellular matrix (ECM) in terms of chemical composition and physical properties. Tissue-engineered scaffolds used in regenerative medicine have micron dimensions that fail to mimic the structure of natural ECM. Biomaterials have been widely used to fabricate tissue-engineered scaffolds because of their good biological activity that current researches appear to indicate for these materials. Remarkably, the recent identification of nanotechnology with enhanced ability to mimic natural ECM proteins such as collagen has led to the discovery of new class of biomaterials with specific properties for tissue engineering applications. One of the common approaches to produce fibers similar to ECM proteins is self-assembling of collagen. This technique is sufficient to cause its rapid transduction into a variety of different tissues *in vitro* as well as *in vivo*. Moreover, this novel technique for proteins and peptides appears to circumvent many problems associated with cells and drug-based methods.

Biomedical Engineering: Materials, Technology, and Applications, First Edition. Hossein Hosseinkhani.

Nanomaterials technology has been approached in two ways: top-down and bottom-up approaches [1]. The top-down approach is miniaturization techniques to construct micro- and nanoscale structures from a macroscopic material or a group of materials by utilizing machining and etching techniques. The best example of this approach is photolithography technique that have been widely used in semiconductor industry to fabricate components of an integrated circuit by etching micro- and nanoscale patterns on a silicon wafer [2]. The bottom-up approach refers to the construction of macromolecular structures from atoms or molecules that have the ability to self-organize or self-assemble to form a macroscopic structure [3]. In other words, this approach has been referred as molecular nanotechnology. These nanomaterials can be effectively applied in the field of regenerative medicine, pharmacotherapeutics, and medical health care to discover novel drugs, drug delivery systems, imaging, and gene therapy. Self-assembly, electrospinning, and phase separation are three different approaches toward the formation of nanofibrous materials [4]. The fabricated materials by each method are very different in terms of physical and chemical characteristics, which lead to its development and biological applications. Self-assembly is a very unique and attractive method that can fabricate small diameter of nanofibers in the range of natural ECM. Electrospinning can only form large diameter nanofibers that are very far from the range of natural ECM. Phase separation, on the other hand, can generate nanofibers in the same range as natural ECM and allows for the design of macroporous structures. Figure 4.1 shows a schematic illustration of an electrospinning machine, and Figure 4.2 shows scanning electron microscope (SEM) image of electrospun poly(ethylene terephthalate) (PET) nanofibers.

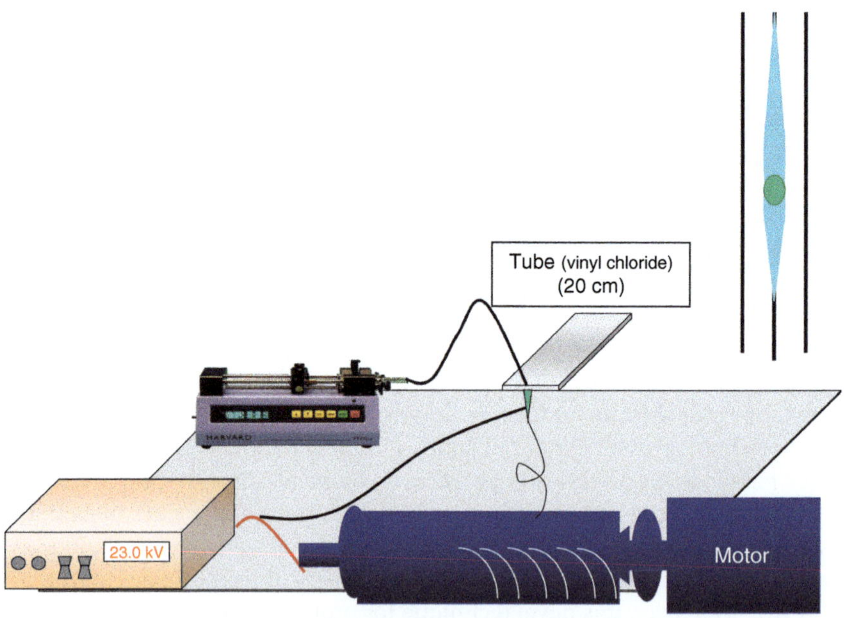

Figure 4.1 Schematic illustration of an electrospinning machine.

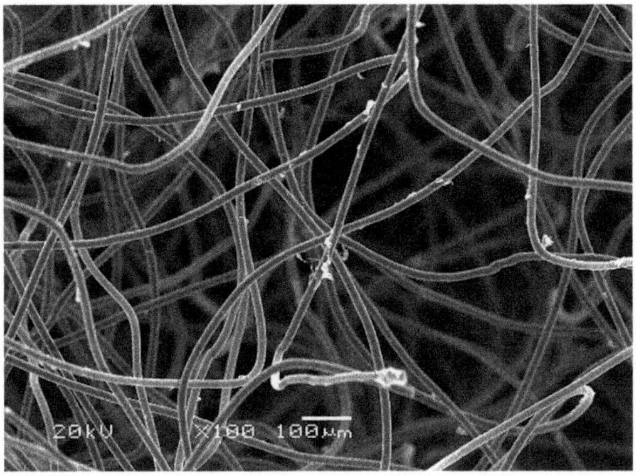

Figure 4.2 Scanning electron microscope (SEM) image of electrospun poly(ethylene terephthalate) (PET) nanofibers.

Self-assembly is the autonomous organization of individual components into patterns or structures without human intervention. Self-assembly has been defined as "the non-covalent interaction of two or more molecular subunits to form an aggregate whose novel structures and properties are determined by the nature and positioning of the individual components" [5]. Self-assembly has also been defined as "the spontaneous assembly of molecules into structured, stable, non-covalently joined aggregates" [6, 7]. Self-assembly is the inherent ability of numerous multimeric biological structures to assemble from their component parts through random movements of molecules and formation of weak chemical bonds between surfaces with complementary shapes. A few examples of self-assembled structures that can be found in biological systems are the phospholipid bilayer of human cell membranes [8] and RNA and DNA complexes [9, 10]. The detergent surfactant molecules exhibit self-assembly phenomenon due to their amphiphilic properties [11]. The molecular building mechanisms underlying the formation of bacteriophage and viral particles are all based on self-assembly [12]. Molecular self-assembly is a powerful phenomenon borrowed from nature by scientists for fabricating novel supramolecular architectures and defined as spontaneous organization of molecules under near-thermodynamic equilibrium conditions into structurally well-defined and stable arrangements through non-covalent interactions [11]. This is mainly governed by weak non-covalent bonds like electrostatic interactions (ionic bonds), hydrogen bonds, hydrophobic and hydrophilic interactions, water-mediated hydrogen bonds, and van der Waals interactions. Although these forces are weak, their collective interactions can produce molecules that are structurally and chemically stable. Frequently, molecular self-assembly relies on chemical complementarily and structural compatibility [13]. The molecular components need complementary properties such as specific surface characteristics, surface charge and polarizability, and mass and surface functionalities to self-assemble into different physiological forms [13]. Biological molecules like proteins, peptides, nucleic acids, lipids, and other cellular components with complementary properties self-assemble to form the fundamental biological unit – the cell. Cellular events like amyloid fibril formation, antigen–antibody recognition, chromatin assembly, and phospholipid membrane self-assembly are excellent examples of molecular self-assembly.

This chapter will consider the self-assembled systems and the recent developments for their potential applications in regenerative medicine. Self-assembling proteins and peptides for designing novel biomaterials and their potential applications in regenerative medicine and biomedical applications will be also discussed in detail from the viewpoint of their biological applications.

4.2 Classification of Self-Assembled Systems

Self-assembly is a native process. It can be classified into two types: static and dynamic [14]. Most research studies on self-assembly have been focused on static type, while the studies on dynamic self-assembly is still poorly understood. Static self-assembly contributes to systems that are at global or local equilibrium and do not dissipate energy [14]. In static self-assembly, formation of the ordered structure may require energy, but it is stable once it is formed. A few examples of static self-assembly phenomenon tailored by nature are lipid molecules forming oil droplets in water, four hemoglobin polypeptides forming a functional hemoglobin protein, and the combination of RNA and ribosomal proteins to form a functional ribosome. Dynamic self-assembly occurs when the formation of an ordered state of equilibrium requires dissipation of energy. In other words, the interactions responsible for the formation of structures or patterns between components occur only if the system dissipates energy [14]. Self-assembly takes place at molecular, mesoscopic, and macroscopic scales. Based on this criteria, self-assembly has been classified as molecular and nanoscale self-assemblies (classical form of self-assembly in chemistry involving atoms, molecules, and crystal formation) and meso- and macroscopic self-assemblies (e.g. engineered microparts). Molecular and nanoscale self-assembly can be further classified as intramolecular and intermolecular self-assembly [11]. In intramolecular self-assembly, molecules are often complex polymers that can assemble from the random coil into a well-defined stable structure (secondary and tertiary structure). Protein folding to form secondary and tertiary amine structures is a good example of intramolecular self-assembly. Intermolecular self-assembly is the ability of molecules to form supramolecular assemblies (quaternary structure). Supramolecular micelle formation by surfactant molecules in solution and self-assembled monolayers (SAMs) on a substrate are classic examples of intermolecular self-assembly. Magnetic, capillary, electrostatic, and gravitational forces play a vital role in self-assembly of structures on the meso and macro scale levels [11].

4.2.1 Proteins and Peptides

Proteins are fundamental components of all living cells. They can be considered as a group of complex organic macromolecules containing carbon, hydrogen, nitrogen, oxygen, and sulfur. They are composed of one or more chains of amino acids. Two or more amino acids linked by a peptide bond form a peptide molecule. Large numbers of peptide molecules arrange themselves in different fashions to make up different kinds of proteins. Enzymes, hormones, and antibodies are few examples of biological substances that are made up of proteins and are required for the proper function in living organism. Structural analysis of protein molecules has revealed that they take up various shapes to form a stable macroscopic structure.

Nature has used proteins to build a vast array of structures like keratin, collagen, coral, pearl, and shell. In the past few decades, numerous researches have been done to understand the structural characteristics that influence the self-assembly of protein and peptide molecules. The potential applications of self-assembling proteins and peptides have been effectively utilized to design novel biomaterials that have tremendous applications in the fields of biotechnology and biomedicine [11–20].

4.2.2 Self-Assembly of Proteins and Peptides

Several peptides and proteins that self-assemble to form various nanostructures like nanotubes, vesicles, helical ribbons, and fibrous scaffolds are discussed in detail. These structures are analyzed to design and fabricate new materials that will have potential applications in biomedical nanotechnology. The findings in the research paper have been discussed and grouped under different headings like self-assembling peptide (SAP) systems, amphiphilic and surfactant peptides, peptides self-assembly into 3D matrix scaffolds, and designed peptide hydrogels for 3D cell culture and regenerative biology [21, 22]. Wang et al. have designed artificial proteins that self-assemble to form hydrogels [23]. The artificial proteins designed by the researchers were made up of ionic self-complementary peptide group that had an alternating polar and nonpolar fashion of arrangement of peptide molecules. These peptides formed a stable β-strand and β-sheet structures that self-assembled to form nanofibers. These nanofibers form interwoven matrices that further give a scaffold hydrogel with high water content. This hydrogel structure has water as its dispersion medium and responds to changes in pH and other environmental factors. These protein hydrogels can be used for advanced wound closure and tissue repair in regenerative medicine and tissue engineering. Biodegradable protein hydrogels can be utilized in drug delivery systems, delivering pharmaceutical protein complexes in treatment of diseases like cancer. Adams et al. have designed peptide nanotubes using surfactant-like peptides [24]. These peptide nanotubes can be used as templates for growing metal nanocrystals and to fabricate nanowires. Peptide nanotubes can also serve as ion channels when incorporated in phospholipid bilayer of the cell membrane. Recent studies also reported biomimetic protein structures and peptide systems that can form complexes with metals and semiconducting elements. Surface-binding peptides can bind covalently with metal surfaces like gold [25]. These peptides can also form complexes with DNA, which in turn can be bound to a metal surface. This property can be exploited to design and fabricate nano-biosensors. Certain peptides with strong dipoles undergo drastic conformational changes between the α-helical and the β-sheet structures. These are called as molecular switch peptides. Gold nanoparticles can be attached to these dipolar peptides to fabricate tiny molecular switches [26].

4.2.3 Amphiphilic and Surfactant-like Peptides

Amphiphilic molecules have a hydrophilic (polar) and a hydrophobic (nonpolar) component. In the presence of water, they self-assemble into distinct structures whose shape is largely determined by the size and shape of the hydrophilic polar head. These molecules have one or two amino acids in the polar region and four or more consecutive hydrophobic amino acids in their nonpolar end. One such example discussed is the V_6D amino acid complex. The V_6D amino acid sequence (VVVVVVD) has six valine (V) residues that are hydrophobic and an aspartic acid (D) residue that is negatively charged [27, 28]. Valine (α-aminoisovaleric acid) is an essential amino acid, whereas aspartic acid is a nonessential amino acid. The V_6D peptide complex formed various nanostructures in aqueous solution like nanotubes and nanovesicles. The samples in aqueous solution were frozen in liquid propane (−180 °C) and surface-coated with a thin layer of platinum and carbon to preserve the structures formed. Zhang et al. observed nanotubes and nanovesicles by transmission electron microscopy (TEM). The nanotubes measured about 30–50 nm [29]. It has been demonstrated that changing the sequence of the amino acids in the peptide chain and the environmental factors can modify the self-assembled structure. These peptide nanotubes can be incorporated into self-assembled membranes to use in bio-nanosensor devices [30]. It is also suggested that these surfactant peptides can be functionally engineered using techniques like biotinylation. Biotinylation is a process of incorporation of biotinyl groups into molecules to visualize specific substrates by incubating with avidin or streptavidin. It is a rapid method of detecting nucleic acids in Western blot technique. When these surfactant peptide nanostructures are made to undergo the process of biotinylation, they can bind to streptavidin-coated inorganic metal surface. Histidine-tagged peptides and proteins can be bound to nickel surfaces. Thus, these nanostructures can be attached to metallic surfaces.

4.2.4 Three-Dimensional Peptides Matrix

A wide variety of self-assembling proteins and peptides have inspired the searchers to fabricate nanoscale fibers and fiber network [16, 17]. The main factor influencing the entire process was the chirality of the individual building components of the peptide complex. Two molecules are said to be chiral if their mirror images do not superimpose on each other. Zhang et al. have studied the KFE_8 peptide complex, which is an eight-residue peptide complex with a FKFEFKFE sequence [31]. This structure represents a group of SAPs that spontaneously assemble under certain physiological conditions. This peptide self-assembles in aqueous solution into left-handed helical ribbons when the peptide backbone is twisted in the opposite direction [31]. When certain amino acids (such as K) in the hydrophobic side chains replaced with another amino acid sequence, little changes were observed. On the other hand, when a positively charged lysine (Lys) was replaced by a positively charged arginine (Arg), likewise when a negatively charged glutamate (Glu) was replaced by a negatively charged aspartate (Asp), very little changes were observed in the nanofibers that were formed [31]. But when the positively charged residue was replaced by a negatively charged residue or vice versa, the peptides did not self-assemble. When the hydrophobic residues replaced with the alanine, there was a greater tendency to self-assemble and form peptide matrices with enhanced strength. This has led the researchers to concentrate their attention toward understanding the basis of protein conformational diseases. Protein conformational diseases are a group of disorders characterized by accumulation of malformed protein structures in cells. Proteins must fold into a proper three-dimensional (3D) structure to carry out their normal functions. But when they do not fold properly, they form malfolded protein structures that accumulate in cells leading to pathological conditions. Alzheimer's disease, prion disease, and Parkinson's disease are a few examples of protein conformational diseases. Thus, by understanding the mechanism of formation of peptide nanofibers and the factors controlling their self-assembly, researchers are aiming to formulate a remedy for protein conformational diseases.

Self-assembled peptide fibers can be classified into three theoretical models. The first model is the molecular model where the β-sheet peptides self-assemble into helical ribbons. The second model is the semi-continuum model where the peptides self-assemble to form elastic tape-like structures composed of brick-like building blocks. The third model is the full continuum model where the peptides self-assemble to form tubules. These approaches have helped to know more about the mechanisms underlying the formation of different structures when peptides and proteins self-assemble and will ultimately guide them to design efficient peptide-based and protein-based biomaterials.

4.3 *In Vitro* Biological Approaches of Self-Assembled Systems

Since several years ago, peptides and nucleotides were considered to be useful building blocks for materials engineering, and molecular self-assembly is known as an important route to produce novel materials that can cover a wide range of biological applications. Many amphiphilic peptides have been used in different applications, thanks to their facilities to form nanostructures [32].

4.3.1 Antimicrobial Activities

Cationic antimicrobial peptides can kill the microbes by interacting and disrupting bacterial cell membranes. After adopting stable secondary structures (α-helices, relaxed coils, β-sheets), the peptide sequences with 6–15 amino acids rich in hydrophilic moieties such as K, R, D, or E and in hydrophobic moieties such as A, V, I, L, F, W, or Y display characteristic amphipathicity [33–37]. Amphiphilic peptide A_mK ($m = 3, 6, 9$) displayed a large antimicrobial activities with permeation and disruption to the bacterial membranes [32]. Mitra et al. have studied lipopeptides-based molecules with P, F, or W as part of the head groups and

C_{14} as tails [36]. These lipopeptides showed high growth inhibition activity on both gram-positive and gram-negative bacteria and fungus. C_{16}-KXK series of lipopeptides (X is A, G, L, or K) have also been reported to be potent to both bacteria and fungi. Among the lipopeptides studied, C_{16}-KKK is the best antimicrobial compound.

Recently, the membrane translocation sequence TAT (YGRKKRRQRRR) has been used to construct an antimicrobial agent [38]: Chol-G_3R_6TAT contains a hydrophobic tail of cholesterol, three glycine residues as spacer, and six arginine to enhance the performance of membrane translocation sequence TAT. The formation of micelles resulted in an increased cationic charge density outside of the nanoparticles, therefore enhancing the antimicrobial activity. The presence of the TAT sequence could also help the nanoparticles to cross the blood–brain barrier (BBB) to the brain, hereby making the molecule a good candidate for the brain infection treatment. The nanoparticles showed high antimicrobial effects against drug-resistant bacteria, yeast, and fungi. Electronic microscopic results revealed the disruption and lysis of cell walls of both bacteria and fungi.

4.3.2 Three-Dimensional Cell Culture

The attractive features of nanostructures associated with gel network formation and antimicrobial effects from peptide amphiphiles (PAs) make them good candidates as cell culture matrixes or scaffolds in tissue engineering and regenerative medicine. Extensive studies have already indicated their good biocompatibility [39].

Hydrogels constructed through self-assembly are receiving increasing attention for a variety of biomedical and biotechnological applications, including scaffolds for regenerative medicine, the controlled release of therapeutics, and defined cell culture matrices [19, 40–44].

β-Sheet fibrillizing peptides and peptide amphiphiles have received particular attention recently as matrices for cells owing to their stimulus-sensitive fibril formation, their ability to form hydrogels, their ease of synthesis, and the availability of many amino acid sequences known to influence cell behavior through the binding of integrins and other receptors [16, 40–51]. These studies have been performed using the RAD_{16} peptide self-assembling sequences from laminin, or self-assembling sequences from transthyretin, demonstrating that several different β-sheet fibril-forming peptides are able of presenting bioavailable ligands on their surfaces [44–49]. The incorporation of ligands within β-sheet fibrillar hydrogels can significantly alter their mechanical properties, complicating the interpretation of cell behavior on these materials [47, 52].

Some reports have described non-immunogenic unfunctionalized fibrillar peptide assemblies and non-inflammatory ligand-bearing assemblies [40, 53–55]. Jung and coworkers [56] investigated the co-assembly, gelation, ligand presentation, technical properties, and immunogenicity of peptide hydrogels based on the sequence QQKFQFQFEQQ (Q_{11}), which has previously been utilized for producing substrates for endothelial cells with adjustable stiffness [57]. They reported a co-assembling set of peptides based on the sequence of Q_{11} that form hydrogels, where the display of multiple ligands may be adjusted simply by mixing different peptides in solution and inducing gelation. RGDS-Q_{11} and IKVAVQ_{11} were quantitatively incorporated into background gels of Q_{11} in a wide range of peptide ratios, and ligand incorporation had no significant impact on fibril morphology or secondary structure. Gel viscoelasticity was minimally changed upon ligand inclusion, allowing ligand incorporation to be adjusted independently of gel mechanics. Both RGDS-Q_{11} and IKVAV-Q_{11} were presented on the surface of co-assemblies with Q_{11}, and these ligands modulated HUVEC behavior *in vitro*. RGDS-Q_{11} significantly affected HUVEC attachment, spreading, and proliferation, while IKVAV-Q_{11} had a small effect on cell attachment and a subtle influence on cell morphology.

The ECM contains a lot of signals that activate various intracellular signaling pathways to control and guide the cell behavior. Recapitulating the ECM regulatory mechanisms is of central importance in fundamental cell studies and cell-based applications such as tissue engineering. Methods have been established to capture the nanotopographical and biochemical characteristics in the natural ECM [58, 59]. Most of these methods, however, entail chemical syntheses that lack an adequate flexibility to generate tunable biochemical patterns for the rational design of extracellular environment. A comprehensive understanding of ECM cell interactions and the screening of biomaterials may involve high-throughput studies that require systematically varying material and biochemical properties in a facile way [60, 61]. Chau et al. [52] explored a peptide-based platform for enriching biological functions in self-assembling 3D scaffolds. A new family of biofunctional materials can potentially be obtained from automated peptide synthesis. SAPs have recently emerged as an attractive class of 3D scaffolding materials, mainly due to their nanoscale fibrous and porous topographies that mimic the natural ECM features [62–64]. Among them, $(RADA)_4$ peptides have been used to form scaffolds *in situ* for tissue engineering applications [19, 53]. This material has been shown to support the growth and differentiation of a variety of cells, including those originated from human, mammals, mouse, and chicken and covering stem cells, progenitor cells, and established cell lines [65]. Zhao et al. investigated a new SAP template with two modules: one for generating ordered secondary structures to enable hydrogel formation and the other for supplying biological cues to elicit specific cellular responses.

Fmoc-RGD mimicking the ECM has also been reported [66]. These molecules self-assembled into nanofibers and bioactive hydrogels through π–π stacking of the Fmoc groups, leaving the RGD groups outside the nanofiber surfaces. The self-assembled hydrogels displayed excellent performance in 3D cell culture using human adult dermal fibroblast cells.

Peptide amphiphiles have also been used to create 3D microscale topographical patterns to study the behavior of human mesenchymal stem cells [67]. The hydrophilic part of the amphiphile was a peptide sequence RGDSKKLLA(K) containing the cell adhesive epitope RGD, while the hydrophobic tail was alkyl chain ($-C_8H_{16}$–diacetylene–$C_{12}H_{25}$) bearing a photosensitive group diacetylene, which could covalently link the self-assembled molecules upon UV irradiation. Cell growth on the patterned peptide amphiphile surfaces demonstrated that cells recognized not only the biomolecular signaling provided by RGDS epitopes but also the physical guidance provided by the topographical patterns. Mata et al. found that cell differentiation was significantly affected by the type of substrate created. Hole microtextures were better for the osteoblast differentiation than all other surfaces. The negatively charged C_{16}–$V_3A_3E_3$ peptide-coated surface was demonstrated to be suitable for the growth of bone marrow mononuclear cells. A binary peptide system containing 10 wt% C_{16}–$V_3A_3K_3$RGDS and 90 wt% C_{16}–$V_3A_3E_3$ lipopeptide molecules was found to promote optimal cell adhesion [68]. *In vivo* delivery of luciferase-expressing cells using the binary lipopeptide nanofiber system into the mouse model revealed the enhanced viability and proliferation of associated bone marrow-derived stem and progenitor cells. Lipopeptides with branched head groups containing RGDS also showed excellent performance as scaffolds for the growth of human bladder smooth muscle cells [69].

4.3.3 Self-Assembled Systems in Drug Delivery

The ways in which drugs are administered have gained increasing attention in the past two decades. Normally, drug is administered in a high dose at a given time only to have to repeat that dose several hours or days later. As a consequence, increasing attention has been focused on methods of giving drugs continually for prolonged time periods and in a controlled fashion. In the pharmaceutical field, in addition to the importance of polymers, an understanding of the physiological barriers in the human body is also critical to developing appropriate controlled release systems.

Self-assembled proteins and peptide may serve as a new class of biomaterials for applications in drug delivery system. Given that the degradation products consist of the drug and amino acid, this drug delivery system has an advantage over polymer-based drug delivery system that generate polymer fragments with heterogeneous chain lengths upon degradation that may present complex toxicity profiles.

Branco and coworker have described in detail the application of self-assembling materials for therapeutic delivery [70]. As described in their article, the first self-assembled materials used as drug carriers were predominantly prepared from lipids. Vesicles (liposomes) and lipid-based micelles have been extensively studied and have been investigated for drug delivery applications as early as the 1970s. As a result, a significant number of liposomal drug formulations are available commercially, and many others are undergoing clinical trials, making these systems one of the leading drug vehicles today. Recently, novel polymeric and peptide self-assembling systems have been developed for drug delivery. These carriers mimic the capabilities of conventional lipid systems and in some cases demonstrate improved drug delivery qualities. Diblock copolymers are commonly used to construct micellar assemblies. In these systems, the hydrophobic blocks assemble, yielding a hydrophobic core shielded by a hydrophilic shell. The self-assembly of block copolymers is often spontaneous and can be performed in the presence of small molecules, affording micelles encapsulated with drugs that have diameters up to hundreds of nanometers. The majority of diblock copolymers have hydrophilic blocks of polyethylene glycol (PEG). Since the therapeutic is present during the self-assembly process, drugs are loaded easily. In addition to micelles, amphiphilic block copolymers can self-assemble into vesicles in water and are referred to as polymersome [70–72]. Polymersomes have been used for the delivery of small hydrophobic drugs and DNA/RNA molecules and possess similar drug delivery characteristics to micelles.

5-Fluorouracil (5-FU) is widely used in clinical practice to inhibit the fibroblasts to proliferate and improve the success rate of glaucoma-filtering surgery, but 5-FU has many toxic effects to normal ocular tissues. Liang et al. have designed a novel targeting drug delivery system based on self-assembled peptide hydrogel that contains the RGD (arginine–glycine–aspartic acid) peptide sequence that act as a specific receptor for the proliferated fibroblasts adhesion [73]. They hypothesized that a novel peptide containing a bioactive RGD sequence was designed and prepared. When dissolving the peptide in distilled water, a supramolecular hydrogel with nanofibers was formed through the self-assembly of the peptide. In addition, this self-assembled peptide hydrogel could integrate 5-FU into its nanofibers during the process of self-assembly of the peptide. The proliferated fibroblasts can be attached in the peptide hydrogel through the recognition of the RGD sequence, and 5-FU is delivered to the fibroblasts by this model of targeting of drugs.

Zhao et al. synthesized an amphiphilic peptide, RATEA16, [CH_3CO]-RATARAEARATARAEA-[NH_2], by solid phase method using standard Fmoc strategy [74]. Peptide RATEA16 and insulin-FITC were mixed in Milli-Q water and Tris HCl buffer (pH 7.5), and the final concentrations of the RATEA16 and the protein in the mixture were 0.3–1.0 wt% and 30–200 μM, respectively. Subsequently, the drug release was investigated *in vitro*, and the release profiles of the insulin-FITC from the peptide hydrogels were examined after the addition of Milli-Q water to the hydrogels containing the insulin-FITC. The hydrogels stably retained these protein molecules, and its release ratio was less than 10 wt% over 48 hours. It is concluded that the protein was very slowly released from the hydrogel that contained water more than 99 wt%. Koutsopoulos et al. reported that the acetyl-(Arg-Ala-Asp-Ala)$_4$-$CONH_2$ [Ac-(RADA)$_4$-$CONH_2$] peptide hydrogel is an efficient slow-delivery carrier of small molecules [75]. In this work, they used a variety of proteins, including lysozyme, trypsin inhibitor, BSA, and IgG with differing physicochemical properties (pI 4.6, 11.4; molecular mass, 14.3–150 kDa) and morphologies and encapsulated them within the Ac-(RADA)$_4$-$CONH_2$ peptide hydrogel. Release kinetics and diffusion coefficients for all systems were determined by using a single-molecule fluorescence correlation spectroscopy (FCS) method. Their results show that large proteins are retained inside the hydrogel scaffold for longer times compared with smaller proteins. Also, they showed that protein release from the Ac-(RADA)$_4$-$CONH_2$ peptide hydrogel was more sensitive to physical size than protein charge. For example, lysozyme, which is the smallest among the proteins studied, was released fastest. Soybean trypsin inhibitor, which is slightly larger and oppositely charged compared with lysozyme, was released slower than lysozyme but faster than BSA, which is larger. IgG, the largest of the proteins examined, was released more slowly than all other proteins.

Curcumin, an antioxidant with anti-inflammatory and anti-tumorigenic properties, has poor water solubility and a relatively low bioavailability that limits its therapeutic use. Altunbas et al. showed a SAP hydrogel to be an effective vehicle for the localized delivery of curcumin over sustained periods of time [76]. The curcumin hydrogel is prepared *in situ* where curcumin encapsulation within the hydrogel network is accomplished concurrently with peptide self-assembly. Physical and *in vitro* biological studies demonstrated the effectiveness of curcumin-loaded β-hairpin hydrogels as injectable agents for localized curcumin delivery. Notably, rheological characterizations of the curcumin-loaded hydrogel before and after shear flow have indicated solid-like properties even at high curcumin payloads. *In vitro* experiments confirmed that the encapsulation of the curcumin within the hydrogel does not have an adverse effect on its bioactivity. Most importantly, the rate of curcumin release and its consequent therapeutic efficacy can be conveniently modulated as a function of the peptide concentration.

Antisense oligonucleotides provide a promising therapeutic approach for several disorders including cancer. Gels composed of nanofibrous peptide network have been used as carriers for controlled delivery of oligonucleotides. A self-assembled peptide nanofibrous system was formed by mixing a cationic peptide amphiphile (PA) with Bcl-2 antisense oligodeoxynucleotide (ODN) through electrostatic interactions [76]. The self-assembly of PA–ODN gel was characterized by circular dichroism, rheology, atomic force microscopy, and scanning electron microscopy. These microscopy images revealed establishment of the nanofibrous PA–ODN network. Due to the electrostatic interactions between PA and ODN, ODN release can be controlled by changing PA and ODN concentrations in the PA–ODN gel. Cellular delivery of the ODN by PA–ODN nanofiber complex was observed by using fluorescently labeled ODN molecule. Cells incubated with PA–ODN complex had enhanced cellular uptake compared to cells incubated with naked ODN. Furthermore, Bcl-2 mRNA amounts were lower in MCF-7 human breast cancer cells in the presence of PA–ODN complex compared to naked ODN and mismatch ODN evidenced by quantitative RT-PCR studies. These results suggest that PA molecules can control ODN release, enhance cellular uptake, and present a novel efficient approach for gene therapy studies and oligonucleotide-based drug delivery.

Recently, a review on molecular self-assembly of peptides, peptide amphiphiles, and peptidomimetics into molecules through nanoarchitectures to hydrogels has been published [77]. Their applications in the field of tissue engineering have been highlighted. The design rules of this rapidly growing field are centered mainly on the construction of peptides in the form of peptide amphiphiles, aromatic short peptide derivatives, all-amino acid peptide amphiphiles, lipidated peptides with single and multiple alkyl chains, and peptide-based block copolymers and polymer peptide conjugates. An overview of the diversity of the patent applications is also provided for self-assembling systems based on nano- and/or microscale such as fibers, fibrils, gels, hydrogels, vesicles, particles, micelles, bilayers, and scaffolds.

Major efforts have been undertaken to develop responsive nanostructures that respond to applied stimuli and dynamically undergo defined changes, thereby producing switchable properties [78]. This introduction of stimuli-responsive functions into aqueous self-assembly provides an attractive approach for the creation of novel nanomaterials that are able of responding to environmental changes.

Kim et al. described the general self-assembly of the rod amphiphiles based on a rigid-flexible molecular architecture in aqueous solution. They then highlight the structural changes and the optical/macroscopic switching that occurs in the aqueous assemblies in response to the external signals. For example, the aqueous nanofibers formed through the self-assembly of the rod amphiphiles respond to external triggers by changing their shape into nanostructures such as hollow capsules, planar sheets, helical coils, and 3D networks. When an external trigger is applied, supramolecular rings laterally associate and merge to form 2D networks and porous capsules with gated lateral pores.

Peptide amphiphiles are excellent candidates for drug delivery due to their trans-membrane capability. The amphiphilic nature of peptide molecules and self-assembled nanostructures can facilitate the internalization of the drugs encapsulated by them. A number of cationic peptide amphiphiles have been reported as drug and gene delivery carriers. Self-assembled cholesterol-conjugated H_5R_{10} and $H_{10}R_{10}$ oligopeptides outperformed PEI in plasmid DNA delivery into both HepG2 and HEK293 cell lines. Increasing the number of histidine residues was found to further enhance gene expression efficiency [79]. It was thought that weak cationically charged histidine served as “proton sponge” and could enhance gene delivery. Peptide amphiphiles with different tails (NH_2–I_5–, NH_2–W_5–, NH_2–F_5–) but with the same head group (–H_4R_8–$CONH_2$) have been compared for their gene delivery capability. Different gene expression efficiencies were observed, indicating the impact of tail hydrophobicity [80].

Peptide amphiphiles $A_{12}H_5K_{10}$ and homologs have shown comparable gene delivery efficiencies to PEI but with better biocompatibility. These peptides, when dissolved in aqueous solution, would form core–shell structured nano-assemblies with diameters around 800 nm. Seow et al. proposed that charge-mediated interaction occurred via the cationically charged micelles and DNA. The increased cationic charge density at the outside of the micellar shell offered better DNA binding capability and protected the DNA from enzymatic degradation. However, because free peptide amphiphile molecules were also present in the solution, complexation with DNA might also proceed through direct molecular interaction. The molecular complexation might produce smaller complexes in greater numbers and might affect the entire transfection efficiency more significantly. Results have shown that the addition of A_{12} tail to H_5K_{10} improved the gene expression without causing significant increase in cytotoxicity [81]. However, due to the weak hydrophobicity of the A_{12} tails, these peptides were not effective at delivering hydrophobic drugs such as doxorubicin and paclitaxel [82]. To increase the hydrophobicity of the tails, six alanine residues inside the tail have been replaced by phenylalanine residues. The new cationic peptide amphiphile Ac–$(AF)_6$–H_5–K_{15}–NH_2 has been evaluated as carrier for co-delivery of drug (doxorubicin) and genes (luciferase reporter gene and p53 gene). The nanostructures could then efficiently encapsulate doxorubicin into the micelles and achieve sustained release without obvious initial burst. Compared with the free doxorubicin, micelles loaded with doxorubicin had better internalization capability into the HepG2 cells. Simultaneous delivery of a model drug (hydrophobic FITC) and gene (rhodamine-labeled DNA) has also been achieved. The co-delivery of doxorubicin and p53-encoding plasmid using the self-assembled nano-micelles synergistically suppressed the proliferation of HepG2 cells. Therefore, designed peptide amphiphiles have great potential as effective carriers for both drugs and genes for therapeutic applications.

4.3.4 Stabilization of Membrane Proteins

Membrane proteins are important natural molecular devices in living cells and involved in many life functionalities such as energy conversions, cell–cell communications, and ion transport. They also work as natural biosensors for sight, hearing, smell, taste, touch, and temperature [16, 83, 84]. Therefore, they are particularly useful in current nanobiotechnology endeavor. However, there is a significant lack of understanding of their structures and functions due to the difficulties in extracting, purifying, and stabilizing them [85]. Simple but improved methods that can obtain membrane proteins without disturbing their structures and biological functions have great potential in nanobiodevice fabrication.

Peptide amphiphiles have been reported as excellent materials to solubilize and stabilize membrane proteins [85, 86]. Peptides V_6D, A_6D A_6K, and V_6K can significantly increase the activity and stabilize diverse membrane proteins including *Escherichia coli* glycerol-3-phosphate dehydrogenase [86], G protein-coupled receptor bovine rhodopsin [87], or multi-domain protein complex photosystem-I (PS–I) on dry surface [88] and in aqueous solution [89]. Peptide A_6K can stabilize the PS–I complex in a dried form at room temperature for at least three weeks. Both the polarity, the number of charges on the head group, and the size and hydrophobicity of the tails have important effects on different membrane proteins [40].

Apart from peptide amphiphiles, other peptides or lipopeptides have also been considered as excellent stabilizers for membrane proteins. Natural lipid bilayers provide not only the surrounding interactions but also suitable pressure to membrane proteins to retain their structures. Many membrane proteins lose their native structures and biological functions during the purification in the surfactant-solubilized state.

Lipopeptides have very similar properties to membrane lipids and can provide suitable environment for membrane proteins. The lipopeptide used by McGregor et al. [90] has an α-helix structure. The polar residues are restricted to one side of the helix while the nonpolar alanine residues are on the opposite side. Two acyl chains are attached at the two ends of the α-helix and interact with the nonpolar alanine residues forming a surfactant feature. The length of the helical structure is similar to the width of a biological membrane. The lipopeptide has the capability to disrupt phospholipid bilayers and has proven to be effective at solubilizing helical membrane proteins including bacteriorhodopsin and lactose (lac) permease, as well as the *E. coli* PagP protein. Experimental studies aiming at elucidating the location and extent of peptide binding in the future will certainly expedite this area of research, leading to the design of more efficient peptides for protein stabilization.

4.4 Future Prospects

As stated previously, there is a large need for novel therapies for the repair and regeneration of a multitude of tissues. Specifically, the primary significance is the development and application of novel technology for the identification of cell/biomaterial/signaling factor systems that promote tissue regeneration with optimal functional properties. There is currently great interest in harnessing these unique properties because, by using stem cells, it may be possible to generate cells in the laboratory that can be used to either replace damaged tissues inside the body or to identify safer and more effective medicines.

Self-assembly of biological molecules forms the basic principle in the formation of complex biological structures. The topic discussed earlier on nanostructured biological materials through self-assembly of peptides and proteins gives us a wide knowledge about the basic principle underlying molecular self-assembly of proteins and peptides. It gives a general overview of different kinds of self-assembling protein and peptide systems. Throughout the topics discussed earlier, it clearly guides the reader about the potential applications of these self-assembled structures that can be applied for fabricating wider range of novel biomaterials for the use in medical applications. With appropriate references and examples, it opens up the reader's mind, incorporating a wider range of knowledge about the self-assembling proteins and peptide systems. These macroscopic structures have inspired the researchers to use them in various areas of science like electronics, biotechnology, nanotechnology, and medicine.

Discussed in this article are a few applications and the advantages of self-assembled proteins and peptides in regenerative medicine therapy. Even though there are certain technological hurdles, these can be overcome by understanding the drawbacks of the individual systems and finding an alternative to overcome this drawbacks [44, 91–103]. However, it is necessary to create new and other alternative methods if we encounter any problems using the current technology in delivering biomolecules, such as proteins, growth factors, and DNA. Such systems have previously been created as one alternative method to enhance the *in vitro* and *in vivo* localization of biomolecules [19, 30, 104–132]. Thus, self-assembling proteins and peptides and its versatile applications have kindled the interests of numerous scientists and researchers to find newer ways to apply it in the field of regenerative medicine. Tissue engineering is evolving rapidly with much more potential impacts in treatment of diseases like heart failures and diabetes. Numerous studies are being conducted worldwide to discover newer solutions that can be effectively applied to treat these diseases using self-assembled proteins and peptides. Results obtained in regenerative medicine are inspiring the scientific community to discover new innovative non-invasive tools at the micro- and nanoscale level for such purposes.

References

1 Wang, Y., Brown, P., and Xia, Y. (2011). *Nat. Mater.* 19: 482.
2 Dvir, T., Timko, B.P., Kohane, D.S., and Langer, R. (2011). *Nat. Nanotechnol.* 6: 13.
3 Farokhzad, O.C. and Langer, R. (2006). *Adv. Drug Delivery Rev.* 58: 1456.
4 Smith, L.A. and Ma, P.X. (2004). *Biointerfaces* 10: 125.
5 Peer, D., Karp, J.M., Hong, S. et al. (2007). *Nat. Nanotechnol.* 2: 751.
6 Mrsny, R.J. (2009). *Adv. Drug Delivery Rev.* 61: 172.
7 Whitesides, G.M. and Boncheva, M. (2002). *Proc. Natl. Acad. Sci. U.S.A.* 99: 4769.
8 Ng, C.C., Cheng, Y.L., and Pennefather, P.S. (2004). *Biophys. J.* 87: 323.
9 Nasalean, L., Baudrey, S., Leontis, N.B., and Jaeger, L. (2006). *Nucleic Acids Res.* 34: 1381.
10 Won, J., Chae, S.K., Kim, J.H. et al. (2005). *J. Membr. Sci.* 249: 113.
11 Duval-Terrié, C., Huguet, J., and Muller, G. (2003). *Colloids Surf., A* 220: 105.
12 Chung, W.J., Oh, J.W., Kwak, K. et al. (2011). *Nature* 478: 364.
13 Whitesides, G.M. and Grzybowski, B. (2002). *Science* 295: 2418.
14 Lee, J., Macosko, C.W., and Urry, D.W. (2001). *Biomacromolecules* 2: 170.
15 Zhang, S., Marini, D.M., Hwang, W., and Santoso, S. (2002). *Curr. Opin. Chem. Biol.* 6: 865.
16 Hartgerink, J.D., Beniash, E., and Stupp, S.I. (2001). *Science* 294: 1684.
17 Niece, K.L., Hartgerink, J.D., Donners, J., and Stupp, S.I. (2003). *J. Am. Chem. Soc.* 125: 7146.
18 Vauthey, S., Santoso, S., Gong, H. et al. (2002). *Proc. Natl. Acad. Sci. U.S.A.* 99: 5355.
19 Kisiday, J., Jin, M., Kurz, B. et al. (2002). *Proc. Natl. Acad. Sci. U.S.A.* 99: 9996.
20 Zhang, S., Gelain, F., and Zhao, X. (2005). *Semin. Cancer Biol.* 15: 413.
21 Hosseinkhani, H. and Hosseinkhani, M. (2008). *Chem. Today* 26: 30.
22 Hosseinkhani, H. and Hosseinkhani, M. (2008). *Chem. Today* 26: 35.
23 Wang, K., Keasling, J.D., and Muller, S.J. (2005). *Int. J. Biol. Macromol.* 36: 232.
24 Adams, D.J., Holtzmann, K., Schneider, C., and Butler, M.F. (2007). *Langmuir* 25: 12729.
25 McMillan, R., Meeks, B., Bensebaa, F. et al. (2001). *J. Biomed. Mater. Res.* 54: 272.
26 Sun, X., Sheardown, H., Tengvall, P., and Brash, J.L. (2000). *Biomed. Mater. Res.* 49: 66.
27 Pilkington, S.M., Roberts, S.J., Meade, S.J., and Gerrard, J.A. (2010). *Biotechnol. Progr.* 26: 93.
28 Zhang, S., Yan, L., Altman, M. et al. (1999). *Biomaterials* 20: 1213.
29 Sharma, J., Chhabra, R., Andersen, C.S. et al. (2008). *J. Am. Chem. Soc.* 130: 7820.
30 Zhang, S., Holmes, T.C., Lockshin, C., and Rich, A.S. (1993). *Proc. Natl. Acad. Sci. U.S.A.* 90: 3334.
31 Zhang, S. (2003). *Mater. Today* 20.

32 Zhao, X., Pan, F., Xu, H. et al. (2010). *Chem. Soc. Rev.* 39: 3480.
33 Brogden, K.A. (2005). *Nat. Rev. Microbiol.* 3: 238.
34 Zasloff, M. (2002). *Nature* 415: 389.
35 Chen, C.X., Pan, F., Zhang, S.Z. et al. (2010). *Biomacromolecules* 11: 402.
36 Mitra, R.N., Shome, A., Paul, P., and Das, P.K. (2009). *Org. Biomol. Chem.* 7: 94.
37 Makovitzki, A., Baram, J., and Shai, Y. (2008). *Biochemistry* 47: 10630.
38 Liu, L., Xu, K., HuayingWang, J.T.P.K. et al. (2009). *Nat. Nanotechnol.* 4: 457.
39 Ghanaati, S., Webber, M.J., Unger, R.E. et al. (2009). *J. Biomater.* 30: 6202.
40 Silva, G.A., Czeisler, C., Niece, K.L. et al. (2004). *Science* 303: 1352.
41 Ramachandran, S. and Yu, Y.B. (2006). *BioDrugs* 20: 263.
42 Law, B., Weissleder, R., and Tung, C.H. (2006). *Biomacromolecules* 7: 1261.
43 Zhang, S., Holmes, T.C., DiPersio, C.M. et al. (1995). *Biomaterials* 16: 1385.
44 Horii, A., Wang, X., Gelain, F., and Zhang, S. (2007). *PLoS One* 2: 190.
45 Haines-Butterick, L., Rajagopal, K., Branco, M. et al. (2007). *Proc. Natl. Acad. Sci. U.S.A.* 104: 7791.
46 Gelain, F., Bottai, D., Vescovi, A., and Zhang, S. (2006). *PLoS One* 1: 119.
47 Genove, E., Shen, C., Zhang, S., and Semino, C.E. (2005). *Biomaterials* 26: 3341.
48 Kasai, S., Ohga, Y., Mochizuki, M. et al. (2004). *Pept. Sci.* 76: 27.
49 Gras, S.L., Tickler, A.K., Squires, A.M. et al. (2008). *Biomaterials* 29: 1553.
50 Storrie, H., Guler, M.O., Abu-Amara, S.N. et al. (2007). *Biomaterials* 28: 4608.
51 Guler, M.O., Hsu, L., Soukasene, S. et al. (1855). *Biomacromolecules* 2006: 7.
52 Chau, Y., Luo, Y., Cheung, A.C. et al. (2008). *Biomaterials* 29: 1713.
53 Holmes, T.C., de Lacalle, S., Su, X. et al. (2000). *Proc. Natl. Acad. Sci. U.S.A.* 97: 6728.
54 Davis, M.E., Hsieh, P.C.H., Takahashi, T. et al. (2006). *Proc. Natl. Acad. Sci. U.S.A.* 103: 8155.
55 Kapadia, M.R., Chow, L.W., Tsihlis, N.D. et al. (2008). *J. Vasc. Surg.* 47: 173.
56 Jangwook, P.J., Arun, K.N., Emily, K.F. et al. (2009). *Biomaterials* 30: 2400.
57 Jung, J.P., Jones, J.L., Cronier, S.A., and Collier, J.H. (2008). *Biomaterials* 29: 2143.
58 Lutolf, M.P. and Hubbell, J.A. (2005). *Nat. Biotechnol.* 23: 47.
59 Stevens, M.M. and George, J.H. (2005). *Science* 310: 1135.
60 Anderson, D.G., Levenberg, S., and Langer, R. (2004). *Nat. Biotechnol.* 22: 863.
61 Flaim, C.J., Chien, S., and Bhatia, S.N. (2005). *Nat. Methods* 2: 119.
62 Zhang, S.G. (2003). *Nat. Biotechnol.* 21: 1171.
63 MacPhee, C.E. and Woolfson, D.N. (2004). *Curr. Opin. Solid State Mater. Sci.* 8: 141.
64 Holmes, T.C. (2002). *Trends Biotechnol.* 20: 16.
65 Zhao, X.J. and Zhang, S.G. (2007). *Macromol. Biosci.* 7: 13.
66 Zhou, M., Smith, A.M., Das, A.K. et al. (2009). *Biomaterials* 30: 2523.
67 Mata, A., Hsu, L., Capito, R. et al. (2009). *Soft Matter* 5: 1228.
68 Webber, M.J., Tongers, J., Renault, M.A. et al. (2010). *Acta Biomater.* 6: 3.
69 Harrington, D.A., Cheng, E.Y., Guler, M.O. et al. (2006). *J. Biomed. Mater. Res. Part A* 78: 157.
70 Branco, M.C. and Schneider, J.P. (2009). *Acta Biomater.* 5: 817.

71 Rajagopal, K. and Schneider, J.P. (2004). *Curr. Opin. Struct. Biol.* 14: 480.
72 Lee, K.Y. and Yuk, S.H. (2007). *Prog. Polym. Sci.* 32: 669.
73 Liang, L., Yang, J., Li, Q. et al. (2011). *Biomater. Nanobiotechnol.* 2: 622.
74 Zhao, Y., Tan, T., Yokoi, H. et al. (2008). *Polym. Sci. Part A: Polym. Chem.* 46: 4927.
75 Koutsopoulos, S., Unsworth, L.D., Nagai, Y., and Zhang, S. (2009). *Proc. Natl. Acad. Sci. U.S.A.* 106: 4623.
76 Altunbas, A., Lee, S.J., Rajasekaran, S.A. et al. (2011). *Biomaterials* 32: 5906.
77 Bulut, S., Erkal, T.S., Toksoz, S. et al. (2011). *Biomacromolecules* 12: 3007.
78 Kumar, P., Pillay, V., Modi, G. et al. (2011). *Recent Pat. Drug Delivery Formulation* 5: 24.
79 Kim, H.J., Kim, T., and Lee, M. (2011). *Acc. Chem. Res.* 44: 72.
80 Guo, X.D., Tandiono, F., Wiradharma, N. et al. (2008). *Biomaterials* 29: 4838.
81 Seow, W.Y. and Yang, Y.Y. (2009). *Adv. Mater.* 21: 86.
82 Wiradharma, N., Khan, M., Tong, Y.W. et al. (2008). *Adv. Funct. Mater.* 18: 943.
83 Wiradharma, N., Tong, Y.W., and Yang, Y.Y. (2009). *Biomaterials* 30: 3100.
84 Dickerson, M.B., Sandhage, K.H., and Naik, R.R. (2008). *Chem. Rev.* 108: 4935.
85 Jiang, H., Guler, M.O., and Stupp, S.I. (2007). *Soft Matter* 3: 454.
86 Sargeant, T.D., Guler, M.O., Oppenheimer, S.M. et al. (2008). *Biomaterials* 29: 161.
87 Spoerke, E.D., Anthony, S.G., and Stupp, S.I. (2009). *Adv. Mater.* 21: 425.
88 Wallin, E. and Von-Heijne, G. (1998). *Protein Sci.* 7: 1029.
89 Loll, P.J. (2003). *J. Struct. Biol.* 142: 144.
90 McGregor, C.-L., Chen, L., Pomroy, N.C. et al. (2003). *Nat. Biotechnol.* 21: 171.
91 Hosseinkhani, H., Hosseinkhani, M., Khademhosseini, A. et al. (2008). *Biomed. Mat. Res. A* 85: 47.
92 Peppas, N.A. and Langer, R. (1994). *Science* 263: 1715.
93 Langer, R. and Peppas, N.A. (1981). *Biomaterials* 2: 201.
94 Hosseinkhani, H. (2006). *Int. J. Nanotechnol.* 3: 416.
95 Mohammad-Taheri, M., Vasheghani-Farahani, E., Hosseinkhani, H. et al. (2012). *Iran. Polym. J.* 21: 239.
96 Hosseinkhani, H. and Tabata, Y.S. (2006). *J. Nanosci. Nanotechnol.* 6: 2320.
97 Langer, R. and Tirrell, D.A. (2004). *Nature* 428: 487.
98 Tysselin-Mattiace, V.M., Sahni, V., Niece, K.L. et al. (2008). *J. Neurosci.* 28: 3814.
99 Kayser, O., Lemker, A., and Trejo, N.H. (2005). *Curr. Pharm. Biotechnol.* 6: 3.
100 Neuberger, T., Schöpf, B., Hofmann, H. et al. (2005). *Magn. Magn. Mat.* 293: 483.
101 Hosseinkhani, H. and Hosseinkhani, M. (2009). *Curr. Drug Saf.* 4: 79.
102 Hosseinkhani, H., Hosseinkhani, M., Vasheghani, E., and Nekoomanesh, M. (2009). *Adv. Sci. Lett.* 2: 70.
103 Tian, F., Hosseinkhani, H., Estrada, G., and Kobayashi, H. (2007). *J. Phys.* 61: 587.
104 Khalaji, S., Ebrahimi, N.G., and Hosseinkhani, H.Y. (2021). *Int. J. Polym. Mater. Polym. Biomater.* 70: 459.

105 Hosseinkhani, H., Hosseinkhani, M., and Khademhosseini, A. (2006). *Drug Discovery* 1: 32.

106 Konishi, M., Tabata, Y., Kariya, M. et al. (2005). *J. Controlled Release* 103: 7.

107 Domb, A.J., Sharifzadeh, G., Nahum, V., and Hosseinkhani, H. (2021). *Pharmaceutics* 13: 1615.

108 Hosseinkhani, H., Azzam, T., Tabata, Y., and Domb, A.J. (2004). *Gene Ther.* 11: 194.

109 Ickowicz, D., Domb, A.J., and Hosseinkhani, H. (2004). *J. Controlled Release* 97: 157.

110 Hosseinkhani, H. and Tabata, Y. (2021). *Int. J. Nanotechnol. Appl.* 11: 1.

111 Hosseinkhani, H., Aoyama, T., Ogawa, O., and Tabata, Y. (2003). *Curr. Pharm. Biotechnol.* 4: 109.

112 Steinman, N.Y., Campos, L.M., Feng, Y. et al. (2020). *Pharmaceutics* 12: 678.

113 Hosseinkhani, H. (2020). *Org. Med. Chem. Int. J.* 10: 1.

114 Hosseinkhani, H. and Domb, A.J. (2019). *Polym. Adv. Technol.* 30: 2647.

115 Abedini, F., Ebrahimi, M., Roozbehani, A.H. et al. (2018). *Polym. Adv. Technol.* 29: 2564.

116 Abedini, F., Ebrahimi, M., and Hosseinkhani, H. (2018). *MicroRNA* 7: 74.

117 Hosseinkhani, H. (2018). *J. Cell Res.* 1: 1.

118 Hosseinkhani, H., Kobayashi, H., and Tabata, Y. (2006). *Pept. Sci.* 2005: 341.

119 Hosseinkhani, H., Kobayashi, H., and Tabata, Y. (2006). *Pept. Sci.* 2005: 63.

120 Hosseinkhani, H., Hosseinkhani, M., Hattori, S. et al. (2010). *J. Biomed. Mater. Res. Part A* 94: 1.

121 Lindstrom, S., Iles, A., Persson, J. et al. (2010). *JBSE* 5: 272.

122 Abdullah, S., Yeo, W.Y., Hosseinkhani, H. et al. (2010). *J. Biomed. Biotechnol.* 2010: 284840.

123 Subramani, K., Hosseinkhani, H., Khraisat, A. et al. (2009). *Curr. Nanosci.* 5: 134.

124 Mahmoudi, M., Hosseinkhani, H., Hosseinkhani, M. et al. (2011). *Chem. Rev.* 111: 253.

125 Mohageri, S., Hosseinkhani, H., Ebrahimi, N.G. et al. (2010). *Tissue Eng. Part A* 16: 3821.

126 Abedini, F., Ismail, M., Hosseinkhani, H. et al. (2010). *J. Cell Anim. Biol.* 4: 170.

127 Abedini, F., Hosseinkhani, H., Ismail, M. et al. (2012). *Int. J. Nanomed.* 7: 4159.

128 Semino, C.E. (2008). *J. Dent. Res.* 87: 606.

129 Aggeli, A., Boden, N., Cheng, Y.L. et al. (1996). *Biochemistry* 35: 16213.

130 Aggeli, A., Bell, M., Boden, N. et al. (1997). *Nature* 386: 259.

131 Kirkham, J., Firth, A., Vernals, D. et al. (2007). *J. Dent. Res.* 86: 426.

132 Zhang, S., Lockshin, C., Cook, R., and Rich, A. (1994). *Biopolymers* 34: 663.

5

Cytotoxicity of Materials

5.1 Introduction

There are enormous growth in nanomaterials research based on a vast range of applications in medicine, photonics, and electronics [1–5]. Any structure, material, or formulation with at least one dimension ranging between 1 and 100 nm is regarded as a nanostructure, nanomaterial, or nanoparticle (NP) [6]. In contrast to their bulk counterparts, nanomaterials present excellent properties including high surface-to-volume ratio; superb electrical, magnetic, mechanical, optical, and thermal behavior; and high surface energy [7]. These nanomaterials exist in diverse structures in which each has its own advantages and applications [8–10]. The application of nanomaterials or nanoscale formulations in medicine and health is generally referred to as nanomedicine, involving biology, chemistry, physics, medical sciences, and engineering [11]. Although this advanced field of medicine has emerged during only recent decades, it has led to several challenges in traditional medicine such as systemic and organ toxicity, impaired target specificity, and poor bioavailability [12]. Thus, a broad range of medical products (more than 200) including diagnostic agents, medical devices, and therapeutics has been clinically approved or is being evaluated due to nanomedicine research [13]. There are, however, several hurdles that eliminate the clinical translocation of nanomaterials [14, 15]. (i) It is difficult to synthesize nanomaterials capable of encapsulating effective therapeutic molecules with activated release. (ii) Difficulty exists to efficiently deliver NPs to the target site for *in vivo* studies. (iii) The efficacy and toxicity of nanomaterials are significant obstacles. (iv) The clinical translocation and industrial production of NPs in a large scale with reasonable cost are challenging. These barriers must be completely overcome to expand the efficacy and application of nanomaterials for clinical purposes.

Due to side effects, rapid development of molecular nanotechnology and nanovaccinology exploiting the improved and often novel physical, chemical, and biological properties of materials at the nanometer scale must be considered before clinical application. Nanotechnology has the potential to enable early detection and prevention and to essentially improve diagnosis, treatment, and follow-up of diseases. It is committed to supporting activities such as theranostics, as well as enabling diagnostic devices and therapeutics to be combined for real benefit

Biomedical Engineering: Materials, Technology, and Applications, First Edition. Hossein Hosseinkhani.

to patients. Growing advances in nanotechnology have stimulated researchers to create innovative technologies to enhance and modify the quality of new devices. On the other hand, little is known about the potential risks of the nanoscale materials. Control of adverse effects is critical in the development of new technologies. The study of side effects of nanomaterials on the ecosystems and living organisms, along with the prevention and improvements of these adverse effects, is termed "nanotoxicology." Nanomedicine, nanotechnology, and nanotoxicology are primarily aimed to improve human life by reducing risks [16]. Nanomaterials have been widely developed and applied in commercial products such as electronic sensors, energy equipment, sun creams, and biomedical devices.

There has not been sufficient evaluation of nanomaterial interfaces with tissue and organs, although several research projects have raised a questions if nanoscale materials should be treated as carcinogens or asbestos. It is important to perform toxicity research on the basic interaction of nanomaterials with natural tissues, since nanotechnology has a high impact on consumer and biomedical products. It is critical to understand the exact role of nanomaterials *in vivo*. When NPs are administrated within the human body, the effect of NP-loaded therapeutics followed by their distribution into tissues (i.e. kidneys, lung, and liver) should be precise. Clearance of nanomaterials can potentially occur through pulmonary, hepatic, and renal systems. They may not degrade or eliminate, which makes predicting long-term behavior particularly difficult. NPs may accumulate in various organs and interact with off-target cells. They may degrade and eliminate from the body due to their very small size [17]. There are few studies addressing whether nanomaterials are (bio)compatible with natural organ, tissues, or cells. This chapter discusses the chemical, physical, and biological properties of various nanomaterials (both natural and synthetic) with specific focus on recent technologies related to the fabrication and process of nanomaterials with different systems, while emphasizing possible toxicity using *in vitro* and *in vivo* assays. More importantly, the *in vivo* fate of nanomaterials, including routes of administration, biodistribution, metabolism, routes of clearance, and blood biocompatibility is discussed. Special emphasis is placed on highlighting the impact of physicochemical properties of nanomaterials and their toxicity. Figure 5.1 indicates several factors that are known as potential risk of nanomaterials in biological application.

5.2 Mechanism of Toxicity

NPs are capable of inducing toxicity both *in vitro* and *in vivo* through various different mechanisms. Some of the most common mechanisms are oxidative stress, cell death mechanisms (apoptosis, autophagy, and necrosis), genotoxicity, and immunological response. Each of these mechanisms can lead to NP toxicity in a different pattern. In this section the different mechanisms of NP toxicity are explained.

5.2.1 Oxidative Stress

Oxidative stress is described as an imbalance between the production of reactive oxygen species (ROS) and antioxidant mechanisms [18]. This imbalance can

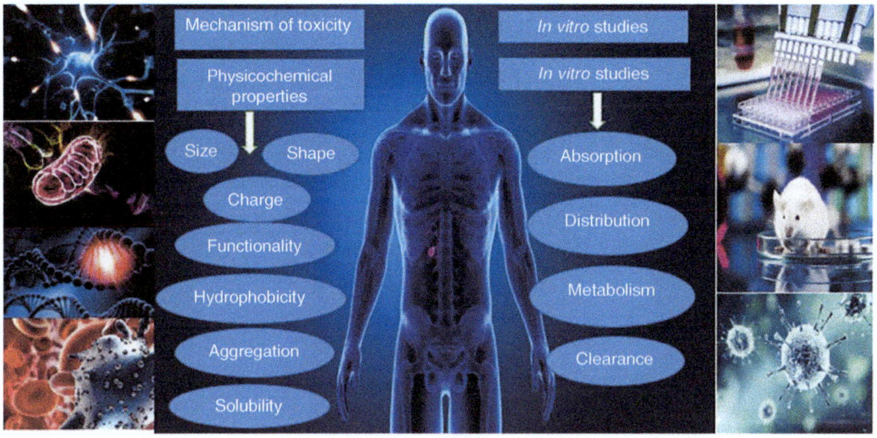

Figure 5.1 Toxicity risk factors of nanomaterials.

be explained as an increase in the generation of ROS or decreased amounts of antioxidants. One of the significant antioxidants found in animals, plants, and fungi is glutathione (GSH), which can stop losses caused by ROS. It is available in oxidized (GSSG) and reduced (GSH) modes, while in healthy situations most of the GSH pool (about 90% of the total) is in its reduced state. An increased GSSG-to-GSH ratio shows oxidative stress [19]. ROS can be triggered through various mechanisms including (i) interaction with oxidative organelles such as mitochondria, (ii) involvement with redox active proteins, (iii) chemical reaction of surface groups or coating from the NPs in the acidic environment, and (iv) activation of different signaling routes via interaction with cell surface receptors [18, 20]. On the contrary, such patterns do not support all NPs such as cerium oxide (CeO_2). CeO_2 NPs do not cause ROS, and they display a preserving effect against ROS damage both *in vitro* and *in vivo* [21, 22]. In recent research, Vitamin C-conjugated NPs protected cells from oxidative stress at micromolar concentration, while it induced cell death at millimolar concentration [23]. The particular molecular target and chemical property of oxidative stress and how it affects modification of distinct biological procedures in cells subjected to engineered NPs (ENPs) is not fully understood. Several findings propose that reversible oxidative posttranslational modifications of protein cysteines via ROS and RNS demonstrate the basic mechanism of cell signaling, which adjusts the protein functions of several cellular activities [24, 25]. Protein S-glutathionylation (SSG) is a vital redox adjustment that alters apoptosis, mitochondrial metabolism, and transcription [26]. SSG alternations are controlled by physiological properties and can be repaired by glutaredoxin (Grx) enzymes [26, 27]. Site-specific profiling of SSG modifications at a proteome-wide scale can be achieved through a quantitative redox proteomics method [28]. Using SSG mechanism, Duan et al. evaluated the modification of macrophage innate immune functions by ENPs via quantitative redox proteomics for site-specific measurement of SSG alternations [29]. To stimulate a disrupt macrophage function and cellular ROS, three ENPs [silicon oxide (SiO_x), Fe_3O_4, and cobalt oxide (II) (CoO)] were applied, causing low, moderate, and high propensity, respectively. SSG regulations indicate specific Cys residues and a wide range of redox sensitive proteins that correspond with the total of cellular oxidative stress ($CoO > Fe_3O_4 \gg SiO_2$). ENPs that cause moderate and serious ROS exhibit different pathways in response to SSG. Pathways modified protein stability and translation representative of ER stress response, but proteins in phagocytosis are highly susceptible to SSG in the presence of Fe_3O_4. ENPs triggers a subcytotoxic degree of redox stress. ENP mitochondrial energetic pathways and classical stress response are notably affected by SSG alternation at higher degrees of redox stress caused by CoO.

5.2.2 Cell Death Mechanisms

This natural process is caused by cell replacement or when old cells are dying. It normally begins through organelle stress [30]. NPs generally induce death-associated signaling via their interactions with various membranes or proteins in organelles. Particular proteins (receptors) act in the cell death triggered by a number of NPs [31]. Cell death is categorized by three major mechanisms: apoptosis, autophagy, and necrosis. Apoptosis is a highly regulated process activated either intrinsically (mitochondria dependent) or extrinsically (death receptor-dependent). It is an indispensable adult tissue remodeling as well as an ontogenetic development. It helps reduce damaged, infected, or tumoral cells [32]. Apoptosis may or may not need the activation of caspases [33]. Autophagy plays a substantial homeostatic role in the cell, but unrestrained autophagy can kill cells. It mediates the elimination of dysfunctional or damaged organelles, thus enabling the recycling of cellular building blocks. Necrosis is a disorganized process that causes cell death through autolysis.

5.2.2.1 Apoptosis

Apoptosis, sometimes referred as type-1 cell death, is a process of programmed cell death vital for development and preservation of tissue homeostasis [34]. It plays a key role in various diseases, since there is a powerful link between apoptosis and oxidative stress in pathophysiology conditions including cancer, Alzheimer's disease, Parkinson's disease, AIDS, diabetes, ischemia of the heart and brain, and autoimmunity [35]. In mammalian cells, apoptosis can be classified as extrinsic (dependent on death receptors) or intrinsic (dependent on/convergent on mitochondria controlled signaling) [34]. Extrinsic apoptosis is caspase-dependent, whereas the intrinsic pathway can be either caspase-dependent or caspase-independent signaling [33]. Wang et al. demonstrate the inherent mechanism of caspase-dependent apoptosis containing both intrinsic and extrinsic pathways on bone tumor impression via selenium [36]. *In vivo* data show that selenium-doped hydroxyapatite NPs (Se-HNPs) cause tumor apoptosis to suppress tumor growth and reduce toxicity. Membrane integrity is inhibited during apoptosis as opposed to necrosis due to the existence of many molecules regarded as "eat me" signals in the plasma membrane of dying cells [37]. Apoptotic bodies are quickly removed via phagocytic cells involving macrophages, dendritic, and epithelial cells because of these signals. Likewise, apoptosis causes an inflammatory immune response, but the mechanism of this reaction is not similar to necrosis. The apoptotic process can be adjusted by two essential molecule groups: (i) caspases consisting of pro-inflammatory (caspase-1, -4, and -5), initiator (caspase-2, -8, -9, and -10), and executioner (caspase-3, -6, and -7) subsets and (ii) Bcl-2 group divided into pro-apoptotic (Bak, Bax, Bid, Bim, Noxa, etc.) or antiapoptotic (Bcl-2, Mcl-1, Bcl-X_L, etc.) proteins [38, 39]. These two classes are linked by the mitochondrial apoptosis route. Pardo and colleagues have elucidated the molecular mechanism of apoptosis in mice for laser PTT [40]. Induced apoptosis via the intrinsic/mitochondrial pathway was regulated through the Bak and Bax by the activation of the BH3-only protein Bid employing laser-irradiated hot gold nanoprisms (NPRs). The NPR-mediated apoptosis obtained was fully related to caspase-3 and caspase-9 proteins. Another molecular mechanism was introduced by Hou et al. for *in vivo* photodynamic therapy through the mitochondrial apoptosis pathway [41]. They applied an NIR light base on TiO_2-modified upconversion NP (UCNP) core/shell nanocomposites (UCNPs@TiO_2 NCs). $NaYF_4$:Yb^{3+} and Tm^{3+}@$NaGdF_4$:Yb^{3+} core/shell UCNPs effectively converted NIR light to UV emission. The UCNPs@TiO_2 NCs produce intracellular ROS in the presence of NIR irradiation, reducing the mitochondrial membrane potential to release cytochrome c in the cytosol, thus leading to activation of caspase 3 and causing apoptosis of the cancer cell. In contrast to UV light, NIR-based UCNPs@TiO_2 exhibits deeper tissue penetration, inhibiting tumor growth more effectively.

5.2.2.2 Autophagy

Autophagy is a vital lysosome-dependent metabolic process to remove dysfunctional organelles and unfold proteins [42]. During autophagy, organelles and cytosol are sequestered by a double-membrane vesicle (autophagosomes). These are the most important markers of autophagy. Eventually, the autophagosomes merge into and integrate with lysosomes, thus forming an autolysosome. They degrade the components through lysosomal hydrolases to provide energy [43]. The key roles of autophagy are cellular homeostasis, antimicrobial, antiaging, and development roles. Autophagy is associated with severe diseases [44, 45]. In recent research, Lin and colleagues report that tetrahedral DNA nanostructures (TDNs) with a concentration of 250 nM enhance autophagy and upregulate various autophagy-related proteins and genes [46]. TDNs internalize through chondrocytes in the absence of other auxiliary agents. They focus on the cytoplasm of chondrocytes, although only few single-strand DNAs (ssDNAs) are internalized by cells. TDNs increase autophagy via the PI3K/AKT/mTOR (mammalian target of rapamycin) pathway.

Several studies reveal that NPs are able to disturb autophagy that can lead to severe toxicological problems [47]. Additional studies are required to clarify the precise role of NPs in the induction of autophagy. Induction of autophagy by a broad range of NPs such as polymer [48], nanogel [49], and iron oxide [44] has been studied. Most of the aforementioned examples are *in vitro* systems based on NP-induced autophagy, which fails to replicate the difficulty of *in vivo* physiological conditions, thereby lacking real-time data. Zhou et al. evaluated QDs-induced autophagy in live organisms using *Caenorhabditis elegans* as the model organism [50]. They triggered autophagy in the intestinal cells through the addition of QDs into the intestinal cells of *C. elegans* via oral feeding or microinjection. The real-time *in situ* tracking of autophagosomes formation in live organisms is presented. This *in vivo* QD-induced autophagy is a defensive strategy of the organism to degrade and recycle damaged cellular organelles. Fan and colleagues developed an autophagy regulation strategy based on folic acid-modified N-doped carbon dots (FA-CDs) for visualized tumor therapy [51]. Stable FN-CDs exist in the autophagic vacuoles in tumor cells, and their specific cellular uptake rate is up to 93.40%. Cellular metabolism is influenced by FN-CDs that cause autophagy. Autophagic vacuoles are opened through autophagy inhibitors, and they release NPs activated signaling pathway, thus killing tumor cells. This strategy shows therapeutic efficacy in 26 types of tumor cell lines. *In vivo* results indicate much higher efficiency (98%) than traditional treatment (68%). Real-time *in situ* imaging tracking of the treatment can be reported through FA-CDs.

5.2.2.3 Necrosis

Necrosis is another type of cell death process that has been considered for many years to be nonprogrammed and unregulated. Recently, however, it was determined that necrosis is able to undergo a programmed process by a number of intracellular signal transductions [52]. Necrosis caused by death receptor (e.g. TNFR1) stimulation usually requires the activation of receptor-interacting protein kinase 1 and 3 (RIPK1 and RIPK3) in response to a series of stresses, such as DNA damage, oxidative stresses, and viral infections [53–55]. RIPK1 and RIPK3 both exist with caspase-8, Fas-associated death domain-containing protein (FADD) and probably TNF receptor-related death domain-containing protein (TRADD) in necrosome. This can trigger necrosis or apoptosis. As mentioned, although caspase-8 activation causes apoptosis, if this caspase is blocked or absent, RIPK1 and RIPK3 phosphorylated can cause necrosis [33]. Liu and colleagues report that GO triggered necrosis to macrophages through activation of a toll-like receptor 4 (TLR4), signaling autocrine TNF-α production [56]. Prevention of TLR4 signaling with a selective inhibitor almost fully inhibits cell death. Also, GO-induced necrosis of macrophages is partially related to the RIPK1–RIPK3 complex necrosis of TNF-α induction. GO treatment leads to massive intracellular ROS generation. GO uptake into macrophages causes reduction in the ability of phagocytosis and disruption of cytoskeleton. Zhang and colleagues report that single-walled carbon-nanohorns (SNH) and CNT induce cell death via necrosis and apoptosis [57]. Their results show that CNT triggers 1.9–4.8-folds of necrosis compared to SNH, indicating that SNH causes much lower necrosis inducibility. As a consequence, SNH and CNT trigger cell death through pyroptosis-mediated necrosis (pyroptosis is a common regulated necrosis pathway) and SNH always induces lower cell death than CNT via apoptosis or necrosis.

Among programmed cell death processes, necrosis is regarded as the most common cellular response during tumor ablation performed through NP-mediated PTT [58]. Generally, uncontrolled necrosis results in the release of intracellular constituents into the extracellular milieu, thus the responses that are induced may generate a harmful knock-on impact through damaging adjacent tissue, triggering tumor growth, etc. [59]. Loss of membrane integrity occurs particularly in necrotic cells. However, considering *in vitro* environments, apoptotic cells also suffer from loss of membrane integrity (secondary necrosis) in the absence of phagocytic cells. This behavior is seen later than during primary necrosis.

5.2.3 Genotoxicity

The destruction of genetic materials of the cell due to the toxicity is typically referred to as genotoxicity [60]. It indicates any DNA damage or chromosomal that contains gene mutations and chromosome rearrangements and breaks [61] using sophisticated signal transduction pathways [62]. Nanomaterials may influence genotoxicity due to their unique properties such as very small size and large surface area. Several factors are involved in the interaction of genetic materials with CNTs, namely, the similarity in size to those to microtubules, the capability of CNTs to penetrate nuclear membrane, and the high affinity of SWCNTs to G–C-rich area in DNA sequences [63]. It has been reported that MWCNTs and SWCNTs trigger DNA damage *in vitro* in mesothelial cells [64]. NPs induce genotoxicity directly (primary genotoxicity) or indirectly (secondary genotoxicity). Secondary genotoxicity arises due to inflammatory responses and oxidative stress, while primary genotoxicity may occur through the interaction of NPs with DNA or the mitotic apparatus [65]. Indirect genotoxicity of NPs can occur by production of ROS through inflammatory cells. The obtained ROS can interact with DNA, resulting in DNA oxidation, breakage, and/or lipid peroxidation-mediated DNA adducts [66].

Genotoxicity can be detected *in vitro* (by the Ames test, comet assay, and chromosome aberration test) and *in vivo* (rodent carcinogenicity, chromosome aberration, and mutation of endogenous genes). These assays are generally based on detection of induced damages. A wide range of NPs has been reported to be genotoxic, such as polymeric NPs and ENPs [67, 68]. Many human cells, including fibroblasts and RAW264.7 macrophages, normal lung epithelial BEAS-2B and SAEC cells, normal and malignant mesothelial cells, and malignant alveolar epithelial A549 cells have been used to evaluate the genotoxicity of CNTs. They show that high doses (>50 to 100 $\mu g\,ml^{-1}$) of MWCNT and SWCNT induce genotoxicity. CNTs disrupt their dynamics and functions by mimicking microtubules [69]. The Ames test indicates sensitivity lower than micronucleus assays for a genotoxicity assessment of AgNPs in human lymphoblastoid (TK6) cells [70]. Such reduction in sensitivity of the Ames test, which employs bacterial cells to evaluate mutagenicity, can be attributed to bacterial cells that are not endocytic. Thus, this test is not suitable for specific NPs [71]. The micronucleus assay has also yielded false results because of reagent interference (i.e. cytochalasin B) with cellular uptake of NPs [72]. DNA damage induced by NPs has also been measured by other assays such as the 8-oxoguanine (8-oxo-dG) ELISA [73]. This assay achieves high throughput with the 96-well plate design, while it measures only a biomarker of DNA oxidation, 8-oxo-dG, it does not show other types of DNA damage such as alkali-sensitive sites, double- and single-strand breaks, and abasic sites. Comet assay is the most common test for assessing the genotoxicity of nanomaterials. This method suffers from low reproducibility and throughput due to sample-to-sample variation (one sample showed different results when tested on two separate glass slides) [74].

Watson et al. introduced a high-throughput screening (HTS) assay based on the developed CometChip approach that provides assessment of abasic sites, alkali-sensitive sites, and single-stranded DNA breaks in TK6 and adherent Chinese hamster ovary (H9T3) cells [75]. The authors observed dose-dependent increases in cytotoxicity and DNA damage through the exposure to ENPs (Ag, CeO_2, Fe_2O_3, SiO_2, and ZnO) at a dose range of 5, 10, and 20 μg ml^{-1}. The genotoxicity of ENPs in TK6 cells at 4 hours and in H9T3 cells at 24 hours was ZnO > Ag > Fe_2O_3 > CeO_2 > SiO_2 and Ag > Fe_2O_3 > ZnO > CeO_2 > SiO_2, respectively.

5.2.4 Immune Response

A wide range of NPs is able to stimulate, suppress, and/or avoid immune responses. Interaction with the immune responses is associated with physiochemical properties including shape, size, and charge [76]. NPs can adsorb, encapsulate, and/or chemically immobilize antigens within their surfaces, resulting in controlled release and delivery [77, 78]. Niikura et al. have demonstrated that different shapes of AuNPs modified with Nile virus envelope (E) protein cause cytokine secretion behavior in dendritic cells. Cubic and spherical AuNPs lead to the secretion of the pro-inflammatory cytokines at high levels, whereas rod AuNPs cause secretion of inflammasome-related cytokines, interleukin (IL) 1β, and IL-18 [79]. Long aspect ratio ENPs, such as CNTs and nanowires, induce activation of inflammasome that is secondary to oxidative stress and induce a shape-dependent effect at the lysosomal level [80]. NPs are also capable of activating cell and humoral immune responses through interacting with cell receptors and plasma proteins, and they can be internalized via mammalian cells. NPs are able to deliver sufficient antigens to targeted sites and enhance immune responses by favorable bioavailability and cellular uptake, thus reducing adverse effects [81].

Nanomaterials can be engineered as vaccines to improve immune response with the ability to reduce unwanted activation or suppression of immune systems. This is appropriate for infectious diseases and immunotherapy of tumors [82]. Nanovaccines are able to regulate their hydrophilicity, shape, antigen loading, surface potential, and scale by various fabrication procedures and materials [82]. They cause long-term immune response by adhesion, mixing, conjugation, and encapsulation of the antigen with NPs for vaccine delivery [83]. Pulendran and coworkers developed synthetic polymer NPs loaded with hemagglutinin (HA) from TLR ligands and avian influenza H5N1. This is effective against swine influenza virus strains and lethal avian in mice, and it causes robust immunity against H1N1 influenza [84]. Despite the aforementioned examples that are NP-based vaccines, non-vaccine nanomaterials can also induce immune response [85]. Zhai and colleagues designed cross-linked (X-NPs) and non-cross-linked self-assembled NPs (NX-NPs) for antiproliferation and antimetastasis of melanoma [86]. The chondroitin sulfate (CS)–adipic dihydrazide (ADH)–chlorin e6 (Ce6)–lipoic acid (LA)-loaded docetaxel (DTX) functions as a redox/enzyme/ultrasound. Responsive conjugates form self-assembled NPs in the presence of aqueous solutions, suitable for combining sonodynamic therapy (SDT) and chemotherapy. X-NPs induce the generation of ROS, causing mitochondrial damage followed by cell apoptosis by the mitochondria–caspase pathway. CD44 nanoscale distribution and targeted properties lead to successful tumor homing ability of NPs, causing potential antimetastasis and antitumor efficiencies with declined tumorous uPA and COX-2 proteins expression. The SDT of X-NPs triggers immunological response via cell death and causes release of tumor-associated antigens.

5.3 Toxicity Assessment of Nanomaterials

5.3.1 *In Vitro* Characterization

In vitro evaluation of their biocompatibility with blood components is a necessary part of early preclinical development. For example, many research papers report NP hemolytic properties. It is likely that the biological activity of NPs will depend on physicochemical parameters not routinely considered in toxicity screening studies. Physicochemical properties that may be important in understanding the toxic effects of test materials include particle size and size distribution, agglomeration state, shape, crystal structure, chemical composition, surface area, surface chemistry, surface charge, and porosity. *In vitro* techniques allow specific biological and mechanistic pathways to be isolated and tested under controlled conditions, in ways not feasible in *in vivo* tests. Tests have been suggested for portal-of-entry toxicity for lungs, skin, and mucosal membranes and to target organ toxicity for endothelium, blood, spleen, liver, nervous system, heart, and kidney. Noncellular assessment of NP durability, protein interactions, complement activation, and pro-oxidant activity has also been considered.

5.3.1.1 Scanning Electron Microscopy

Scanning electron microscopy (SEM) can provide significant information with images regarding *in vitro* NP composition and topography through interactions of an electron beam. A sample is examined by detection of backscattered electrons from its surface/subsurface or secondary electrons emitted from its surface [87]. SEM is able to produce high-resolution images on the order of 1 nm. Horvath et al. investigated the *in vitro* cytotoxicity of boron nitride nanotubes (BNNTs) using SEM [88]. Nanotubes attached to the surface of exposed cells can cause toxicity via activating signaling pathways. SEM, therefore, has been applied to study the distribution of BNNTs on the surface of macrophages (RAW 264.7) and epithelial cells (A549). BNNT-treated A549 cells reveal no significant morphological changes, while many nanotubes bind to cell surface in BNNT-treated macrophages. SEM indicates ultrastructural modification of the A549 epithelial cell surface in the response of ENMs. In another example, the biocompatibility and dynamical processes of interaction of nanomaterials with macrophage cells dosed with metal oxide NPs (CeO_2, TiO_2, and ZnO) were examined through field emission scanning electron microscopy (FE-SEM) [89]. FE-SEM confirms the interaction and exact details of adsorption, internalization, and ultrastructural location of NPs via main constituents of the innate immune system. FE-SEM offers new insights into biocompatibility and biodistribution of NPs with immune cells. SEM is also used for determination of tumor margins based on the actual visualization of anti-AuNR-epidermal growth factor receptors (EGFRs) [90]. The distribution of EGFRs in and around tumor cells was evaluated by air SEM (airSEM) as a highly sensitive microscopy for biological samples. AirSEM provides localized and actual visualization of the AuNR on tissue. The distance between healthy and tumor regions is in a radius of ~1 mm.

5.3.1.2 Transmission Electron Microscopy

Transmission electron microscopy(TEM) utilizes electrons transmitted through a sample to form images. By visualization of NP locations in a cell/tissue and together with other techniques, TEM reveals considerable information regarding *in vitro* NP localization and uptake [91]. Both SEM and TEM are highly suitable for characterization of nanomaterials before and after administration within materials. Several drawbacks limit their application: long sample preparation times with a low throughput, limited use of biological samples with thickness of 50–100 nm, and structural damage to the sample during preparation. Nevertheless, TEM has been applied as a method in cellular localization of various NPs [91, 92]. It is also capable of detecting morphological changes on the cells caused by NPs. Ghandehari and coworkers utilized TEM for cellular internalization of SNPs composed of nonporous spherical (Stöber) SNPs with diameters of 46 ± 4.9 nm (Stöber50), 432 ± 18.7 nm (Stöber500), and mesoporous spherical (MSN) with a diameter of 466 ± 86 nm (MSN500) [93]. The SNPs were localized inside vesicles situated in the cytosol of macrophages. TEM imaging demonstrates that the number of Stöber500 particles attached to the outer part of cell is higher than MSN500. Compared to Stöber500 SNPs, the higher internalization of MSN500 is sub-cytotoxic and causes enlargement of macrophage cells. Both nonporous and mesoporous spherical SNPs were applied to study the gene expression of cell. Stöber SNPs have a minor effect on gene expression, whereas MSN can change gene transcription without exhibiting any acute toxicity. Scanning transmission electron microscopy (STEM) is a modified type of TEM that forms images by using the combination of suitable detectors and a device that raster scans the beam. Valiyaveettil and colleagues used STEM to investigate the biodistribution of silver NPs, where 100 mg ml^{-1} of silver NPs was treated with cancer cells [94]. The presence of silver NPs within the mitochondria and nucleus was imaged by STEM; the direct effect of silver NPs in DNA damage and mitochondrial toxicity was confirmed. Peckys and de Jonge applied the STEM to provide information about the intracellular uptake of 30 nm AuNPs in live fibroblast cells (COS-7) in a microfluidic chamber [95]. Information regarding AuNPs distribution in pristine cells was obtained, since the cells are alive at the onset of STEM. That information included the number of vesicles in cluster, vesicle diameter, and the number of detected AuNPs per 0.15×0.15 μm^2 of the image.

5.3.1.3 Atomic Force Microscopy

Atomic force microscopy (AFM) at high resolution (below 10 nm) can provide considerable information regarding mechanical, topographical, and chemical properties based on scanning a cantilever with a very sharp tip to measure interactions between a sample and a scanning probe tip [96]. AFM is applicable for imaging soft/biological materials such as living cells and biomolecules in their natural environments, because they do not require vacuum. Electron microscopes such as SEM and TEM are mostly suitable for imaging metallic NPs. For example, Chrzanowski and coworkers used infrared spectroscopy AFM (AFM-IR) to study the role of NDs on fibronectin-FN as extracellular proteins and those proteins inside liver cell as intracellular proteins [97]. Conformational changes including protein aggregations and formation of antiparallel β-turns, β-sheet, and intermolecular β-sheet can be observed for both intracellular and extracellular proteins at the ND–protein interface, implying an adverse effect of individual NPs on the protein structure. AFM also elucidates how curcumin-conjugated gold clusters (CUR-AuNCs) change cell morphology of HeLa cancer cells [98]. Pore formation and surface roughness of cells increase by CUR-AuNCs. Cancer cells treated with CUR-AuNCs indicate uneven, damaged surfaces at 24 hours and breakage at 48 hours. Such morphological damage in cancer cells confirms the antitumorigenic activity of the CUR-AuNCs, suggesting their potential for cancer therapy. Passeri and colleagues utilized contact resonance AFM (CR-AFM) as a nondestructive subsurface nanomechanical characterization tool for the detection of stiff (magnetic) NPs internalized in cells [99]. Magnetic NPs were internalized within microglial cells through a phagocytosis process. By increasing contact stiffness in the biological network, magnetic NPs agglomerate in the cell, showing mechanical inhomogeneities that can be visualized via CR-AFM nanomechanical imaging. These results are confirmed by other methods including HarmoniX^{TM}, light microscopy, and magnetic force microscopy, revealing the profound implication of CR-AFM for imaging of NPs into biological tissues.

5.3.1.4 Confocal Laser Scanning Microscopy

Confocal laser scanning microscopy (CLSM) is an important class of optical imaging, enabling lateral (*xy*) and axial (*z*) images for visualization of living specimens in a short time. Compared with other imaging procedures, CLSM can reject light from outside the focal plane (out-of-focus flare) through point scanning. This method has been exclusively applied in various types of cancer studies to clarify the internalization and distribution of NPs that affect cell morphology, resulting in toxicity. In a notable example, DNA-carrying NPs transported leukemia-targeting CAR genes into T-cell nuclei; they were rapidly internalized (120 minutes) within cytoplasm, likely due to receptor-induced endocytosis [100]. Further research illustrates that two types of alginate NPs consisting of mannose modified alginate (MAN-ALG) and an antigen model [ovalbumin (OVA) conjugate alginate (ALG=OVA)] were internalized into mouse bone marrow dendritic cells (BMDCs) [101]. Results indicate that MAN-ALG/ALG=OVA NPs enhance antigen uptake, delivery, and cytosolic release of the antigen in BMDCs. Xu et al. report that the cellular uptake of coumarin-6 (C6) in 4T1 mammary tumor cells delivered via shrapnel NPs greatly increases with matrix metalloproteinases (MMP-9) [102]. These results are promising for inhibition of metastasis and the growth of breast cancer. CLSM is also able to provide nanoscale information on the transportation of NPs in intestinal epithelial cells and related mechanisms that enhance the bioavailability of orally administrated drugs. Both polymer and pure drug NPs were employed to evaluate endocytosis, and intracellular trafficking in Caco-2 epithelial cells [103, 104]. CLSM colocalization and distribution methods illustrate the key role of actin in internalization of polymer NPs. Endocytosis inhibitors lead to considerable suppression on the internalization of C6-polymer NPs. Intracellular trafficking increases by raising incubation time. CLSM further confirms that polymer NPs transport to cells through both lysosome- and endoplasmic reticulum (ER)-trafficking pathways [103], while good colocalization with ER has been reported for drug NPs [104]. These findings may shed new light on treatment of various types of cancers and provide important information for oncologists.

5.3.1.5 Dark-Field Microscopy

Dark-field microscopy uses scattered light from a sample by rejection of excitation light. This method enables a large signal-to-noise ratio. The scattered light from a sample can be detected on a dark background [105]. Although powerful, it is limited to highly scattering objects. For example, Chan and colleagues utilized 3D dark-field microscopy based on NP scattering probes for three distinct biomedical applications: (i) molecular imaging of whole tissues, (ii) tracking of nanodrug carriers in mouse tumor models, and (iii) characterization of lesions in a mouse model of multiple sclerosis [106]. 3D dark-field microscopy enabled the authors to observe the AuNP scattering signal in intestine and kidney tissues. They also visualized both diffuse and localized signals from AuNPs in cleared tissues with respect to gold deposition. Dark-field imaging can also be used in combination with hyperspectral imaging, which enables enhanced quantitative intensity for the detection, composition, and location of plasmonic NPs in biological samples. One example of hyperspectral dark-field microscopy (HSDFM) is with plasmonic nanoprobes to map and quantify multiple epigenetic marks at single-modification resolution in single cells [107]. Plasmonic nanoprobes with strong LSP enable HSDFM to collect key information about cytosine modification in a single nucleus including colocalization, dynamic abundance, distribution, and density. By spectrally differentiable AuNPs and AuNRs, HSDFM can also shed light on multiplex detection of proteins and oncogenic mRNAs for a single gene [108]. The intracellular formation of gold particulates at a spatiotemporal resolution can be achieved with this single-cell optical clearing method.

5.3.1.6 Light-Scattering Microscopy

Light-scattering microscopy employs an incident laser beam that scatters light based on the movement and size of the molecule. This technique is divided into static and dynamic light scattering (DLS). The technique provides important information regarding hydrodynamic radius and diameter, uniform size distribution, formation of RNA cubes for applications in influenza, and drug delivery [109–111]. Zhai et al. used DLS to investigate the hydrodynamic diameter distributions and stability of non-cross-linked micelles (NCMs) and core cross-linked micelles (CCM), which are remarkable in tumor combination chemotherapy [112]. Upon modification with fetal bovine serum, CCM shows narrow size distribution and uniformity in the initial 30 minutes during the process, while the size of NCM increases to 176 nm, it then swells dramatically to over 600 nm with a broad distribution at 24 hours, exhibiting the formation of large aggregates. Using DLS, the hydrodynamic radii (R_h) of hexameric DNA and RNA nanocubes were 6.2 and 6.4 nm, respectively [110]. DLS, cryogenic electron microscopy (cryo-EM), and native poly-acrylamide gel electrophoresis (PAGE) results confirm the formation of compact, closed molecules with different strands.

5.3.2 *In Vitro* Cell-Based Cytotoxicity Assay

In vitro biological assays study biological properties of NPs, identify interferences with *in vitro* tests, and provide the first comprehensive insight into potential sources of this interference, demonstrate the usefulness of including NP-only controls, and illustrate the importance of physicochemical characterization of NP formulations. They also visually monitor test samples to avoid false-positive or false-negative results. There are several concerns regarding the safety of nanomaterials, particularly for biological applications. There has been a call for more studies on nano-biocompatibility. *In vitro* assays of material in any given matrix allow the examination of specific biological response(s) and/or mechanism of action under controlled conditions not easily studied in complex *in vivo* situations. *In vitro* assays are important, since they are cost-effective and provide quick results. There is no need to use animals in these assays, thus positively addressing ethical issues. Nonetheless, original physiological results (in *in vivo* research) may differ from *in vitro* assays, since *in vitro* tests are not able to completely replicate the sophisticated interactions that take place in the body mainly due to the use of one cell type or a mixture of several cells.

5.3.2.1 Cell Viability Assays

Cell viability assay provides valuable information regarding the biocompatibility of materials by determining the number of survived or death cells in a sample. It is, thus, vital in the assessment of nanomaterials for biomedical applications. A variety of cell types can be applied in cell viability assays, including but not limited to cancer cell lines, endothelial, epithelial, fibroblast, hepatic, neural, phagocytic, and red blood cells [113, 114]. Typically, cell viability is evaluated by various cellular processes, such as DNA synthesizers, metabolic activity, and proliferation. Common *in vitro* assays are 3-(4,5-dimethylthiazol-2-yl)-2,5-diphenyltetrazolium bromide (MTT), 4-[3-(4-Iodophenyl)-2-4-(nitrophenyl)-2H-5-tetrazolio]-1,3-benzene disulfonate (WST-1), 3-(4,5-dimethylthiazol-2-yl)-5-(3-carboxymethoxyphenyl)-2-(4-sulfophenyl)-2H-tetrazolium (MTS), 2,3-bis[2-methoxy-4-nitro-5-sulfophenyl]-2H-tetrazolium-5-carboxanilide (XTT), and adenosine triphosphate (ATP). These assays evaluate the activity of mitochondrial enzymes (e.g. succinate dehydrogenase). MTT is positively charged to easily enter cells, while MTS, WST-1, and XTT are negatively charged molecules that cannot readily penetrate cells. The latter require an intermediate electron acceptor. MTT is an assay for measuring *in vitro* viability and cell interactions with nanomaterials. MTT is a positively charged molecule converted into an insoluble formazan product through mitochondrial enzyme succinate dehydrogenase. It easily enters viable cells [115]. Formazan that precipitates in MTT assays require solubilization, which hampers the probability to mix various assays on the same well [116]. Using MTT assay, Sokolova et al. show that calcium phosphate NPs at low concentration ($2.4 \times 109\ ml^{-1}$) exhibit no evidence of cytotoxicity on differentiated THP-1 cells after 1 or 24 hours of incubation [117]. By increasing the concentration of NP to $2.4 \times 10^9\ ml^{-1}$, the viability of THP-1 cells was about 50%. Such cytotoxic effect for higher concentration of NPs is possibly attributed to the increased concentration of PEI applied for the colloidal stabilization of the NPs. MTS is a negatively charged molecule converted into a soluble formazan product in the presence of phenazine methosulfate (PMS). Compared with MTT, this assay allows sequential analysis of treated cells with other assays. The content of tetrazolium salt can be measured photometrically at 490 nm. Similar to MTS, WST-1 assay is a negatively charged molecule that is a water-soluble tetrazolium salt. WST-1 assay is more accurate than MTT for evaluation of CNT toxicity, because insoluble MTT formazan crystals can bind to CNTs and produce inaccurate absorbance values [118]. WST-1 assays resolve this problem by producing a water-soluble formazan product. ATP is linked to the metabolic activity of viable cells. It is, therefore, employed as a marker to evaluate viability and proliferation of cultured cells [119, 120]. Its intracellular level is modified accurately in healthy cells due to the key role of ATP in cellular metabolism. This luminescence assay measures the metabolic activity of cells through the enzymatic transformation of intracellular ATP and D-luciferin into oxiluciferin and light.

5.3.2.2 Cell Cytotoxicity Assays

Mean cytotoxicity/reactivity grades related to the concentration and toxicity of soluble components diffuse from the test sample. *In vitro* testing using primary cells can quickly screen for new chemical entities with serious toxicology consequences. Nanomaterials have similar properties hazardous to other physical forms of the substance. Cell cytotoxicity assays are generally involved with cell death (13 types of cell death mechanisms have been introduced) [121]. Cell death is a natural biological process that occurs as a result of old cell dying or cell replacement as well as localized injury and disease. There are three main cell death mechanisms: necrosis, apoptosis, and autophagy. These are activated by nanomaterials [33]. There are several assays to measure cell membrane damage, including lactate dehydrogenase (LDH) enzyme (measures the release of cytosolic molecules), propidium iodide, and trypan blue (measures the entrance of molecular probes).

LDH assays measure release of lactate dehydrogenase enzyme into the extracellular media as a result of either cell death or membrane damage [122]. Released LDH enzymes catalyze the conversion of lactate into pyruvate, followed by reduction of nicotinamide adenine dinucleotide (NADH) into NAD^+ [123]. Reduction of NADH absorption peaks is attributed to the growth of extracellular LDH concentrations [123]. By interacting with basic tetrazolium salts, released NADH forms a colored formazan compound that results in quantification of cells [124]. Kostarelos and coworkers used modified LDH assay to study the effect of different MWNTs on neuronal and glial cultures isolated from fetal rat frontal cortex (FCO) and striatum (ST) [125]. LDH results indicate no important reduction from both FCO and ST induced by MWNTs. However, non-neuronal (mixed glial) cell cultures were reduced in ST-derived mixed glial cultures due to the higher number of microglial cells in this brain region as opposed to FCO-derived ones. These findings confirm that FCO-derived mixed glial cultures are less sensitive to MWNT exposure than ST-derived mixed glial cultures.

Propidium iodide is a charged dye with the capability of penetrating damaged membranes excluded from viable cells [122]. Generally, propidium iodide is applied to measure cell death after the breakdown of the cell membrane. It enters cells with damaged membranes and binds into double-stranded RNA and DNA of non-viable cells. This leads to formation of a fluorescent complex. Trypan blue is a diazo dye employed as a colorimetric assay to measure cell cytotoxicity. Similar to propidium iodide, trypan blue can enter damaged membranes and exclude from viable cells [122]. Cells are modified with trypan blue in this test and visually reveal whether they exclude or uptake the dye [126]. Although a dead (nonviable) cell should have blue cytoplasm, the viable cell should possess a clear cytoplasm. The main advantage of this assay is ease of estimation of the actual number of viable cells. Neutral red uptake (NRU) assay was involved in the first nongenotoxicity *in vitro* test accepted for regulatory assessment of chemicals [127]. It is based on the ability of viable cells to incorporate and bind dye neutral red. NRU enters cell membranes by nonionic passive diffusion and is retained in the lysosomes. It is then extracted from the viable cells using an acidified ethanol solution [127]. ATP is necessary to maintain the pH gradients in lysosomes. Under physiological conditions neutral red is an uncharged dye able to penetrate the cell membrane. There is a proton gradient in the lysosomes that retains a pH lower than cytoplasm, hence the neutral red dye becomes charged. When the pH gradient is reduced or the cell dies, the dye cannot be maintained, and the amount of retained dye shows cytotoxicity [127].

5.3.3 Validity of *In Vitro* Assays

A wide range of suitable standard controls should be taken into consideration to validate the results of *in vitro* assays. These standard controls are divided into positive or negative soluble components (e.g. released components or ions, vehicle/dispersant, or excipients) [128]. Positive controls are reagents able to induce peak changes in the endpoint of the *in vitro* assay [128]. They are associated with the hypothesis being assessed, for example, use of *tert*-butyl hydroperoxide for testing the cytotoxicity effect of nanomaterials through oxidative stress [128, 129]. There are two positive controls for assessing cell viability: sodium deoxycholate and acetaminophen [130, 131]. Negative controls are typically demonstrated via incubation buffers, such as Hank's Balanced Salt Solution (HBSS) for Caco-2 cell transport experiments or cell culture medium [131, 132]. Soluble components of nanomaterials are significant controls to recognize the contribution of free constitutive molecules and/or elements as well as bioavailable released ions and/or impurities in inducing the biological response [128]. Vehicle-only treatment plays a key role in studying the influence of dispersion media on measured endpoints [128]. Since vehicle-only treatment indicates the starting point of dose–response curves, it should not cause any cytotoxic impact on the cells. Excipient controls are suitable to examine the possible effect of formulation excipients on cell viability [131]. Nanomaterials are also able to be considered as negative and positive controls, as are polystyrene NPs with either carboxyl (PS-COOH) or amine (PS-NH_2) functionalization. Negatively charged polystyrene NPs (PS-COOH) are employed as a potential negative control for NP-induced cytotoxicity, whereas positively charged NPs (PS-NH_2) are used as a positive control [133].

5.3.4 In Silico Assay

In silico procedures are cost-effective and rapid experiments performed by computers to predict the toxicity of materials. Such computational/simulation methods are important in the design of new materials and can positively reduce the length of the assays as well as the risk assessment of chemicals. One of the main *in silico* methods for toxicity is quantitative structure–activity relationships (QSARs), enabling the toxicity prediction of nanomaterials based on their biological and physicochemical properties [134]. Yoon and colleagues employed nano-SAR models to investigate the influence of quality screening and data gap filling approaches [135]. They performed a comprehensive analysis of 216 publications on oxide NPs with 14 attributes of toxicological, physicochemical, and quantum-mechanical features. Results reveal that the mixture of toxicological and physicochemical attributes has a higher effect on toxicity than quantum-mechanical attributes. In other research nano-QSAR was utilized to predict the toxicity of 17 different metal oxide NPs [136]. Theoretical methods and experimental data confirm the relationship between cytotoxicity to *Escherichia coli* cells and NP structures.

5.3.5 *In Vivo* Studies

In vivo experiments provide information regarding biological interactions in living organisms through the use of animal models as well as non-human primates. These experiments are the principal assays in assessing toxicity of nanomaterials. They are able to determine the cytotoxicity and organotoxicity of NPs on a specific organism. Applying this assay, the distribution, cellular uptake, metabolism, and clearance of nanomaterials can be obtained. They remain the most reliable methods to evaluate toxicity of nanomaterials despite extensive time and cost. Typically, however, *in vivo* interaction of the biological system and NPs is sophisticated, and these methods are limited compared with *in vitro* and *in silico* assays. There are various animal models commonly employed to study *in vivo* nanotoxicity. These include mice, rats, zebrafish, frogs, and nonhuman primates. In this section, we focus in detail on principal factors that affect the *in vivo* toxicity of nanomaterials.

5.3.5.1 Blood Circulation

The initial encounter between the organisms and NPs occurs in the circulatory system; thus blood compatibility of NPs is fundamental for *in vivo* applications. The absence of blood compatibility may cause adsorption of plasma proteins, platelet adhesion, and activation of complement cascades that leads to hemolysis and clot formation (thrombogenicity) [137]. Although blood compatibility is the first requisite for NPs that are proposed for clinical applications, few studies have evaluated the hematological and blood coagulation effects of NPs *in vivo* [138, 139]. There are non-precise/standard procedures for the *in vitro* evaluation of these NPs. Hemolysis is the disintegration of red blood cells in light of the changes in their cell membranes. During hemostasis, clot formation naturally occurs to stop bleeding from injured blood cells. Several negatively charged porous NPs (including MSNPs and zeolites) are able to accelerate hemostasis of the blood by triggering a coagulation cascade. For example, Liu and colleagues developed ordered sphere-shaped MSNPs doping with silver and calcium through one-step based-catalyzed self-assembly and subsequent ion-exchange for hemorrhage control [140]. These 3.2 nm NPs with a pore volume of $0.74\,m^3\,g^{-1}$ and surface area of $919\,m^2\,g^{-1}$ exhibit suitable antibacterial and degradation abilities. The authors have concluded that the optimal formulation of MSNPs can activate the intrinsic pathway of coagulation cascade, cause platelet adherence, enhance blood clotting, and obtain hemorrhage control in rabbits using liver injury and femoral artery models. However, clot formation inside blood vessels through the injection of MSNPs may result in several unfavorable effects by blocking the blood vessels. This can cause strokes and death [141, 142]. One example is the use of generation-7 (G7) poly(amidoamine) dendrimer in zebrafish embryos that adversely affects various blood components to generate rapid coagulation, notably different from normal coagulation mechanisms [143]. Interaction of positively charged dendrimers with negatively charged blood proteins (albumin and fibrinogen) cause coagulopathies (bleeding disorders). As a consequence, the clot formation of NPs should be examined prior to *in vivo* applications. Likewise, protein adsorption, usually regarded as the initial step in the interaction of NPs with blood, should be determined, since it leads to coagulation and other problems. Materials such as carbon and silver NPs trigger platelets and promote vascular thrombosis as well as venous thrombus formation [144, 145]. AgNPs (0.05–$0.1\,mg\,kg^{-1}$ i.v. or 5–$10\,mg\,kg^{-1}$ intratracheal administration) improve platelet aggregation and venous thrombus formation in rats. The results are consistent with *in vitro* analyses [145]. Carbon-based particles, including carbon black, CNT, and urban dusts induce platelet aggregation and accelerate the rate of vascular thrombosis in rats [146]. Although all of applied particles lead to upregulation of GPIIb/IIIa in human platelets, dissimilar particles influence the release of platelet granules and the activity of matrix metalloproteinase-, thromboxane-, ADP-, and protein kinase C-dependent pathways of aggregation.

Several analyses evaluate the plasma blood coagulation, such as thrombin time (TT), prothrombin time (PT), activated clotting time (ACT), and activated partial thromboplastin time (APTT), as well as hematotoxicity properties of NPs including complete blood count and the possible genotoxic effect for clinical applications [147, 148]. Ruiz et al. applied magnetite NPs with a narrow size distribution (mean size = 7 nm; SD < 0.15) functionalized with both PEG and dimercaptosuccinic acid (DMSA) to assess hematotoxicity of these NPs using *in vitro* and *in vivo* tests [148].

5.3.5.2 Pharmacokinetics

Pharmacokinetics (PK) demonstrate how organisms affect particular materials (e.g. nanostructures) after entering into the body via several processes (e.g. absorption, distribution, metabolism, and elimination). Absorption occurs after administration by interactions of nanomaterials with biological substances such as cells. They can then distribute within many living organs in which they may become altered or remain unchanged [149]. Nanomaterials may remain in an organ for an unspecified time before leaving to other cells or being excreted. A comprehensive analysis of PK is required to understand the toxicity of NPs and their potential effects in living organisms. Such biological information may influence the design/development of NPs with the highest efficiency in the field of biomedical areas.

Absorption NPs can be administrated to the body through three routes: airways, skin, and ingestion [150]. They can translocate to other organs, but only in small quantities [151]. Current *in vivo* results have been obtained mostly from intratracheal instillation and inhalation in rodents including the bulk of toxicity information for diverse NPs [152]. In the airways route, NPs can access the body via the respiratory tract. The nasopharyngeal region absorbs particles of 10–20 nm [153]. The health effects of NPs are related to the residence time in the respiratory system and lung burden [154]. It has been reported that smaller particles induce higher toxicity and lead to inflammatory reaction greater than larger particles of the same compounds [155]. At high concentrations ultrafine TiO_2 (<100 nm) and fine CNTs (<2.5 μm) cause tumors in the respiratory tract [156–158].

One of the major concerns about translocation is the ability of NPs to pass air–blood barrier in the lungs to further allow them access to other organs of the body [159]. Mucus and mucociliary escalators are known as the defense mechanisms against particles in the body [160]. NPs successfully cross these barriers and translocate from the lung into the spleen, liver, heart, and perhaps other organs [161]. Gold particles introduced intratracheally into rats penetrate into the body by passing and translocating to other organs including blood, urine, and different tissues as well as being retained in the skeleton [162]. The translocation mechanisms of NPs via the air–blood barrier have not been clarified, while the importance of transcellular routes has been suggested [163]. Elimination of NPs from the lung is maintained mainly through alveolar macrophages (AMs) [164]. Inhaled NPs such as Ni, Mn, Co, and Cd can also gain access to the nervous system via the olfactory nerves and/or BBB [154]. Poorly soluble and insoluble NPs are able to enter the brain through the nasal route and uptake via the BBB from systemic circulation [165].

NPs are capable of entering the human body via the skin through hair follicles and flexed or broken skin [152, 166, 167]. For example, TiO_2 NPs, often applied in commercially available sunscreen products, may enter into the body, but their translocation and dermal absorption are not completely understood and need further study [168]. Other NPs such as QDs [150] and fullerenes [169] enter the dermis [170, 171] based on size and surface coatings [172]. It is important to note that skin penetration of NPs depends both on properties of NPs (e.g. size, dose, and surface charge) and skin conditions (e.g. skin barrier and age). In the case of QDs, the animal species plays a key role in the results. The skin of mice and porcine animals was penetrated by QDs to pass the dermal barrier, whereas such penetration was not reported for rat skin [170, 173, 174].

The gastrointestinal tract (GIT) is another route of administration of NPs into the body. Even though GIT exposure may lead to the translocation and uptake of NPs to various organs, there has been less research than on other routes. There are growing concerns related to the possible access of NPs into the blood stream via GIT mainly due to the increasing use of NPs in food packaging and processing or as food additives [175]. The kinetics of NPs in the GIT depend on the charge in which negatively charged particles pass across the mucus layer, whereas positively charged latex particles stay in the negatively charged mucus [176]. Some diseases such as diabetes may induce greater particle absorption in the GIT [176]. Although high concentrations of most NPs are quickly removed from the body through feces, small proportions are absorbed by the gut, showing size-dependent behavior [177, 178].

Distribution After absorption of NPs in the body, they distribute to diverse cells, organs, and tissues. More *in vivo* biodistribution studies with precise characterization of NPs are required. Various factors affect the fate of nanomaterials in a human body including their size, solubility, and surface charge. Optical imaging, post-mortem NP determination, and isotopic labeling with radioactive elements have been utilized to analyze the biodistribution of nanostructures [179–181]. For example, Rojas et al. report *in vivo* biodistribution NDs via ^{18}F Radionuclide Emission. They demonstrate the effect of surfactant agents as well as kinetic particle size on biodistribution [182]. Radiolabeled NDs are largely retained in the lung, spleen, and liver, and they are eliminated by the urinary tract. The introduction of Tween 80 and PEG_{8000} as surfactant agents shows a small impact on this pattern, while they slightly reduce the urinary excretion rate. In contrast, the kinetic size completely impedes uptake of radiolabeled NDs in the both lung and spleen and decreases in the liver. Distribution of NPs of larger size faces difficulties with lower penetration rates in solid tumors studies. These NPs are centralized heterogeneously in the perivascular region [14, 183]. Matsumoto et al. intravitally imaged the transport of fluorescent-labeled polymeric NPs (30 and 70 nm) at 10 minutes intervals to investigate eruption and distribution in hypovascular human pancreatic BxPC3-GFP tumors implanted in BALB/c nu/nu mice by applying intravital CLSM [184]. They introduced dynamic vent and eruption as the time-limited formation of an opening in the vessel wall and vigorous extravasation through the vent. Eruptions occurred stochastically but continuously during the 10 hours observation period.

The existence of eruptions was observed for all modified groups in which 30 nm NPs indicated shorter-lived eruption plumes as opposed to 70 nm NPs. The size of the eruption plumes made by the 30 nm NPs was gradually smaller than those prepared by 70 nm NPs. Vessel dynamics probably did not change according to NP size. Duration and eruption plume size are not similar due to the faster dispersion of smaller NPs. When NPs disperse sufficiently, fluorescence intensity declines under the detection threshold. The use of 30 nm NPs rather than 70 nm NPs shows quicker grow of background fluorescence, indicating that smaller NPs can distribute more easily in a tumor. As the eruption starts during the early phase, the rate of plume growth in both sets reveals that smaller NPs transfer more rapidly than larger ones. This is because during the initiation phase NP movement is essentially convection driven with bigger NPs having a little more movement resistance from the tissue structures. When complete, plume dispersal is controlled through diffusion. This is very difficult for larger NPs. Therefore, vascular eruptions are responsible for deep penetration of NPs as well as restricted dispersion from the early plume. Campbell et al. provide mechanistic insights of NP–liver interactions in zebrafish at the molecular level, demonstrating an essential role of the stabilin-2 scavenger receptor (stab-2) in the uptake of anionic liposomes via sinusoidal (or scavenger) endothelial cells (SECs) [185]. They employed confocal microscopy to observe real-time distribution of fluorescently labeled NPs in the body of zebrafish as a transparent model. The authors utilized liposomes to evaluate the influence of surface charge on distribution. They intravenously injected three types of liposomes including Myocet (neutral), EndoTAG-1 (positively charged), and Ambisome (negatively charged) into the duct of Cuvier of zebrafish embryos. Since each of these liposomes is formed from a combination of phospholipids and cholesterol, their cholesterol content and chemical structure (level of saturation and length of alkyl chains in fatty acids) are both responsible for physicochemical properties, hence their *in vivo* fate. To reduce possible variation of chemical composition of liposome on the distribution, the researchers also used 1,2-dioleoyl-*sn*-glycero-3-phosphocholine (DOPC), 1,2-distearoyl-*sn*-glycero-3-phosphocholine (DSPC), and 1,2-dioleoyl-*sn*-glycero-3-phospho-(1′-rac-glycerol) (DOPG). DOPG (a negatively charged, unsaturated lipid) and DSPC (a neutral saturated lipid) liposomes were strongly associated with venous ECs, while DOPC (a neutral, unsaturated lipid) circulated freely. NP–SEC interactions were inhibited by a competitive inhibitor of stab2, dextran sulfate, and other receptors. The authors finally applied these specific interactions for targeted intracellular drug delivery. This significant finding in zebrafish may open a new avenue for researchers in biomedical engineering to solve the delivery issue of *in vivo* NPs.

Metabolism *In vivo* metabolism of NPs in tissues after systemic administration is an important milestone for the rational design of nanostructure materials for clinical applications. This not only enables a better understanding of *in vivo* NP dynamics but also can effectively minimize the toxicity risk of NPs. Degradation of NPs is essential for the successful delivery of therapeutics into clinical areas, since NP degradation is directly associated with the delivery and release of drugs at their desired location. It should be noted that *in vivo* degradation of NPs is related to NP composition.

For instance, NPs made from Au, MSN, and QDs are generally not degraded *in vivo*, while other materials such as Ag, CdSe, SPIONs, and ZnO release metal ions [186]. Considering the case of SPIONs, several analyses are often applied to quantify the degradation of NPs over time in the body, including inductively coupled plasma mass spectrometry (ICP-MS), temperature-dependent susceptibility measurements, electron paramagnetic resonance (EPR), MRI, and radioactive labeling [187–189]. Distinguishing between the NPs and their degradation products is fundamental. Taking SPIONs as an example, an ICP-MS test is not suitable because of the high amount of endogenous iron forms, whereas EPR provides information regarding the biotransformation of the superparamagnetic iron core in both short and long time [190]. Gazeau and colleagues employed electron spin resonance (ESR) to quantify superparamagnetic iron oxide from the spleen and liver during one year after intravenous injection in mice [191]. Gold/iron oxide nanoheterostructures were coated with PEG and an amphiphilic polymer, and then they degraded *in vivo*. NPs coated with the amphiphilic polymer exhibited higher concentrations in the liver and spleen as opposed to PEG-coated NPs. Additionally, the initial coating showed a long-lasting influence on the degradation/clearance of NPs.

Apart from the quantification of NPs at the organ level, observing the NPs at the nanoscale in biological environments is indispensable for assessing the degradation of nanomaterials. Hence, the use of high-resolution imaging techniques (probe the atomic-structure of NP cores), as well as electron energy loss spectroscopies and energy dispersive X-ray (EDX), enables local biodistribution of exogenous materials. These techniques determined that gold/iron oxide heterostructures and iron oxide nanocubes degraded in the spleen and liver seven days after injection [191, 192].

Clearance From a toxicological perspective, nanomaterials should be cleared from the body after administration, because they may react with organs, cells, or tissues that can cause severe health problems. The environment of target tissues as well as physicochemical properties include size, shape, coating, and surface charge. These have a significant impact on the clearance of nanomaterials from the body. There are three main pathways for elimination of nanomaterials: hepatic, pulmonary, and renal clearance that use liver, lung, and kidney organs for removing nanomaterials.

Hepatic Clearance The liver is the second largest human organ after the skin and the major organ for elimination of a wide range of drugs. Both the spleen and the liver are regarded as the main biological barriers to translate NPs, since they are responsible for sequestering most of the administrated NPs and prevent delivery to unhealthy organs. The liver consists of two lobes: parenchymal cells (i.e. hepatocytes) and nonparenchymal cells (i.e. sinusoidal endothelial cells, Kupffer cells, sinusoidal endothelial cells, stellate, and intrahepatic lymphocytes). These cells deal with hepatic clearance of nanomaterials from the body [193]. The two lobes are separated by the middle hepatic vein. Blood is transferred to the liver through the portal vein and hepatic artery. Hepatocytes and phagocytic Kupffer cells are important routes for hepatic elimination of NPs. Several studies indicate that hepatic clearance is influenced by the physicochemical characteristics of nanomaterials. For instance, larger NPs are cleared predominately through the liver [194–196]. Kupffer cells as specific macrophages inside the liver are capable of internalizing administrated NPs up to several hundreds of nanometers in size either via endocytotic uptake or phagocytosis mechanisms [197, 198]. Typically, NPs bigger than ~200 nm are efficiently eliminated by Kupffer cells due to the slow blood flow in liver sinusoids. This provides enough time for micropinocytosis and phagocytosis of NPs. However, smaller-sized particles may pass the endothelium into the Disse space or undergo transcytosis to be taken up by hepatocytes [199, 200]. Hepatic clearance is also affected by the surface charge of NPs. For example, hepatocytes play a key role in the internalization of cationic NPs, whereas Kupffer cells are efficient for anionic NPs [200–202]. *In vivo* clearance of nanomaterials is also associated with the interaction of NPs with biological components (i.e. enzymes, proteins, and biological fluids). A wide variety of Phase I and Phase II enzymes, including epoxide hydrolase, monooxygenase, esterases, and transferases, exists in the liver. This may affect the hepatic clearance of NPs [203].

Pulmonary Clearance Compared to the liver, the lung is involved in the clearance of inhaled NPs that reach the body through the upper airways, because it is directly exposed to the environment. The fate of NPs is related to lung compartment deposition, which is size dependent. For example, NPs smaller than 100 nm have a preference for deposition in the peripheral lung regions (i.e. alveoli), while larger particles with sizes of 1–10 μm deposit in conducting airways with ciliated epithelial cells [204]. The AMs and airway are located at the front line of lung defense. The original mechanisms for clearance of materials are mucociliary transport (as the most rapid elimination pathway for inhaled components) and phagocytic uptake (a slow process) [205]. Intranasal administrated NPs enter the alveoli areas in the lungs [206]. AMs phagocytize the IONPs, digest them and their by-products, are swallowed, or are released within pulmonary lymphatics [207, 208]. Research reveals that the number of lung macrophages can grow in the presence of IONPs via the transportation of monocytes into the lung. This leads to quicker ingestion of NPs [209]. Other physicochemical properties of NPs, such as coating and surface charges, can affect their clearance from the body. For instance, Cho et al. utilized negatively charged cross-linked IONPs. The researchers demonstrate the clearance of most percentage of NPs from the lung 3 hours after penetration. This is because of increased macrophage uptake in the lung [210].

The clearance kinetic of nanosized drug carriers after pulmonary administration in the lung has not yet been fully understood. Kaminskas and coworkers compared lung clearance kinetics and the pulmonary pharmacokinetic pattern of two inhaled NPs (anionic liposomes and solid lipid NPs) with sizes of ~150 nm in rats. They used ^{3}H-labeled structural lipids (phosphatidylcholine and tristearin) [211]. Although both liposomes and solid lipid NPs were deposited differently into the lung, they showed similar clearance rates through the mucociliary escalator. Generally, however, prolonged lung exposure after an initial rapid clearance was observed for both NPs. Phospholipids from the liposomes were better absorbed from the lung, while the mucociliary clearance was more pronounced for solid lipid NPs. Lipid NPs were largely incorporated into the lung tissue and plasma proteins.

Renal Clearance The kidney is involved in the effective elimination of foreign substances from the body, such as nanomaterials. The main structural and functional parts of the kidney are called nephrons, which are abundantly available. They assist the kidney to maintain the homeostasis of electrolytes and body fluids in extravascular, extracellular, and intracellular compartments. They are included in the specific filtering of proteins and saccharides from the blood, as well as NPs. Intravenously administrated NPs are able to enter the blood vessels of the nephrons via the renal hilum and are then excreted in urine through the ureter and then by the urinary bladder [212]. Renal clearance is regarded as a suitable route for clearance of NPs, because it needs minimal interaction and degradation within the body, causing fewer health hazards [213]. In this clearance route the NPs in the kidney first reach the glomerular capillary wall as the blood filtration site in the nephrons, which is highly shape-, size-, and charge-dependent [214]. Other factors, such as particle dose and molecular weight, strongly affect the renal clearance of NPs [215, 216]. The glomerular capillary wall has three layers: the endothelium with fenestrae (70–90 nm), glomerular basement membrane (2–8 nm pores), and the epithelium with filtration slits (4–11 nm) between the podocyte extensions [217]. Therefore, only NPs with specific hydrodynamic sizes (below 6 nm) can be easily removed by the kidneys through the glomerular structure, while NPs with sizes above 6 nm are eliminated from the blood stream via the RES (liver and spleen) that requires much longer time [218–220]. Choi et al. demonstrate effective clearance of <6 nm zwitterionic QDs NPs by the urinary system, where a small reduction of the hydrodynamic size from 5.64 to 4.64 nm leads to a considerable increase in renal elimination efficiency from 45% to 75% injection dose (%ID) (4 hours post injection) [221]. Similar studies with remarkable increases in renal clearance are also reported for silica and AuNPs [222, 223]. These studies show that the glomerular filtration barrier allows proteins or NPs with sizes below than 6 nm to be readily cleared by the kidneys. Smaller particles eliminate more effectively than larger particles. Zheng and colleagues discovered that the glomerular wall in the sub-nanometer size range (~1 nm) acts as an atomically precise bandpass filter to notably slow the elimination of AuNCs with various sizes (Au_{18}, Au_{15}, and $Au_{10\text{-}11}$) but with the same surface ligands (GSH) [224]. In Au_{25} (~1.0 nm), renal elimination of AuNCs improves significantly through addition number of atoms in a particle, since smaller AuNCs are more readily retained via the glomerular glycocalyx than larger particles. This interaction slows the extravasation of very small-sized AuNCs from normal blood vessels and improves their accumulations in cancerous tissues by an enhanced permeability and retention (EPR) effect. As opposed to small drug molecules, NPs are often passively accumulated with high concentrations for longer times due to EPR [225, 226]. When larger NPs are compared with small, renal-clearable molecules [227, 228], they intensely accumulate in RES organs (liver and spleen) [229], causing low targeting specificity and potential toxicity [230, 231]. To resolve this problem for tumor targeting, NPs should be retained in the blood plasma at a relatively high concentration for >6 hours; the sizes of NPs should be smaller than 6 nm, allowing them to successfully pass kidney filtration [226, 232]. For example, Zheng and colleagues compared the passive tumor targeting of two renal-clearable materials, organic fluorophore (IRDye 800CW) and ~2.5 nm GSH-coated luminescent AuNPs (GS-AuNPs) [233]. In the initial stage of tumor targeting the AuNPs acted like small molecules with much longer tumor retention time than the dye, showing the EPR effect and effective renal clearance of NPs. Results reveal that GS-AuNPs is retained in blood plasma for a long clearance half-life of 8.5 hours and is eliminated from normal tissues about three times faster than the dyes.

5.4 Physicochemical Properties of NPs Affecting Toxicity

The physicochemical properties of NPs include but are not limited to size, shape, surface functionality, surface charge, composition, hydrophobicity, aggregation, and solubility. They are essentially tied with their interactions with cells, thus affecting toxicity. It is accepted that controlling these parameters should be optimal for biomedical/biological applications of NPs. In this section we explain in detail the influence of these factors on the toxicity of nanomaterials.

5.4.1 Effect of Size

The size of NPs plays a critical role in pharmacokinetics, *in vivo* biodistribution, therapeutic effect, and tumor accumulation [234–236]. NPs can passively accumulate in tumors based on their size due to the EPR effect and different pore size of tumor vessels. In compared healthy tissues, the EPR effect results in more than a 50-fold rise in the accumulation of NPs [237]. NPs are cleared via the mononuclear phagocytic cells in the RES of the spleen and liver, causing inadequate accumulation at the target site. Another crucial barrier for drug delivery is the fast cleaning of NPs and drugs from the tumor tissue [238]. For example, the *in vivo* distribution of AgNPs with the sizes of 20, 80, and 110 nm demonstrate that larger particles (80, 100 nm) are distributed mainly in the spleen, followed by the liver and the lung, while 20 nm sized NPs are primarily distributed to the liver, followed by the kidneys and spleen [239]. Several modifications have been employed to enhance biodistribution and drug accumulation in tumors such as addition of zwitterionic polymer membrane and folic acid- and dopamine-decorated hyaluronan to the surface of NPs [240–242]. Jiang and colleagues investigated the impact of size and surface chemistry on the accumulation of boronic acid-rich BSA NPs in tumors [243]. Particle residence time and drug accumulation were increased by the addition of the boronic acid group into the NPs. When doxorubicin accumulation at tumor site reaches above 12% injected dose per gram of tumor, a 16-fold drug is injected.

Many efforts have been applied in cancer research to determine optimum size NPs and to enhance tumor treatment and diagnosis [244, 245]. Tang and coworkers studied the therapeutic effect and tumor accumulation of micelles with sizes between 30 and 200 nm. Smaller micelles indicate higher penetration and better therapeutic efficacy, while larger micelles show greater tumor accumulation [246]. The 12.1 and 27.3 nm PEG-coated AuNPs have stronger sensitization impacts than 4.8 and 46.6 nm particles via necrosis and apoptosis [247]. *In vivo* results indicate that all sizes of NPs remarkably reduce tumor weight and volume after 5 Gy radiation. Small NPs cause a considerable decrease in tumors compared to larger ones. Small seed NPs apparently accumulate with high concentrations at the tumor site. *In vivo* toxicity tests of blood biochemistry and immune response confirm that PEG-coated NPs do not cause kidney and spleen damages except for slight toxicity of the liver. Chen and colleagues tested [64Cu]-labeled perylene diimide (PDI) NPs with sizes of 30, 60, 100, and 200 nm as PTT agents, as well as photoacoustic (PA) and positron emission tomography (PET) imaging probes [248]. NPs with sizes 100 nm are ideal for differentiating popliteal and sciatic LNs. NPs migrate from popliteal LNs to sciatic LNs during an interval of about 60 minutes. PDI NPs of 60 nm are optimum for photothermal cancer therapy and tumor imaging because of their high tumor accumulation efficiency.

Small-sized NPs can penetrate deeper into tumors, but they are removed quickly with lower blood circulation time, whereas larger NPs cannot penetrate deeply into tumors and are retained in tumorous tissue for a longer time [249–252]. Size-reducible NPs based on temperature, pH, enzyme, and UV light as stimuli were synthesized. They exhibited higher antitumor effects and better homogeneous drug distribution in tumors [253, 254]. There are, however, several factors that restrict the extravasation of large-to-small transformable NPs comprising dense tumor matrix, solid stress, and high interstitial fluid pressure (IFP). These factors weaken the effect of size-reducible NPs. Gao and coworkers designed a tumor microenvironment responsive and adjustable NP consisting of a laser-responsive nitric oxide (NO) donor, degradable HA, small-sized dendrimeric prodrug (IDD) of DOX, and indocyanine green (ICG) as a photothermal agent [255]. HAase-induced size-reduction occurred in a HAase-rich tumor microenvironment. This enabled the exposure of IDD with deep penetration. NPs improved the NO release for enhancing the EPR effect in the presence of NIR laser irradiation and triggered potent hyperthermia for PTT. The Gao Group further developed a size-reducible nanoplatform (AuNC@CBSA@HA) composed of degradable HA, cationic BSA (CBSA), and AuNCs for breast cancer and lung metastasis [256]. The HN shell protects the positive charge on AuNC@CBSA, reducing toxicity and increasing circulating time of NPs. HA shields NPs with an active targeting ability that can degrade in tumors to decrease the size of NP, thus improving tumor penetration. Results reveal that 200 nm is best with an ideal EPR effect, which is then loaded with ICG and PTX for adjusting tumor microenvironment, drug delivery, and chemo-photothermal therapy. AuNC@CBSA-PTX-ICG@HA-NO_3 with size-reducible properties displays homogeneous intra-tumor distribution. It has high accumulation in breast cancer, impedes 88.4% of lung metastasis growth, and prevents 95.3% of *in situ* tumor growth. The size of NPs can be regulated in the presence of different stimuli, changing from smaller to higher sizes in tumor microenvironments. Such an ability can combine the positive effects of both larger and smaller NPs, thus leading to optimal efficacy for tumor treatment.

5.4.2 Effect of Shape

The shape of NPs is also considered a critical factor that can significantly affect their role in biomedical applications. Although most laboratory-scale studies and clinical trials employ spherical-shaped NPs, nonspherical shape particles have attracted attention in recent decades [257]. Such shapes as nanowires, nanorod, nanocube, nanobelt, nanostar, and nanothron can be formed mainly via postsynthesis manipulation or templates for polymeric NPs and wet chemical methods for metal NPs. Generally, the shape of nanomaterials affects biocompatibility, biodistribution, cellular uptake, and blood circulation [258]. Nonspherical NPs, such as disc and rod, have displayed higher tumor inhibition rate and cellular uptake efficacy. They can easily pass through the spleen [259]. Compared with other NP shapes, nanodiscs are internalized more effectively, and the mechanism of uptake, efficiency, and kinetics display shape-dependent behavior [259]. There have been continuous efforts to study bio–nano interactions of NPs. Elongated shapes of NPs possess better properties than spherical NPs [260]. However, the interaction between the outcomes of suicide gene therapy and magnetic MSNPs (M-MSNPs) has not been clarified. Dong and coworkers addressed the efficiency of shaped-controlled M-MSNPs in magnetic hyperthermia therapy, MRI, and suicide gene therapy of hepatocellular carcinoma (HCC) [261]. Two different shapes of NPs consisting of rod and sphere shapes were fabricated. Rod NPs show faster release behavior, greater loading capacity, better magnetically improved gene delivery, and higher magnetic hyperthermia properties. The magnetic ability of MSNPs has led to significantly efficient dual magnetically improved suicide gene therapy *in vivo* with the ability to observe therapeutic effect through MRI and with reduced toxicity. The author suggests that the potential of both shaped M-MSNPs be applied as safe and effective for HCC theranostics. This group gives insight into the shape effect of M-MSNPs on the isolation and detection of circulating tumor cells (CTCs) [262]. They conjugated fluorescent M-MSNPs of rod and spherical shapes with the antibody of EpCAM. They report good sensitivity of both NPs on CTCs. Rod shape M-MSNPs offer higher effective detection and faster enrichment of CTCs in real clinical blood samples and spiked cells as opposed to their sphere shape counterparts. Recently, molecular bottlebrushes as unimolecular micelles with shapes of rod, sphere, and worm were synthesized by click chemistry and controlled/living polymerization for cancer therapy [263]. IR780 as a photothermal agent was loaded into these core–shell NPs. Maximum loading content of c. 25% does not result in aggregation and morphological changes of these molecular NPs in cell culture media. Unimolecular rod shape NPs exhibit the best penetration *in vitro* and *in vivo,* as well as cellular uptake, which efficiently inhibits tumors. Although shape notably affects cellular uptake, circulation time, tumor accumulation, etc., it is not clear how NP shape influences loading capacity and drug release. Rampersaud et al. compared drug release and efficiency of solid spherical and cage shapes of IONPs-loaded riluzole as anticancer drugs [264]. The cytotoxic effect of nanocages was two times greater, and drug delivery of spherical NPs was three times less effective than their counterparts. Such difference in drug delivery can be attributed to the charge-screening impact on the NP surface. When loaded with NPs, the charge of drugs can affect overall charges of nanostructures (NP loaded drugs). Thus, the surface area of NPs plays an important role in how NP shape influences drug release and efficiency.

Shape can also affect the pharmacokinetic and biodistribution of NPs *in vivo*, which makes them ideal candidates in cancer therapy. This is because NPs can target tumors via active and passive targeting mechanisms [265]. For instance, Davis and coworkers employed fluorescence-assisted cell sorting (FACS) to evaluate subtumoral cellular distribution of PEG-based polymer NPs (rod shape of 80 × 320 nm and cuboidal 55 × 70 nm) fabricated by polymer replication in nonwetting templates (PRINT) [266]. NPs administrated intravenously by the tail vein accumulated mainly in RES organs. Based on the cellular distribution of particles in the tumor, NPs are better related to cancer cells than macrophages, fibroblasts, and neutrophils endothelial tumor-associated cells. Cuboidal NPs are related to cancer cells more than rod-shaped particles. NPs also enhanced pharmacokinetic and tumor localization for drug delivery. Cylindrical PLGA NPs with 200 nm diameter × 200 nm length or 80 nm diameter × 320 nm length loaded with DOX exhibit 20 times higher drug plasma exposure than free drugs [267]. NPs with the longest aspect ratio or smallest diameter cause the maximum concentration of DOX in tumors, as well as lower amounts in the liver, spleen, and lung. The authors postulate that the smaller diameter allows NPs to escape more easily from the vascularized RES, which agrees with nanorods and QDs studies [268].

Together with targeting ligand density, the shape effect of NPs is employed in tumor targeting. Rod and cuboidal shaped NPs, similar to those used by Davis and colleagues [266] were targeted for EGFR through an anti-EGFR affibody [269]. Different shaped NPs of various targeting ligand densities were applied in EGFR-overexpressing epidermoid carcinoma (A431) bearing mice. Cuboidal NPs show higher accumulation in tumor xenografts than rod shape NPs in the absence of a targeting ligand. The cuboidal particles also indicate higher accumulation in the liver than rod-shaped NPs; however more accumulation in the spleen is seen with rod-shaped NPs. This agrees with previous findings [270, 271]. Surprisingly, with rod-shaped NP, circulation time decreases when increasing targeting ligand density. Compared to the PEG control, the elimination of NPs in the experiments was between 5- and 30-fold greater for low-to-high ligand densities. The nonspecific adsorption of proteins to the targeting ligand-bearing NPs is responsible for low NP pharmacokinetics. Both ligand density and liver accumulation targets of the rod-shaped NP directly relate to each other, apparently due to the native expression of EGFR in the liver. Similarly, nanoworms of iron oxide target tumors through a PEG-linked short peptide sequence. This exhibits an ideal peptide concentration that tumor-targeting efficiency optimizes [272].

5.4.3 Effect of Surface Charge

Surface charge can substantially influence the efficiency and the *in vivo* fate of NPs [273, 274]. Surface densities and charges (e.g. negative, neutral, or positive) are different for each NP. Typically, negatively and neutrally charged NPs possess higher biocompatibility, longer circulation half-lives, enhanced stealth-like effect, and reduced undesirable clearance [275, 276]. Due to the negative charge of cell membranes, both negatively and neutrally charged NPs display lower degrees of internalization compared to positively charged (cationic) counterparts [240]. Cationic NPs interact with negatively charged groups on the cell surface and transfer through the plasma membrane. The degree of positive charge is related to cellular internalization [277]. Cationic polymers, including PEI, poly(L-lysine) (PLL), DEAE-dex, and chitosan have been applied in cellular uptake. The highest density of charged groups has been reported for PEI, which significantly enhances membrane permeability [278, 279]. Several cationic materials without fusogenic activity can disrupt endosomal membranes via their buffering capacity [280]. Plasma proteins are able to interact with positive surface of NPs and improve cellular uptake [281]. Cationic NPs appear to be important in mucosal drug delivery by improving *in vivo* association to the cornea, nasal mucosae, and GIT mucosae [280]. They apparently pass the BBB to a higher extent, thus improving brain penetration [282]. Cationic NPs, however, are not essential for effective endocytosis. Neutral or negatively charged NPs indicate effective cellular uptake when conjugated with targeting ligands. Dante et al. report the quick localization of negatively charged magnetic and cadmium-based semiconductor NPs (administered at low concentration of 10 nM) on neuronal membranes. This causes electrophysiological manipulations [283]. The same neutral or positively charged NPs display low or no nonspecific interaction with the cellular membrane. The interaction of negative NPs appears to be selective for excitable neuronal cells and dependent on neuronal activity. The results of the pharmacological alternation of electrical neuronal activity suggest impact of neuronal activity on the neuron–NP interaction. The role of different charged NPs on *in vitro* and *in vivo* toxicity can be different. For example, negatively charged QDs have been reported to be considerably lower cytotoxic than cationic QDs [284]. Although highly toxic cationic QDs cause alternations in genes related to mitochondrial functions, negatively charged QDs improve gene expression of those related to DNA damage and proinflammatory cytokines.

The surface charge of NPs can enable them to be deployed effectively in tumor tissues [285]. The ability to bind to tumors is associated with electrostatic attraction between cationic NPs and negatively charged cell membranes inside tumor sites [286]. However, the particle surface has to maintain a neutral or negative charge out of the target site to avoid nonspecific binding during circulation [287]. It is, therefore, important to design a surface coating ligand that has negatively charged and nonspecific protein adsorption in the blood and modifies its surface charge to positive to be distinguishable from tumor cells [240, 288, 289]. Developing pH-responsive surface properties with the ability of charge switching under tumor microenvironment is desirable, because the tumor extracellular environment is more acidic (pH 6.5–6.8) compared to normal tissues and blood (pH $\approx$ 7.4). Such smart nanomaterials are able to adjust NP interaction with tumor cells by dynamically switching the surface charge in the microenvironment. This reduces unwanted side effects and improves therapeutic efficacy [290, 291]. Thus, the particular intracellular uptake can be increased through regulating the blocking/deblocking process of nonspecific ligands [290, 292]. Using a nonspecific D-galactose building block as the switching unit, Shen et al. synthesized a dual signal responsive nanocarrier consisting of PTX-loaded M-NPs-PLGA-L-lysine-D-galactose (PTX-MNP-PLGA-Lys-Gal) NPs for tumor targeted delivery [293]. The use of oleic acid (OA)-coated M-NPs results in a high M-NP/drug payload. Such particles show switching the surface charge of NPs from negatively charged above pH 7 to positively charged below pH 6.5. Under NIR irradiation approximately half of the drug is released within 30 minutes due to the presence of magnetic hyperthermia and PLGA. Although blocking NPs indicates safety during circulation because of a low hemolysis effect, deblocking enables better targeting to tumor cells. pH charge-switchable surfaces can be cationic at tumor sites and negatively charged and corona-free in blood [294]. This feature enhances circulation of manganese oxide nanocubes (MOCs) as a T_1 contrast agent in blood and improves cancer cell uptake. Recently, Feng et al. prepared a rigid pH-sensitive micellar NP with acid-switchable surface charge and tunable rigidity for MDR through improving lysosome escape and cellular uptake [295]. This NP was formed by a pH-sensitive shell and a PLGA core that could be continuously produced within microfluidic chips in a one-step procedure. When internalized via cancer cells, the rigid and positively charged core ameliorates lysosomal escape, while the pH-responsive shell dissociates inside the acidic lysosomes. The cellular and nuclear uptake of DOX is greater for pH-sensitive NPs. Results from *in vivo* experiments of Dox-resistant cancer model (MCF-7/ADR) xenograft tumors in mice confirm that DOX-loaded NPs considerably inhibited tumor growth by 92%, whereas DOX-loaded pH-sensitive micelles exhibit only 36% inhibition of tumor growth.

5.5 Future Prospects

Increasing interest in bio-nanotechnology has stimulated researchers to scrutinize biological elements and learn from nature. Nanostructured biological materials through NPs, nanofibers, nanowires, and nanorods, as well as self-assembly of peptides and proteins, provide knowledge about basic principles underlying the application of nanotechnology. The potential applications of nanostructured systems can be applied for fabricating a wider range of novel biomaterials for the use in bio-nanotechnology. With appropriate references and examples, the reader can benefit from broader knowledge about potential health risk of nanotechnology. These macroscopic structures have inspired researchers to apply this knowledge to electronics, biotechnology, nanotechnology, and medicine. This review has discussed some applications of nanotechnology and the advantages and limitations in nature and medicine.

Marketing nanomaterials has developed rapidly. It is, therefore, necessary to make available courses together with unbiased consulting (e.g. strategic decision making, commercialization lifecycles, and risk management) to aid decision making. This opportunity is available for government, manufacturers using nano-products, nano-research laboratories, investors, lawyers, and professional media. Private nano-companies need to provide education, courses, and consulting services for nano-research laboratories specifically aimed at nanotoxicological issues. Using standard services, one can achieve immense cost savings and risk reduction by proactively addressing potential toxicology concerns. Consulting and training services are based on current reliable reports and that connect the fields of nanotechnology to nanotoxicology. The following can be offered:

- Risk assessments
- Toxicity evaluations for NPs
- Essential equipment for nanotoxicology
- Art of nanotoxicological technical reports
- Design of safe nano-labs

Although nanoscience and technology can offer new commercial and entrepreneurial opportunities, they also offer great potential for severe toxicities. Research institutes can provide comprehensive services on nanotoxicological issues in the international community. The following services must be available:

- Reports on toxicological issues
- Nano-laboratory safe design
- Biocompatibility enhancement of nano-products
- Tools development for assessment and handling of nano-products in both laboratories and markets
- Nanotoxicology standards implementation
- Importance of nano–bio interfaces

References

1 Yun, Q., Li, L., Hu, Z. et al. (2020). Layered transition metal dichalcogenide-based nanomaterials for electrochemical energy storage. *Adv. Mater.* 32: 1903826.

2 Lizundia, E., Puglia, D., Nguyen, T.D., and Armentano, I. (2020). Cellulose nanocrystal based multifunctional nanohybrids. *Prog. Mater Sci.* 112: 100668. https://doi.org/10.1016/j.pmatsci.2020.100668.

3 Joo, J.I.I., Choi, M., Jang, S.H. et al. (2020). Realizing cancer precision medicine by integrating systems biology and nanomaterial engineering. *Adv. Mater.* https://doi.org/10.1002/adma.201906783.

4 Ding, B., Zheng, P., Ma, P., and Lin, J. (2020). Manganese oxide nanomaterials: synthesis, properties, and theranostic applications. *Adv. Mater.* 32: 1905823.

5 Xu, G., Zeng, S., Zhang, B. et al. (2016). New generation cadmium free quantum dots for biophotonics and nanomedicine. *Chem. Rev.* 116: 12234–12327.

6 D'Mello, S.R., Cruz, C.N., Chen, M.L. et al. (2017). The evolving landscape of drug products containing nanomaterials in the United States. *Nat. Nanotechnol.* 12: 523–529.

7 Wong, X.Y., Sena-Torralba, A., Álvarez-Diduk, R. et al. (2020). Nanomaterials for nanotheranostics: tuning their properties according to disease needs. *ACS Nano* 14: 2585–2627.

8 Elder, B., Neupane, R., Tokita, E. et al. (2020). Nanomaterial patterning in 3D printing. *Adv. Mater.* 32: 1907142.

9 Liu, P.S., Hu, M., Zhang, H. et al. (2020). Peroxidase-like nanozymes induce a novel form of cell death and inhibit tumor growth *in vivo*. *Adv. Funct. Mater.* 30: 200647.

10 Sharifzadeh, G. and Hosseinkhani, H. (2017). Biomolecule-responsive hydrogels in medicine. *Adv. Healthcare Mater.* 6: 1700801.

11 Wagner, V., Dullaart, A., Bock, A.K., and Zweck, A. (2006). The emerging nanomedicine landscape. *Nat. Biotechnol.* 24: 1211–1217.

12 Brusini, R., Varna, M., and Couvreur, P. (2020). Advanced nanomedicines for the treatment of inflammatory diseases. *Adv. Drug Delivery Rev.* https://doi.org/10.1016/j.addr.2020.07.010.

13 Etheridge, M.L., Campbell, S.A., Erdman, A.G. et al. (2013). The big picture on nanomedicine: the state of investigational and approved nanomedicine products. *Nanomedicine* 9: 1–14.

14 Jain, R.K. and Stylianopoulos, T. (2010). Delivering nanomedicine to solid tumors. *Nat. Rev. Clin. Oncol.* 7: 653–664.

15 Sanhai, W.R., Sakamoto, J.H., Canady, R., and Ferrari, M. (2008). Seven challenges for nanomedicine. *Nat. Nanotechnol.* 3: 242–244.

16 Oberdörster, G. (2010). Safety assessment for nanotechnology and nanomedicine: concepts of nanotoxicology. *J. Intern. Med.* 267: 89–105.

17 Poon, W., Zhang, Y.N., Ouyang, B. et al. (2019). Elimination pathways of nanoparticles. *ACS Nano* 13: 5785–5798.

18 Nel, A., Xia, T., Mädler, L., and Li, N. (2006). Toxic potential of materials at the nanolevel. *Science* 311: 622–627.

19 Halprin, K.M. and Ohkawara, A. (1967). The measurement of glutathione in human epidermis using glutathione reductase. *J. Invest. Dermatol.* 48: 149–152.

20 Soenen, S.J., Rivera-Gil, P., Montenegro, J.M. et al. (2011). Cellular toxicity of inorganic nanoparticles: common aspects and guidelines for improved nanotoxicity evaluation. *Nano Today* 6: 446–465.

21 Pagliari, F., Mandoli, C., Forte, G. et al. (2012). Cerium oxide nanoparticles protect cardiac progenitor cells from oxidative stress. *ACS Nano* 6: 3767–3775.

22 Niu, J., Azfer, A., Rogers, L.M. et al. (2007). Cardioprotective effects of cerium oxide nanoparticles in a transgenic murine model of cardiomyopathy. *Cardiovasc. Res.* 73: 549–559.

23 Chakraborty, A. and Jana, N.R. (2017). Vitamin C conjugated nanoparticle protects cells from oxidative stress at low dose but induces oxidative stress and cell death at high dose. *ACS Appl. Mater. Interfaces* 9: 41807–41817.

24 Anathy, V., Roberson, E.C., Guala, A.S. et al. (2012). Redox-based regulation of apoptosis: S-glutathionylation as a regulatory mechanism to control cell death. *Antioxid. Redox Signal.* 16: 496–505.

25 Dalle-Donne, I., Rossi, R., Colombo, G. et al. (2009). Protein S-glutathionylation: a regulatory device from bacteria to humans. *Trends Biochem. Sci.* 34: 85–96.

26 Shelton, M.D., Chock, P.B., and Mieyal, J.J. (2005). Glutaredoxin: role in reversible protein S-glutathionylation and regulation of redox signal transduction and protein translocation. *Antioxid. Redox Signal.* 7: 348–366.

27 Allen, E.M. and Mieyal, J.J. (2012). Protein-thiol oxidation and cell death: regulatory role of glutaredoxins. *Antioxid. Redox Signal.* 17: 1748–1763.

28 Guo, J., Gaffrey, M.J., Su, D. et al. (2014). Resin-assisted enrichment of thiols as a general strategy for proteomic profiling of cysteine-based reversible modifications. *Nat. Protoc.* 9: 64–75.

29 Duan, J., Kodali, V.K., Gaffrey, M.J. et al. (2016). Quantitative profiling of protein S-glutathionylation reveals redox-dependent regulation of macrophage function during nanoparticle-induced oxidative stress. *ACS Nano* 10: 524–538.

30 Galluzzi, L., Pedro, J.M.B., and Kroemer, G. (2014). Organelle-specific initiation of cell death. *Nat. Cell Biol.* 16: 728–736.

31 Nel, A.E., Mädler, L., Velegol, D. et al. (2009). Understanding biophysicochemical interactions at the nano–bio interface. *Nat. Mater.* 8: 543–557.

32 Kerr, J.F., Wyllie, A.H., and Currie, A.R. (1972). Apoptosis: a basic biological phenomenon with wide-ranging implications in tissue kinetics. *Br. J. Cancer* 26: 239–257.

33 Andón, F.T. and Fadeel, B. (2013). Programmed cell death: molecular mechanisms and implications for safety assessment of nanomaterials. *Acc. Chem. Res.* 46: 733–742.

34 Fadeel, B. and Orrenius, S. (2005). Apoptosis: a basic biological phenomenon with wide ranging implications in human disease. *J. Intern. Med.* 258: 479–517.

35 Kannan, K. and Jain, S.K. (2000). Oxidative stress and apoptosis. *Pathophysiology* 7: 153–163.

36 Wang, Y., Wang, J., Hao, H. et al. (2016). *In vitro* and *in vivo* mechanism of bone tumor inhibition by selenium-doped bone mineral nanoparticles. *ACS Nano* 10: 9927–9937.

37 Ravichandran, K.S. (2010). Find-me and eat-me signals in apoptotic cell clearance: progress and conundrums. *J. Exp. Med.* 207: 1807–1817.

38 Turk, B. and Stoka, V. (2007). Protease signalling in cell death: caspases versus cysteine cathepsins. *FEBS Lett.* 581: 2761–2767.

39 Adams, J.M. and Cory, S. (2007). Bcl-2-regulated apoptosis: mechanism and therapeutic potential. *Curr. Opin. Immunol.* 19: 488–496.

40 Perez-Hernandez, M., Del Pino, P., Mitchell, S.G. et al. (2015). Dissecting the molecular mechanism of apoptosis during photothermal therapy using gold nanoprisms. *ACS Nano* 9: 52–61.

41 Hou, Z., Zhang, Y., Deng, K. et al. (2015). UV-emitting upconversion-based TiO_2 photosensitizing nanoplatform: near-infrared light mediated *in vivo* photodynamic therapy via mitochondria-involved apoptosis pathway. *ACS Nano* 9: 2584–2599.

42 Autophagy, M.N. (2007). Process and function. *Genes Dev.* 21: 2861–2873.

43 Xie, Z. and Klionsky, D.J. (2007). Autophagosome formation: core machinery and adaptations. *Nat. Cell Biol.* 9: 1102–1109.

44 Xie, Y., Jiang, J., Tang, Q. et al. (2020). Iron oxide nanoparticles as autophagy intervention agents suppress hepatoma growth by enhancing tumoricidal autophagy. *Adv. Sci.* 7: 1903323.

45 Shi, Y., Wang, J., Liu, J. et al. (2020). Oxidative stress-driven DR5 upregulation restores TRAIL/Apo2L sensitivity induced by iron oxide nanoparticles in colorectal cancer. *Biomaterials* 233: 119753.

46 Shi, S., Lin, S., Li, Y. et al. (2018). Effects of tetrahedral DNA nanostructures on autophagy in chondrocytes. *Chem. Commun. (Camb.)* 54: 1327–1330.

47 Peynshaert, K., Manshian, B.B., Joris, F. et al. (2014). Exploiting intrinsic nanoparticle toxicity: the pros and cons of nanoparticle-induced autophagy in biomedical research. *Chem. Rev.* 114: 7581–7609.

48 Wang, S.F., Li, Y.B., Fan, J.J. et al. (2014). The role of autophagy in the neurotoxicity of cationic PAMAM dendrimers. *Biomaterials* 35: 7588–7597.

49 Zhang, H., Ren, Y., Hou, L. et al. (2020). Positioning remodeling nanogels mediated codelivery of antivascular drug and autophagy inhibitor for cooperative tumor therapy. *ACS Appl. Mater. Interfaces* 12: 6978–6990.

50 Zhou, Y., Wang, Q., Song, B. et al. (2015). A real-time documentation and mechanistic investigation of quantum dots-induced autophagy in live *Caenorhabditis elegans*. *Biomaterials* 72: 38–48.

51 Li, J., Yang, S., Deng, Y. et al. (2018). Emancipating target-functionalized carbon dots from autophagy vesicles for a novel visualized tumor therapy. *Adv. Funct. Mater.* 28: 1800881.

52 Feoktistova, M. and Leverkus, M. (2015). Programmed necrosis and necroptosis signalling. *FEBS J.* 282: 19–31.

53 Vandenabeele, P., Galluzzi, L., Vanden Berghe, T., and Kroemer, G. (2010). Molecular mechanisms of necroptosis: an ordered cellular explosion. *Nat. Rev. Mol. Cell Biol.* 11: 700–714.

54 Cho, Y.S., Challa, S., Moquin, D. et al. (2009). Phosphorylation-driven assembly of the RIP1-RIP3 complex regulates programmed necrosis and virus-induced inflammation. *Cell* 137: 1112–1123.

55 Challa, S. and Chan, F.K.M. (2010). Going up in flames: necrotic cell injury and inflammatory diseases. *Cell. Mol. Life Sci.* 67: 3241–3253.

56 Qu, G., Liu, S., Zhang, S. et al. (2013). Graphene oxide induces toll-like receptor 4 (TLR4)-dependent necrosis in macrophages. *ACS Nano* 7: 5732–5745.

57 He, B., Shi, Y., Liang, Y. et al. (2018). Single-walled carbon-nanohorns improve biocompatibility over nanotubes by triggering less protein-initiated pyroptosis and apoptosis in macrophages. *Nat. Commun.* 9: 2393.

58 Fernandez Cabada, T., de Pablo, C.S., Serrano, A.M. et al. (2012). Induction of cell death in a glioblastoma line by hyperthermic therapy based on gold nanorods. *Int. J. Nanomed.* 7: 1511–1523.

59 Grivennikov, S.I., Greten, F.R., and Karin, M. (2010). Immunity, inflammation, and cancer. *Cell* 140: 883–899.

60 Cao, Y., Gong, Y., Liu, L. et al. (2017). The use of human umbilical vein endothelial cells (HUVECs) as an *in vitro* model to assess the toxicity of nanoparticles to endothelium: a review. *J. Appl. Toxicol.* 37: 1359–1369.

61 Genetic, Z.E. (2010). Toxicology testing. In: *Comprehensive Toxicology* (ed. C.A. McQueen), 139–158. Oxford: Elsevier.
62 Lord, C.J. and Ashworth, A. (2012). The DNA damage response and cancer therapy. *Nature* 481: 287–294.
63 Sargent, L.M., Reynolds, S.H., and Castranova, V. (2010). Potential pulmonary effects of engineered carbon nanotubes: *in vitro* genotoxic effects. *Nanotoxicology* 4: 396–408.
64 Lindberg, H.K., Falck, G.C., Singh, R. et al. (2013). Genotoxicity of short single-wall and multi-wall carbon nanotubes in human bronchial epithelial and mesothelial cells *in vitro*. *Toxicology* 313: 24–37.
65 Lindberg, H.K., Falck, G.C., Suhonen, S. et al. (2009). Genotoxicity of nanomaterials: DNA damage and micronuclei induced by carbon nanotubes and graphite nanofibres in human bronchial epithelial cells *in vitro*. *Toxicol. Lett.* 186: 166–173.
66 Moller, P., Jacobsen, N.R., Folkmann, J.K. et al. (2010). Role of oxidative damage in toxicity of particulates. *Free Radical Res.* 44: 1–46.
67 Burgum, M.J., Clift, M.J.D., Evans, S.J. et al. (2020). *In vitro* primary-indirect genotoxicity in bronchial epithelial cells promoted by industrially relevant few-layer graphene. *Small* 17: 2002551.
68 Rubio, L., Barguill, I., Domenech, J. et al. (2020). Biological effects, including oxidative stress and genotoxic damage, of polystyrene nanoparticles in different human hematopoietic cell lines. *J. Hazard. Mater.* 398: 122900.
69 Toyokuni, S. (2013). Genotoxicity and carcinogenicity risk of carbon nanotubes. *Adv. Drug Delivery Rev.* 65: 2098–2110.
70 Li, Y., Chen, D., Yan, J. et al. (2012). Genotoxicity of silver nanoparticles evaluated using the ames test and *in vitro* micronucleus assay. *Mutat. Res. Genet. Toxicol. Environ. Mutagen.* 745: 4–10.
71 Landsiedel, R., Kapp, M.D., Schulz, M. et al. (2009). Genotoxicity investigations on nanomaterials: methods, preparation and characterization of test material, potential artifacts, and limitations—many questions, some answers. *Mutat. Res. Rev. Mutat. Res.* 681: 241–258.
72 Magdolenova, Z., Collins, A., Kumar, A. et al. (2013). Mechanisms of genotoxicity: a review of *in vitro* and *in vivo* studies with engineered nanoparticles. *Nanotoxicology* 8: 233–278.
73 Ma, P., Luo, Q., Chen, J. et al. (2012). Intraperitoneal injection of magnetic Fe_3O_4-nanoparticle induces hepatic and renal tissue injury via oxidative stress in mice. *Int. J. Nanomed.* 7: 4809–4818.
74 García, O., Mandina, T., Lamadrid, A.I. et al. (2004). Sensitivity and variability of visual scoring in the comet assay. Results of an inter-laboratory scoring exercise with the use of silver staining. *Mutat. Res.* 556: 25–34.
75 Watson, C., Ge, J., Cohen, J. et al. (2014). High-throughput screening platform for engineered nanoparticle-mediated genotoxicity using CometChip technology. *ACS Nano* 8: 2118–2133.

76 Hassan, H.A.F.M., Smyth, L., Rubio, N. et al. (2016). Carbon nanotubes' surface chemistry determines their potency as vaccine nanocarriers *in vitro* and *in vivo*. *J. Controlled Release* 225: 205–216.

77 Liu, H. and Irvine, D.J. (2015). Guiding principles in the design of molecular bioconjugates for vaccine applications. *Bioconjugate Chem.* 26: 791–801.

78 Sexton, A., Whitney, P.G., Chong, S.F. et al. (2009). A protective vaccine delivery system for *in vivo* T cell stimulation using nanoengineered polymer hydrogel capsules. *ACS Nano* 3: 3391–3400.

79 Niikura, K., Matsunaga, T., Suzuki, T. et al. (2013). Gold nanoparticles as a vaccine platform: influence of size and shape on immunological responses *in vitro* and *in vivo*. *ACS Nano* 7: 3926–3938.

80 Ji, Z., Wang, X., Zhang, H. et al. (2012). Designed synthesis of CeO_2 nanorods and nanowires for studying toxicological effects of high aspect ratio nanomaterials. *ACS Nano* 6: 5366–5380.

81 Liu, X., Feng, Z., Wang, C. et al. (2020). et al., Co-localized delivery of nanomedicine and nanovaccine augments the postoperative cancer immunotherapy by amplifying T-cell responses. *Biomaterials* https://doi.org/10.1016/j.biomaterials.2019.119649.

82 Bhardwaj, P., Bhatia, E., Sharma, S. et al. (2020). Advancements in prophylactic and therapeutic nanovaccines. *Acta Biomater.* 108: 1–21.

83 Zhou, W., Moguche, A.O., Chiu, D. et al. (2014). Just-in-time vaccines: biomineralized calcium phosphate core-immunogen shell nanoparticles induce long-lasting $Cd8^+$ T cell responses in mice. *Nanomedicine* 10: 571–578.

84 Kasturi, S.P., Skountzou, I., Albrecht, R.A. et al. (2011). Programming the magnitude and persistence of antibody responses with innate immunity. *Nature* 470: 543–547.

85 Feng, B., Zhou, F., Hou, B. et al. (2018). Binary cooperative prodrug nanoparticles improve immunotherapy by synergistically modulating immune tumor microenvironment. *Adv. Mater.* 30: 1803001.

86 Liu, M., Khan, A.R., Ji, J. et al. (2018). Crosslinked self-assembled nanoparticles for chemo-sonodynamic combination therapy favoring antitumor, antimetastasis management and immune responses. *J. Controlled Release* 290: 150–164.

87 Narayan, K. and Subramaniam, S. (2015). Focused ion beams in biology. *Nat. Methods* 12: 1021–1031.

88 Horvath, L., Magrez, A., Golberg, D. et al. (2011). *In vitro* investigation of the cellular toxicity of boron nitride nanotubes. *ACS Nano* 5: 3800–3810.

89 Plascencia-Villa, G., Starr, C.R., Armstrong, L.S. et al. (2012). Imaging interactions of metal oxide nanoparticles with macrophage cells by ultra-high resolution scanning electron microscopy techniques. *Integr. Biol.* 4: 1358–1366.

90 Ankri, R., Ashkenazy, A., Milstein, Y. et al. (2016). Gold nanorods based air scanning electron microscopy and diffusion reflection imaging for mapping tumor margins in squamous cell carcinoma. *ACS Nano* 10: 2349–2356.

91 Guarnieri, D., Sabella, S., Muscetti, O. et al. (2014). Transport across the cell-membrane dictates nanoparticle fate and toxicity: a new paradigm in nanotoxicology. *Nanoscale* 6: 10264–10273.

92 Love, S.A., Liu, Z., and Haynes, C.L. (2012). Examining changes in cellular communication in neuroendocrine cells after noble metal nanoparticle exposure. *Analyst* 137: 3004–3010.

93 Yazdimamaghani, M., Moos, P.J., and Ghandehari, H. (2018). Global gene expression analysis of macrophage response induced by nonporous and porous silica nanoparticles. *Nanomedicine* 14: 533–545.

94 AshaRani, P.V., Low Kah Mun, G., Hande, M.P., and Valiyaveettil, S. (2009). Cytotoxicity and genotoxicity of silver nanoparticles in human cells. *ACS Nano* 3: 279–290.

95 Peckys, D.B. and de Jonge, N. (2011). Visualizing gold nanoparticle uptake in live cells with liquid scanning transmission electron microscopy. *Nano Lett.* 11: 1733–1738.

96 Brown, C.P. (2013). Advancing musculoskeletal research with nanoscience. *Nat. Rev. Rheumatol.* 9: 614–623.

97 Khanal, D., Kondyurin, A., Hau, H. et al. (2016). Biospectroscopy of nanodiamond-induced alterations in conformation of intra- and extracellular proteins – a nanoscale IR study. *Anal. Chem.* 88: 7530–7538.

98 Govindaraju, S., Rengaraj, A., Arivazhagan, R. et al. (2018). Curcumin-conjugated gold clusters for bioimaging and anticancer applications. *Bioconjugate Chem.* 29: 363–370.

99 Reggente, M., Passeri, D., Angeloni, L. et al. (2017). Detection of stiff nanoparticles within cellular structures by contact resonance atomic force microscopy subsurface nanomechanical imaging. *Nanoscale* 9: 5671–5676.

100 Smith, T.T., Stephan, S.B., Moffett, H.F. et al. (2017). *In situ* programming of leukaemia-specific T cells using synthetic DNA nanocarriers. *Nat. Nanotechnol.* 12: 813–820.

101 Zhang, C., Shi, G., Zhang, J. et al. (2017). Targeted antigen delivery to dendritic cell via functionalized alginate nanoparticles for cancer immunotherapy. *J. Controlled Release* 256: 170–181.

102 Xu, P., Meng, Q., Sun, H. et al. (2015). Shrapnel nanoparticles loading docetaxel inhibit metastasis and growth of breast cancer. *Biomaterials* 64: 10–20.

103 He, B., Lin, P., Jia, Z. et al. (2013). The transport mechanisms of polymer nanoparticles in Caco-2 epithelial cells. *Biomaterials* 34: 6082–6098.

104 Xia, D., He, Y., Li, Q. et al. (2018). Transport mechanism of lipid covered saquinavir pure drug nanoparticles in intestinal epithelium. *J. Controlled Release* 269: 159–170.

105 Harutyunyan, H., Palomba, S., Renger, J. et al. (2010). Nonlinear dark-field microscopy. *Nano Lett.* 10: 5076–5079.

106 Syed, A.M., Sindhwani, S., Wilhelm, S. et al. (2017). Three-dimensional imaging of transparent tissues via metal nanoparticle labeling. *J. Am. Chem. Soc.* 139: 9961–9971.

107 Wang, X., Cui, Y., and Irudayaraj, J. (2015). Single-cell quantification of cytosine modifications by hyperspectral dark-field imaging. *ACS Nano* 9: 11924–11932.

108 Cui, Y., Wang, X., Ren, W. et al. (2016). Optical clearing delivers ultrasensitive hyperspectral dark-field imaging for single-cell evaluation. *ACS Nano* 10: 3132–3143.

109 Kwon, S.J., Na, D.H., Kwak, J.H. et al. (2017). Nanostructured glycan architecture is important in the inhibition of influenza a virus infection. *Nat. Nanotechnol.* 12: 48–54.

110 Sharifzadeh, G., Wahit, M.U., Soheilmoghaddam, M. et al. (2016). Kappa-carrageenan/halloysite nanocomposite hydrogels as potential drug delivery systems. *J. Taiwan Inst. Chem. Eng.* 67: 426–434.

111 Sharifzadeh, G., Hezaveh, H., Muhamad, I.I. et al. (2020). Montmorillonite-based polyacrylamide hydrogel rings for controlled vaginal drug delivery. *Mater. Sci. Eng., C* 110: 110609.

112 Zhai, S., Hu, X., Hu, Y. et al. (2017). Visible light-induced crosslinking and physiological stabilization of diselenide-rich nanoparticles for redox-responsive drug release and combination chemotherapy. *Biomaterials* 121: 41–54.

113 Jones, C.F. and Grainger, D.W. (2009). *In vitro* assessments of nanomaterial toxicity. *Adv. Drug Delivery Rev.* 61: 438–456.

114 Rastogi, S.K., Raghavan, G., Yang, G., and Cohen-Karni, T. (2017). Effect of graphene on nonneuronal and neuronal cell viability and stress. *Nano Lett.* 17: 3297–3301.

115 Ciappellano, S.G., Tedesco, E., Venturini, M., and Benetti, F. (2016). *In vitro* toxicity assessment of oral nanocarriers. *Adv. Drug Delivery Rev.* 106: 381–401.

116 Riss, T., Moravec, R., Niles, A. et al. (2016). Cell Viability Assays.

117 Sokolova, V., Shi, Z., Huang, S. et al. (2017). Delivery of the TLR ligand poly(I:C) to liver cells *in vitro* and *in vivo* by calcium phosphate nanoparticles leads to a pronounced immunostimulation. *Acta Biomater.* 64: 401–410.

118 Wörle-Knirsch, J.M., Pulskamp, K., and Krug, H.F. (2006). Oops they did it again! Carbon nanotubes hoax scientists in viability assays. *Nano Lett.* 6: 1261–1268.

119 Marquis, B.J., Love, S.A., Braun, K.L., and Haynes, C.L. (2009). Analytical methods to assess nanoparticle toxicity. *Analyst* 134: 425–439.

120 Costa, C., Brandão, F., Bessa, M.J. et al. (2016). *In vitro* cytotoxicity of superparamagnetic iron oxide nanoparticles on neuronal and glial cells. Evaluation of nanoparticle interference with viability tests. *J. Appl. Toxicol.* 36: 361–372.

121 Galluzzi, L., Vitale, I., Abrams, J.M. et al. (2017). Molecular definitions of cell death subroutines: recommendations of the nomenclature committee on cell death 2012. *Cell Death Differ.* 19: 107–120.

122 Love, S.A., Maurer-Jones, M.A., Thompson, J.W. et al. (2012). Assessing nanoparticle toxicity. *Annu. Rev. Anal. Chem.* 5: 181–205.

123 Han, X., Gelein, R., Corson, N. et al. (2011). Validation of an LDH assay for assessing nanoparticle toxicity. *Toxicology* 287: 99–104.

124 Marchesi, E., Rota, C., Fann, Y.C. et al. (1999). Photoreduction of the fluorescent dye 2′-7′-Dichlorofluorescein: a spin trapping and direct electron spin resonance study with implications for oxidative stress measurements. *Free Radical Biol. Med.* 26: 148–161.

125 Bussy, C., Al-Jamal, K.T., Boczkowski, J. et al. (2015). Microglia determine brain region-specific neurotoxic responses to chemically functionalized carbon nanotubes. *ACS Nano* 9: 7815–7830.

126 Brun, E., Barreau, F., Veronesi, G. et al. (2014). Titanium dioxide nanoparticle impact and translocation through *ex vivo, in vivo* and *in vitro* gut epithelia. *Part. Fibre Toxicol.* 11: 13.

127 Repetto, G., del Peso, A., and Zurita, J.L. (2008). Neutral red uptake assay for the estimation of cell viability/cytotoxicity. *Nat. Protoc.* 3: 1125–1131.

128 Stone, V., Johnston, H., and Schins, R.P. (2009). Development of *in vitro* systems for nanotoxicology: methodological considerations. *Crit. Rev. Toxicol.* 39: 613–626.

129 Fisichella, M., Berenguer, F., Steinmetz, G. et al. (2012). Intestinal toxicity evaluation of TiO_2 degraded surface-treated nanoparticles: a combined physico-chemical and toxicogenomics approach in Caco-2 cells. *Part. Fibre Toxicol.* 9: 18.

130 Yin, L., Ding, J., He, C. et al. (2009). Drug permeability and mucoadhesion properties of thiolated trimethyl chitosan nanoparticles in oral insulin delivery. *Biomaterials* 30: 5691–5700.

131 Buyukozturk, F., Benneyan, J.C., and Carrier, R.L. (2010). Impact of emulsion-based drug delivery systems on intestinal permeability and drug release kinetics. *J. Controlled Release* 142: 22–30.

132 Bimbo, L.M., Mäkilä, E., Laaksonen, T. et al. (2011). Drug permeation across intestinal epithelial cells using porous silicon nanoparticles. *Biomaterials* 32: 2625–2633.

133 Anguissola, S., Garry, D., Salvati, A. et al. (2014). High content analysis provides mechanistic insights on the pathways of toxicity induced by amine-modified polystyrene nanoparticles. *PLoS One* 9: e108025.

134 Cohen, Y., Rallo, R., Liu, R., and Liu, H.H. (2013). *In silico* analysis of nanomaterials hazard and risk. *Acc. Chem. Res.* 46: 802–812.

135 Ha, M.K., Trinh, T.X., Choi, J.S. et al. (2018). Toxicity classification of oxide nanomaterials: effects of data gap filling and PChem score-based screening approaches. *Sci. Rep.* 8: 3141.

136 Puzyn, T., Rasulev, B., Gajewicz, A. et al. (2011). Using nano-QSAR to predict the cytotoxicity of metal oxide nanoparticles. *Nat. Nanotechnol.* 6: 175–178.

137 Sharifi, S., Behzadi, S., Laurent, S. et al. (2012). Toxicity of nanomaterials. *Chem. Soc. Rev.* 41: 2323–2343.

138 Fernández-Pacheco, R., Marquina, C., Valdivia, J.G. et al. (2007). Magnetic nanoparticles for local drug delivery using magnetic implants. *J. Magn. Magn. Mater.* 311: 318–322.

139 Zhu, M.T., Feng, W.Y., Wang, B. et al. (2008). Comparative study of pulmonary responses to nano- and submicron-sized ferric oxide in rats. *Toxicology* 247: 102–111.

140 Dai, C., Yuan, Y., Liu, C. et al. (2009). Degradable, antibacterial silver exchanged mesoporous silica spheres for hemorrhage control. *Biomaterials* 30: 5364–5375.

141 Dobrovolskaia, M.A., Aggarwal, P., Hall, J.B., and McNeil, S.E. (2008). Preclinical studies to understand nanoparticle interaction with the immune system and its potential effects on nanoparticle biodistribution. *Mol. Pharm.* 5: 487–495.

142 Cavadas, M., Gonzalez-Fernandez, A., and Franco, R. (2011). Pathogen-mimetic stealth nanocarriers for drug delivery: a future possibility. *Nanomedicine* 7: 730–743.

143 Jones, C.F., Campbell, R.A., Brooks, A.E. et al. (2012). Cationic PAMAM dendrimers aggressively initiate blood clot formation. *ACS Nano* 9: 9900–9910.

144 Delfino, R.J., Sioutas, C., and Malik, S. (2005). Potential role of ultrafine particles in associations between airborne particle mass and cardiovascular health. *Environ. Health Perspect.* 113: 934–946.

145 Jun, E.A., Lim, K.M., Kim, K. et al. (2011). Silver nanoparticles enhance thrombus formation through increased platelet aggregation and procoagulant activity. *Nanotoxicology* 5: 157–167.

146 Radomski, A., Jurasz, P., Alonso-Escolano, D. et al. (2005). Nanoparticle-induced platelet aggregation and vascular thrombosis. *J. Pharmacol.* 146: 882–893.

147 Sahu, S.C. and Casciano, D.A. (2009). *Nanotoxicity: From in vivo and in vitro Models to Health Risks*. Laurel, MD: Wiley.

148 Ruiz, A., Ali, L.M.A., Cáceres-Vélez, P.R. et al. (2015). Hematotoxicity of magnetite nanoparticles coated with polyethylene glycol: *in vitro* and *in vivo* studies. *Toxicol. Res.* 4: 1555–1564.

149 Borm, P., Klaessig, F.C., Landry, T.D. et al. (2006). Research strategies for safety evaluation of nanomaterials, part V: role of dissolution in biological fate and effects of nanoscale particles. *Toxicol. Sci.* 90: 23–32.

150 Stern, S.T. and McNeil, S.E. (2008). Nanotechnology safety concerns revisited. *Toxicol. Sci.* 101: 4–21.

151 Shen, A.M. and Minko, T. (2020). Pharmacokinetics of inhaled nanotherapeutics for pulmonary delivery. *J. Controlled Release* 326: 222–244.

152 Oberdorster, G., Maynard, A., Donaldson, K. et al. (2005). Principles for characterizing the potential human health effects from exposure to nanomaterials: elements of a screening strategy. *Part. Fibre Toxicol.* 2: 8.

153 Oberdörster, G. (2000). Pulmonary effects of inhaled ultrafine particles. *Int. Arch. Occup. Environ. Health* 74: 1–8.

154 Peters, A., Veronesi, B., Calderón-Garcidueñas, L. et al. (2006). Translocation and potential neurological effects of fine and ultrafine particles a critical update. *Part. Fibre Toxicol.* 3: 1–13.

155 Oberdörster, G., Ferin, J., and Lehnert, B.E. (1994). Correlation between particle size, *in vivo* particle persistence, and lung injury. *Environ. Health Perspect.* 102: 173.

156 Tang, Y., Wang, F., Jin, C. et al. (2013). Mitochondrial injury induced by nanosized titanium dioxide in A549 cells and rats. *Environ. Toxicol. Pharmacol.* 36: 66–72.

157 Wang, J., Xu, Y., Yang, Z. et al. (2013). Toxicity of carbon nanotubes. *Curr. Drug Metab.* 14: 891–899.

158 Wang, Y., Chen, Z., Ba, T. et al. (2013). Susceptibility of young and adult rats to the oral toxicity of titanium dioxide nanoparticles. *Small* 9: 1742–1752.

159 Bennett, W.D. (2002). Rapid translocation of nanoparticles from the lung to the bloodstream? *Am. J. Respir. Crit. Care Med.* 165: 1671–1672.

160 Wu, L., Shan, W., Zhang, Z., and Huang, Y. (2018). Engineering nanomaterials to overcome the mucosal barrier by modulating surface properties. *Adv. Drug Delivery Rev.* 124: 150–163.

161 Choi, H.S., Ashitate, Y., Lee, J.H. et al. (2010). Rapid translocation of nanoparticles from the lung airspaces to the body. *Nat. Biotechnol.* 28: 1300–1303.

162 Kreyling, W.G., Hirn, S., Moller, W. et al. (2014). Air-blood barrier translocation of tracheally instilled gold nanoparticles inversely depends on particle size. *ACS Nano* 8: 222–233.

163 Mühlfeld, C., Gehr, P., and Rothen-Rutishauser, B. (2008). Translocation and cellular entering mechanisms of nanoparticles in the respiratory tract. *Swiss Med. Wkly.* 138: 387–391.

164 Semmler-Behnke, M., Takenaka, S., Fertsch, S. et al. (2007). Efficient elimination of inhaled nanoparticles from the alveolar region: evidence for interstitial uptake and subsequent reentrainment onto airways epithelium. *Environ. Health Perspect.* 115: 728–733.

165 Borm, P.J. and Müller-Schulte, D. (2006). Nanoparticles in drug delivery and environmental exposure: same size, same risks? *Nanomedicine (Lond.)* 1: 235–249.

166 Tinkle, S.S., Antonini, J.M., Rich, B.A. et al. (2003). Skin as a route of exposure and sensitization in chronic beryllium disease. *Environ. Health Perspect.* 111: 1202.

167 Toll, R., Jacobi, U., Richter, H. et al. (2004). Penetration profile of microspheres in follicular targeting of terminal hair follicles. *J. Invest. Dermatol.* 123: 168–176.

168 Tsuji, J.S., Maynard, A.D., Howard, P.C. et al. (2006). Research strategies for safety evaluation of nanomaterials, part IV: risk assessment of nanoparticles. *Toxicol. Sci.* 89: 42–50.

169 Sperling, R.A., Gil, P.R., Zhang, F. et al. (2008). Biological applications of gold nanoparticles. *Chem. Soc. Rev.* 37: 1896–1908.

170 Ryman-Rasmussen, J.P., Riviere, J.E., and Monteiro-Riviere, N.A. (2006). Penetration of intact skin by quantum dots with diverse physicochemical properties. *Toxicol. Sci.* 91: 159–165.

171 Rouse, J.G., Yang, J., Ryman-Rasmussen, J.P. et al. (2006). Effects of mechanical flexion on the penetration of fullerene amino acid-derivatized peptide nanoparticles through skin. *Nano Lett.* 7: 155–160.

172 Ryman-Rasmussen, J.P., Riviere, J.E., and Monteiro-Riviere, N.A. (2007). Surface coatings determine cytotoxicity and irritation potential of quantum dot nanoparticles in epidermal keratinocytes. *J. Invest. Dermatol.* 127: 143–153.

173 Chu, M.Q., Wu, Q., Wang, J.X. et al. (2007). *In vitro* and *in vivo* transdermal delivery capacity of quantum dots through mouse skin. *Nanotechnology* 18: 455103.

174 Mortensen, L.J., Oberdorster, G., Pentland, A.P., and DeLouise, L.A. (2008). *In vivo* skin penetration of quantum dot nanoparticles in the murine model: the effect of UVR. *Nano Lett.* 8: 2779–2787.

175 Elsaesser, A. and Howard, C.V. (2012). Toxicology of nanoparticles. *Adv. Drug Delivery Rev.* 64: 129–137.

176 Hoet, P.H., Brüske-Hohlfeld, I., and Salata, O.V. (2004). Nanoparticles-known and unknown health risks. *J. Nanobiotechnol.* 2: 12.

177 Florence, A.T. (2005). Nanoparticle uptake by the oral route: fulfilling its potential? *Drug Discovery Today* 2: 75–81.

178 Jani, P., Halbert, G.W., Langridge, J., and Florence, A.T. (1990). Nanoparticle uptake by the rat gastrointestinal mucosa: quantitation and particle size dependency. *J. Pharm. Pharmacol.* 42: 821–826.

179 Balasubramanian, S.K., Jittiwat, J., Manikandan, J. et al. (2010). Biodistribution of gold nanoparticles and gene expression changes in the liver and spleen after intravenous administration in rats. *Biomaterials* 31: 2034–2042.

180 Texier, I. and Josser, V. (2009). *In vivo* imaging of quantum dots. *Methods Mol. Biol.* 544: 393–406.

181 Shokeen, M., Fettig, N.M., and Rossin, R. (2008). Synthesis, *in vitro* and *in vivo* evaluation of radiolabeled nanoparticles. *Q. J. Nucl. Med. Mol. Imaging* 52: 267–277.

182 Rojas, S., Gispert, J.D., Martín, R. et al. (2011). Biodistribution of amino-functionalized diamond nanoparticles. *In vivo* studies based on ^{18}F radionuclide emission. *ACS Nano* 5: 5552–5559.

183 Cabral, H., Matsumoto, Y., Mizuno, K. et al. (2011). Accumulation of Sub-100 nm polymeric micelles in poorly permeable tumours depends on size. *Nat. Nanotechnol.* 6: 815–823.

184 Matsumoto, M., Nichols, J.W., Toh, K. et al. (2016). Vascular bursts enhance permeability of tumour blood vessels and improve nanoparticle delivery. *Nat. Nanotechnol.* 11: 533–538.

185 Campbell, F., Bos, F.L., Sieber, S. et al. (2018). Directing nanoparticle biodistribution through evasion and exploitation of Stab2-dependent nanoparticle uptake. *ACS Nano* 12: 2138–2150.

186 Soenen, S.J., Parak, W.J., Rejman, J., and Manshian, B. (2015). (Intra)cellular stability of inorganic nanoparticles: effects on cytotoxicity, particle functionality, and biomedical applications. *Chem. Rev.* 115: 2109–2135.

187 Levy, M., Luciani, N., Alloyeau, D. et al. (2011). Long term *in vivo* biotransformation of iron oxide nanoparticles. *Biomaterials* 32: 3988–3999.

188 Mejias, R., Gutierrez, L., Salas, G. et al. (2013). Long term biotransformation and toxicity of dimercaptosuccinic acid-coated magnetic nanoparticles support their use in biomedical applications. *J. Controlled Release* 171: 225–233.

189 Freund, B., Tromsdorf, U.I., Bruns, O.T. et al. (2012). A simple and widely applicable method to ^{59}Fe-radiolabel monodisperse superparamagnetic iron oxide nanoparticles for *in vivo* quantification studies. *ACS Nano* 6: 7318–7325.

190 Feliu, N., Docter, D., Heine, M. et al. (2016). *In vivo* degeneration and the fate of inorganic nanoparticles. *Chem. Soc. Rev.* 45: 2440–2457.

191 Kolosnjaj-Tabi, J., Javed, Y., Lartigue, L. et al. (2015). The one year fate of iron oxide coated gold nanoparticles in mice. *ACS Nano* 9: 7925–7939.

192 Lartigue, L., Alloyeau, D., Kolosnjaj-Tabi, J. et al. (2013). Biodegradation of iron oxide nanocubes: high-resolution *in situ* monitoring. *ACS Nano* 7: 3939–3952.

193 Tsoi, K.M., MacParland, S.A., Ma, X.Z. et al. (2016). Mechanism of hard-nanomaterial clearance by the liver. *Nat. Mater.* 15: 1212–1221.

194 Abdelhalim, M. and Jarrar, B. (2011). Gold nanoparticles administration induced prominent inflammatory, central vein intima disruption, fatty change and Kupffer cells hyperplasia. *Lipids Health Dis.* 10: 133–138.

195 Wang, B., Feng, W., Wang, M. et al. (2008). Acute toxicological impact of nano- and submicro-scaled zinc oxide powder on healthy adult mice. *J. Nanopart. Res.* 10: 263–276.

196 Chen, Z., Meng, H., Xing, G. et al. (2006). Acute toxicological effects of copper nanoparticles *in vivo*. *Toxicol. Lett.* 163: 109–120.

197 He, Q., Zhang, Z., Gao, F. et al. (2011). *In vivo* biodistribution and urinary excretion of mesoporous silica nanoparticles: effects of particle size and PEGylation. *Small* 7: 271–280.

198 Ogawara, K., Yoshida, M., Higaki, K. et al. (1999). Hepatic uptake of polystyrene microspheres in rats: effect of particle size on intrahepatic distribution. *J. Controlled Release* 59: 15–22.

199 Schluep, T., Hwang, J., Hildebrandt, I. et al. (2009). Pharmacokinetics and tumor dynamics of the nanoparticle IT-101 from PET imaging and tumor histological measurements. *Proc. Natl. Acad. Sci. U.S.A.* 106: 11394–11399.

200 Yang, Y., Yuan, S.X., Zhao, L.H. et al. (2015). Ligand-directed stearic acid grafted chitosan micelles to increase therapeutic efficacy in hepatic cancer. *Mol. Pharm.* 12: 644–652.

201 Wang, L., Li, Y.F., Zhou, L. et al. (2011). Characterization of gold nanorods *in vivo* by integrated analytical techniques: their uptake, retention, and chemical forms. *Anal. Bioanal.Chem.* 396: 1105–1114.

202 Poelstra, K., Prakash, J., and Beljaars, L. (2012). Drug targeting to the diseased liver. *J. Controlled Release* 161: 188–197.

203 Wang, B., He, X., Zhang, Z. et al. (2013). Metabolism of nanomaterials *in vivo*: blood circulation and organ clearance. *Acc. Chem. Res.* 46: 761–769.

204 Heyder, J., Gebhart, J., Rudolf, G. et al. (1986). Deposition of particles in the human respiratory tract in the size range 0.005–15 μm. *J. Aerosol Sci.* 17: 811–825.

205 Knowles, M.R. and Boucher, R.C. (2002). Mucus clearance as a primary innate defense mechanism for mammalian airways. *J. Clin. Invest.* 109: 571–577.

206 Faraj, A.A., Shaik, A.P., and Shaik, A.S. (2015). Effect of surface coating on the biocompatibility and *in vivo* MRI detection of iron oxide nanoparticles after intrapulmonary administration. *Nanotoxicology* 9: 825–834.

207 Heilig, E.A., Thompson, K.J., Molina, R.M. et al. (2006). Manganese and iron transport across pulmonary epithelium. *Am. J. Physiol. Lung Cell. Mol. Physiol.* 290: L1247–L1259.

208 Turi, J.L., Yang, F., Garrick, M.D. et al. (2004). The iron cycle and oxidative stress in the lung. *Free Radical Biol. Med.* 36: 850–857.

209 Valois, C.R.A., Braz, J.M., Nunes, E.S. et al. (2010). The effect of DMSA-functionalized magnetic nanoparticles on transendothelial migration of monocytes in the murine lung via a β2 integrin-dependent pathway. *Biomaterials* 31: 366–374.

210 Cho, W.S., Cho, M., Kim, S.R. et al. (2009). Pulmonary toxicity and kinetic study of Cy5.5-conjugated superparamagnetic iron oxide nanoparticles by optical imaging. *Toxicol. Appl. Pharmacol.* 239: 106–115.

211 Haque, S., Whittaker, M., McIntosh, M.P. et al. (2018). A comparison of the lung clearance kinetics of solid lipid nanoparticles and liposomes by following the ^{3}H-labelled structural lipids after pulmonary delivery in rats. *Eur. J. Pharm. Biopharm.* 125: 1–12.

212 Arami, H., Khandhar, A., Liggitt, D., and Krishnan, K.M. (2015). *In vivo* delivery, pharmacokinetics, biodistribution and toxicity of iron oxide nanoparticles. *Chem. Soc. Rev.* 44: 8576–8607.

213 Yu, M. and Zheng, J. (2015). Clearance pathways and tumor targeting of imaging nanoparticles. *ACS Nano* 9: 6655–6674.

214 Longmire, M., Choyke, P.L., and Kobayashi, H. (2008). Clearance properties of nano-sized particles and molecules as imaging agents: considerations and caveats. *Nanomedicine (Lond.)* 3: 703–717.

215 Xu, J., Yu, M., Peng, C. et al. (2018). Dose dependencies and biocompatibility of renal clearable gold nanoparticles: from mice to non-human primates. *Angew. Chem. Int. Ed.* 57: 266–271.

216 Cheng, L., Jiang, D., Kamkaew, A. et al. (2017). Renal-clearable PEGylated porphyrin nanoparticles for image-guided photodynamic cancer therapy. *Adv. Funct. Mater.* 27: 1702928.

217 Liu, C.J. (2008). Urine formation and excretion. In: *Physiology*, 7e (ed. D.N. Zhu), 212–239. People's Medical Publishing House.

218 Ye, L., Yong, K.T., Liu, L. et al. (2012). A pilot study in non-human primates shows no adverse response to intravenous injection of quantum dots. *Nat. Nanotechnol.* 7: 453–458.

219 Haraldsson, B., Nyström, J., and Deen, W.M. (2008). Properties of the glomerular barrier and mechanisms of proteinuria. *Physiol. Rev.* 88: 451–487.

220 Hall, J.E. (2015). *Guyton and Hall Textbook of Medical Physiology*, 13e. Elsevier Health Sciences.

221 Choi, H.S., Liu, W., Misra, P. et al. (2007). Renal clearance of quantum dots. *Nat. Biotechnol.* 25: 1165–1170.

222 Burns, A.A., Vider, J., Ow, H. et al. (2009). Fluorescent silica nanoparticles with efficient urinary excretion for nanomedicine. *Nano Lett.* 9: 442–448.

223 Zhou, C., Long, M., Qin, Y. et al. (2011). Luminescent gold nanoparticles with efficient renal clearance. *Angew. Chem. Int. Ed.* 123: 3226–3230.

224 Du, B., Jiang, X., Das, A. et al. (2017). Glomerular barrier behaves as an atomically precise bandpass filter in a sub-nanometre regime. *Nat. Nanotechnol.* 12: 1096–1102.

225 Wang, Y., Liu, Y., Luehmann, H. et al. (2012). Evaluating the pharmacokinetics and *in vivo* cancer targeting capability of Au nanocages by positron emission tomography imaging. *ACS Nano* 6: 5880–5888.

226 Iyer, A.K., Khaled, G., Fang, J., and Maeda, H. (2006). Exploiting the enhanced permeability and retention effect for tumor targeting. *Drug Discovery Today* 11: 812–818.

227 Kobayashi, H., Sato, N., Hiraga, A. et al. (2001). 3D-micro-MR angiography of mice using macromolecular MR contrast agents with polyamidoamine dendrimer core with reference to their pharmacokinetic properties. *Magn. Reson. Med.* 45: 454–460.

228 Liu, R.S., Chou, T.K., Chang, C.H. et al. (2009). Biodistribution, pharmacokinetics and PET imaging of [^{18}F]FMISO, [^{18}F]FDG and [^{18}F]FAc in a sarcoma- and inflammation-bearing mouse model. *Nucl. Med. Biol.* 36: 305–312.

229 Zamboni, W.C. (2005). Liposomal, nanoparticle, and conjugated formulations of anticancer agents. *Clin. Cancer Res.* 11: 8230–8234.

230 Kareem, H., Sandstrom, K., Elia, R. et al. (2010). Blocking EGFR in the liver improves the tumor-to-liver uptake ratio of radiolabeled EGF. *Tumor Biol.* 31: 79–87.

231 Yang, R.H., Chang, L.W., Wu, J.P. et al. (2007). Persistent tissue kinetics and redistribution of nanoparticles, quantum dot 705, in mice: ICP-MS quantitative assessment. *Environ. Health Perspect.* 115: 1339–1343.

232 Kobayashi, H., Longmire, M.R., Ogawa, M., and Choyke, P.L. (2011). Rational chemical design of the next generation of molecular imaging probes based on physics and biology: mixing modalities, colors and signals. *Chem. Soc. Rev.* 40: 4626–4648.

233 Liu, J., Yu, M., Zhou, C. et al. (2013). Passive tumor targeting of renal-clearable luminescent gold nanoparticles: long tumor retention and fast normal tissue clearance. *J. Am. Chem. Soc.* 135: 4978–4981.

234 Sindhwani, S., Syed, A.M., Ngai, J. et al. (2020). The entry of nanoparticles into solid tumours. *Nat. Mater.* 19: 566–575.

235 Haque, S., Pouton, C.W.P., McIntosh, M.P. et al. (2020). The impact of size and charge on the pulmonary pharmacokinetics and immunological response of the lungs to PLGA nanoparticles after INTRATRACHEAL administration to rats. *Nanomed. Nanotechnol. Biol. Med.* 102291. https://doi.org/10.1016/j.nano.2020.102291.

236 Ye, W., Chen, X., Li, X. et al. (2020). Structure-switchable DNA programmed disassembly of nanoparticles for smart size tunability and cancer-specific drug release. *ACS Appl. Mater. Interfaces* 12: 22560–22571.

237 Cheng, Z.L., Zaki, A.A., Hui, J.Z. et al. (2012). Multifunctional nanoparticles: cost versus benefit of adding targeting and imaging capabilities. *Science* 338: 903–910.

238 Dreher, M.R., Liu, W., Michelich, C.R. et al. (2006). Tumor vascular permeability, accumulation, and penetration of macromolecular drug carriers. *J. Natl. Cancer Inst.* 98: 335–344.

239 Lankveld, D.P.K., Oomen, A.G., Krystek, P. et al. (2010). The kinetics of the tissue distribution of silver nanoparticles of different sizes. *Biomaterials* 31: 8350–8361.

240 Yuan, Y.Y., Mao, C.Q., Du, X.J. et al. (2012). Surface charge switchable nanoparticles based on zwitterionic polymer for enhanced drug delivery to tumor. *Adv. Mater.* 24: 5476–5480.

241 Cong, Z., Zhang, L., Ma, S.Q. et al. (2020). Size-transformable hyaluronan stacked self-assembling peptide nanoparticles for improved transcellular tumor penetration and photo–chemo combination therapy. *ACS Nano* 14: 1958–1970.

242 Penga, S., Ouyang, B., Men, Y. et al. (2020). Biodegradable zwitterionic polymer membrane coating endowing nanoparticles with ultra-long circulation and enhanced tumor photothermal therapy. *Biomaterials* 231: 119680. https://doi.org/10.1016/j.biomaterials.2019.119680.

243 Wang, J., Wu, W., Zhang, Y. et al. (2014). The combined effects of size and surface chemistry on the accumulation of boronic acid-rich protein nanoparticles in tumors. *Biomaterials* 35: 866–878.

244 Guo, X., Wei, X., Jing, Y., and Zhou, S. (2015). Size changeable nanocarriers with nuclear targeting for effectively overcoming multidrug resistance in cancer therapy. *Adv. Mater.* 27: 6450–6456.

245 Sykes, E.A., Chen, J., Zheng, G., and Chan, W.C.W. (2014). Investigating the impact of nanoparticle size on active and passive tumor targeting efficiency. *ACS Nano* 8: 5696–5706.

246 Wang, J., Mao, W., Lock, L.L. et al. (2015). The role of micelle size in tumor accumulation, penetration, and treatment. *ACS Nano* 9: 7195–7206.

247 Zhang, X.D., Wua, D., Shen, X. et al. (2012). Size-dependent radiosensitization of PEG-coated gold nanoparticles for cancer radiation therapy. *Biomaterials* 33: 6408–6419.

248 Yang, Z., Tian, R., Wu, J. et al. (2017). Impact of semiconducting perylene diimide nanoparticle size on lymph node mapping and cancer imaging. *ACS Nano* 11: 4247–4255.

249 Popović, Z., Liu, W., Chauhan, V.P. et al. (2010). Nanoparticle size series for *in vivo* fluorescence imaging. *Angew. Chem.* 49: 8649–8652.

250 Hickey, J.W., Santos, J.L., Williford, J.M., and Mao, H.Q. (2015). Control of polymeric nanoparticle size to improve therapeutic delivery. *J. Controlled Release* 219: 536–547.

251 Kibria, G., Hatakeyama, H., Ohga, N. et al. (2015). The effect of liposomal size on the targeted delivery of doxorubicin to integrin αvβ3-expressing tumor endothelial cells. *Biomaterials* 34: 5617–5627.

252 Tang, L., Yang, X., Yin, Q. et al. (2014). Investigating the optimal size of anticancer nanomedicine. *Proc. Natl. Acad. Sci. U.S.A.* 111: 15344–15349.

253 Ruan, S., Hu, C., Xian, T. et al. (2016). Increased gold nanoparticle retention in brain tumors by *in situ* enzyme-induced aggregation. *ACS Nano* 10: 10086–10098.

254 Ruan, S., He, Q., and Gao, H. (2015). Matrix metalloproteinase triggered size-shrinkable gelatin-gold fabricated nanoparticles for tumor microenvironment sensitive penetration and diagnosis of glioma. *Nanoscale* 7: 9487–9496.

255 Hu, C., Cun, X., Ruan, S. et al. (2018). Enzyme-triggered size shrink and laser-enhanced NO release nanoparticles for deep tumor penetration and combination therapy. *Biomaterials* 168: 64–75.

256 Liu, R., Xiao, W., Hu, C. et al. (2018). Theranostic size-reducible and no donor conjugated gold nanocluster fabricated hyaluronic acid nanoparticle with optimal size for combinational treatment of breast cancer and lung metastasis. *J. Controlled Release* 278: 127–139.

257 Zhu, X., Vo, C., Taylor, M., and Smith, B.R. (2019). Non-spherical micro- and nanoparticles in nanomedicine. *Mater. Horiz.* 6: 1094–1121.

258 Yang, L., Zhou, Z., Song, J., and Chen, X. (2019). Anisotropic nanomaterials for shape-dependent physicochemical and biomedical applications. *Chem. Soc. Rev.* 48: 5140–5176.

259 Agarwala, R., Singhb, V., Jurneyb, P. et al. (2013). Mammalian cells preferentially internalize hydrogel nanodiscs over nanorods and use shape-specific uptake mechanisms. *Proc. Natl. Acad. Sci. U.S.A.* 110: 17247–17252.

260 Toy, R., Peiris, P.M., Ghaghada, K.B., and Karathanasis, E. (2014). Shaping cancer nanomedicine: the effect of particle shape on the *in vivo* journey of nanoparticles. *Nanomedicine* 9: 121–134.

261 Wang, Z., Chang, Z., Lu, M. et al. (2018). Shape-controlled magnetic mesoporous silica nanoparticles for magnetically-mediated suicide gene therapy of hepatocellular carcinoma. *Biomaterials* 154: 147–157.

262 Chang, Z.M., Wang, Z., Shao, D. et al. (2018). Shape engineering boosts magnetic mesoporous silica nanoparticles-based isolation and detection of circulating tumor cells. *ACS Appl. Mater. Interfaces* 10: 10656–10663.

263 Li, H., Liu, H., Nie, T. et al. (2018). Molecular bottlebrush as a unimolecular vehicle with tunable shape for photothermal cancer therapy. *Biomaterials* 178: 620–629.

264 Rampersaud, S., Fang, J., Wei, Z. et al. (2016). The effect of cage shape on nanoparticle-based drug carriers: anticancer drug release and efficacy via receptor blockade using dextran-coated iron oxide nanocages. *Nano Lett.* 16: 7357–7363.

265 Peer, D., Karp, J.M., Hong, S. et al. (2007). Nanocarriers as an emerging platform for cancer therapy. *Nat. Nanotechnol.* 2: 751–760.

266 Roode, L.E., Brighton, H., Bo, T. et al. (2016). Subtumoral analysis of PRINT nanoparticle distribution reveals targeting variation based on cellular and particle properties. *Nanomedicine* 12: 1053–1062.

267 Chu, K.S., Hasan, W., Rawal, S. et al. (2013). Tumor and tissue pharmacokinetics of docetaxel delivered via nanoparticles of different sizes and shapes in mice bearing Skov-3 human ovarian carcinoma xenograft. *Nanomedicine* 9: 686–693.

268 Chauhan, V.P., Popović, Z., Chen, O. et al. (2011). Fluorescent nanorods and nanospheres for real-time *in vivo* probing of nanoparticle shape-dependent tumor penetration. *Angew. Chem. Int. Ed.* 50: 11417–11420.

269 Reuter, K.G., Perry, J.L., Kim, D. et al. (2015). Targeted print hydrogels: the role of nanoparticle size and ligand density on cell association biodistribution, and tumor accumulation. *Nano Lett.* 15: 6371–6378.

270 Huang, X., Li, L., Liu, T. et al. (2011). The shape effect of mesoporous silica nanoparticles on biodistribution, clearance, and biocompatibility *in vivo*. *ACS Nano* 5: 5390–5399.

271 Kolhar, P., Anselmo, A.C., Gupta, V. et al. (2013). Using shape effects to target antibody-coated nanoparticles to lung and brain endothelium. *Proc. Natl. Acad. Sci. U.S.A.* 110: 10753–10758.

272 Park, J.H., von Maltzahn, G., Zhang, L. et al. (2009). Systematic surface engineering of magnetic nanoworms for *in vivo* tumor targeting. *Small* 5: 694–700.

273 Kankala, R.K., Han, Y.H., Na, J. et al. (2020). Nanoarchitectured structure and surface biofunctionality of mesoporous silica nanoparticles. *Adv. Mater.* 32: 1907035.

274 Ding, L., Lyu, Z., Louis, B. et al. (2020). Surface charge of supramolecular nanosystems for *in vivo* biodistribution: a MicroSPECT/CT imaging study. *Small* 2003290. https://doi.org/10.1002/smll.202003290.

275 Xiao, K., Li, Y., Luo, J. et al. (2011). The effect of surface charge on *in vivo* biodistribution of PEG-oligocholic acid based micellar nanoparticles. *Biomaterials* 32: 3435–3446.

276 Liu, S., Wang, L., Zhang, M. et al. (2019). Tumor microenvironment-responsive nanoshuttles with sodium citrate modification for hierarchical targeting and improved tumor theranostics. *ACS Appl. Mater. Interfaces* 11: 25730–25739.

277 Verma, A. and Stellacci, F. (2010). Effect of surface properties on nanoparticle–cell interactions. *Small* 6: 12–21.

278 Ramos, J., Forcada, J., and Hidalgo-Alvarez, R. (2014). Cationic polymer nanoparticles and nanogels: from synthesis to biotechnological applications. *Chem. Rev.* 114: 367–428.

279 Hühn, D., Kantner, K., Geidel, C. et al. (2013). Polymercoated nanoparticles interacting with proteins and cells: focusing on the sign of the net charge. *ACS Nano* 7: 3253–3263.

280 Hillaireau, H. and Couvreur, P. (2009). Nanocarriers' entry into the cell: relevance to drug delivery. *Cell. Mol. Life Sci.* 66: 2873–2896.

281 Fleischer, C.C. and Payne, C.K. (2012). Nanoparticle surface charge mediates the cellular receptors used by protein-nanoparticle complexes. *J. Phys. Chem. B* 116: 8901–8907.

282 Jallouli, Y., Paillard, A., Chang, J. et al. (2007). Influence of surface charge and inner composition of porous nanoparticles to cross blood-brain barrier *in vitro*. *Int. J. Pharm.* 344: 103–109.

283 Dante, S., Petrelli, A., Petrini, E.M. et al. (2017). Selective targeting of neurons with inorganic nanoparticles: revealing the crucial role of nanoparticle surface charge. *ACS Nano* 11: 6630–6640.

284 Nagy, A., Steinbruck, A., Gao, J. et al. (2012). Comprehensive analysis of the effects of CdSe quantum dot size, surface charge, and functionalization on primary human lung cells. *ACS Nano* 6: 4748–4762.

285 Farokhzad, O.C. and Langer, R. (2009). Impact of nanotechnology on drug delivery. *ACS Nano* 3: 16–20.

286 Shen, Y.Q., Zhan, Y.H., Tang, J.B. et al. (2008). Multifunctioning pH-responsive nanoparticles from hierarchical self-assembly of polymer brush for cancer drug delivery. *AlChE J.* 54: 2979–2989.

287 Yoo, J.W., Doshi, N., and Mitragotri, S. (2011). Adaptive micro and nanoparticles: temporal control over carrier properties to facilitate drug delivery. *Adv. Drug Delivery Rev.* 63: 1247–1256.

288 Chen, S., Rong, L., Lei, Q. et al. (2016). A surface charge-switchable and folate modified system for co-delivery of proapoptosis peptide and p53 plasmid in cancer therapy. *Biomaterials* 77: 149–163.

289 Qiu, N., Liu, X., Zhong, Y. et al. (2016). Esterase-activated charge-reversal polymer for fibroblast-exempt cancer gene therapy. *Adv. Mater.* 28: 10613–10622.

290 Zhao, B.X., Zhao, Y., Huang, Y. et al. (2012). The efficiency of tumor-specific pH-responsive peptide-modified polymeric micelles containing paclitaxel. *Biomaterials* 33: 2508–2520.

291 Min, K.H., Kim, J.H., Bae, S.M. et al. (2010). Tumoral acidic pH-responsive MPEG-poly(β-amino ester) polymeric micelles for cancer targeting therapy. *J. Controlled Release* 144: 259–266.

292 Kaminskas, L.M., Kelly, B.D., McLeod, V.M. et al. (2011). Characterisation and tumour targeting of PEGylated polylysine dendrimers bearing doxorubicin via a pH labile linker. *J. Controlled Release* 152: 241–248.

293 Shen, J.M., Yin, T., Tian, X.Z. et al. (2013). Surface charge-switchable polymeric magnetic nanoparticles for the controlled release of anticancer drug. *ACS Appl. Mater. Interfaces* 5: 7014–7024.

294 Lei, M., Fu, C., Cheng, X. et al. (2017). Activated surface charge-reversal manganese oxide nanocubes with high surface-to-volume ratio for accurate magnetic resonance tumor imaging. *Adv. Funct. Mater.* 27: 1700978.

295 Feng, Q., Liu, J., Li, X. et al. (2017). One-step microfluidic synthesis of nanocomplex with tunable rigidity and acid-switchable surface charge for overcoming drug resistance. *Small* 13: 1603109.

6

Drug-Delivery Technology

6.1 Introduction

The development over the past decade of methods for delivering genes to mammalian cells has stimulated great interest in the possibility of treating human diseases by gene-based therapies. However, despite substantial progress, a number of key technical issues need to be resolved before gene therapy can be safely and effectively applied in the clinic. The major aim of gene therapy is to effectively deliver the genetic materials into cells, genetically modifying and repairing cell functions, which may induce therapeutic healing of disease conditions. While most current gene therapy clinical trials are based on cell- and viral-mediated approaches, nonviral gene medicines are emerging as potentially safe and effective in the treatment of a wide variety of genetic and acquired diseases. Viral vectors have been mainly used because of their inherently high transfection efficiency of gene. However, there are some problems to be resolved for the clinical applications, such as the pathogenicity and immunogenicity of viral vectors themselves. Therefore, many research trials with nonviral vectors have been performed to enhance their efficiency to a level comparable to the viral vector. Nonviral technologies consist of plasmid-based expression systems containing a gene-encoding therapeutic protein and synthetic gene delivery systems. Gene therapy with nanoparticles technologies aims to control the *in vivo* location and function of administered genes.

Gene transfection is a powerful and promising technique that involves the *in vitro* or *in vivo* introduction of exogenous genes into cells for experimental and therapeutic purposes. Whichever is the final goal for experimental biology and gene therapy, the first key issue to be dealt with is enabling the gene to internalize into the cell as efficiently as possible and to facilitate the expression for a long or short time period. The genetic material involves DNA, RNA, antisense, decoy DNA, and ribozymes. Based on basic and scientific knowledge, the delivery technology of genetic material is applicable to produce various proteins of pharmaceutical value (e.g. cytokines, growth factors, and antibodies) and seeds resistant to harmful insects and cold weather damage. In other words, the cells would be enhanced to produce valuable pharmaceutical and food products.

Biomedical Engineering: Materials, Technology, and Applications, First Edition. Hossein Hosseinkhani.

However, the idea underlying gene therapy that human disease might be treated by transfer of genetic material into specific cells of a patient rather than by conventional drugs has yet to make its mark in medicine. Although the concept may appear to be elegantly straightforward and the most direct application of DNA technology, research has indicated that successful implementation of gene transfer in the clinic will require the coordinated development of a variety of new technologies and the establishment of unique interactions between investigators from divergent medical and basic science disciplines.

Although several reviews of gene therapy research have been published [1–9], few have focused on the technical issues that continue to impede the translation of preclinical studies of gene therapy into effective clinical protocols. This chapter gives an overview of the enhanced gene expression of plasmid DNA with different systems and suggests new areas of investigation that may help to resolve them.

6.2 Biodegradable Polymers in Drug Delivery

A number of methods for drug delivery to a patient's cells have been exported [10–15]. Each of these methods offers distinct clinical opportunities and risks, as well as potentials for commercialization. They include the use of natural polymers' delivery, viral and nonviral vectors for gene delivery, and physical methods for drug delivery.

6.2.1 Gene Delivery

Viral-mediated gene therapy involves the use of attenuated or defective viruses, the so-called viral vectors that are created by genetic engineering. These viral vectors are designed to transfer therapeutic genes into patient's cells without causing viral disease, and they function on the premise that they optimally utilize the highly evolved pathways of infection to deliver and express genes in the body. Such pathways provide viral vectors with access to a number of target cells, uptake via receptor-mediated endocytosis, and efficient intracellular trafficking from the endosomes to the nucleus. In addition, certain viruses are capable of either permanently integrating their genomes into the host chromosomal DNA or replicating and persisting as an extrachromosomal element (episome) within the nucleus. Due to their natural ability to infect cells, several viruses, such as retrovirus, adenovirus, adeno-associated virus, and herpes virus, have been investigated for *in vivo* gene delivery [16–20]. In general, viral vectors are highly advantageous in terms of their inherently high efficiency of gene transfection as compared with the nonviral vectors. However, several deficiencies must be remedied before they are employed in clinical applications. One of them is an inherent nature to the induce immune and/or toxic reactions. In fact, one cannot inject an identical viral vector more than twice into an individual. At present, some replication-deficient viral constructs are being used for most gene therapies, but it is difficult to completely rule out the possibility that recombination of the viral gene will generate a pathogenic, replication-competent virus *de novo* [21]. In the case of retroviral vectors applicable

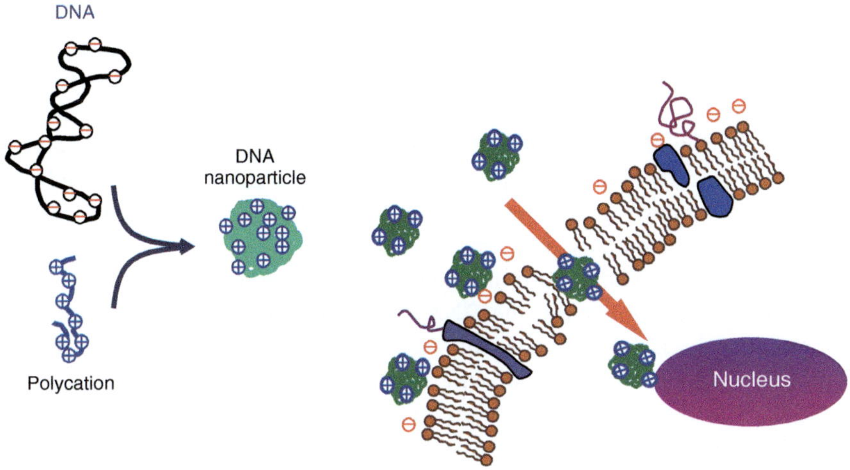

Figure 6.1 Schematic illustration of the formation of the DNA nanoparticles via electrostatic interaction of negatively charged DNA with positively charged polycation.

only to cells in the mitotic phase, their integration into undefined regions of genome may induce undesired activation of potential oncogenic genes. There is also a limitation in the molecular size of gene that can be introduced into cells by the viral vector.

Nonviral vectors have several advantages over viral vectors since they are chemically based materials, although the efficiency of gene transfection is much lower than that of a viral system. The nonviral vectors have low toxicity and inducibility of immune responses compared with the viral vectors, and there is no possibility of the genetic integration into the genome. In addition, the nonviral vector system is not limited by the molecular size of the gene to be introduced. Moreover, it is practically easy to produce and chemically modify nonviral vectors on a large scale. There are at least four categories by which the research approach of nonviral vectors design can be classified, which will be discussed in the following text.

Polycations used for gene delivery are considered to be promising candidates for nonviral gene delivery in part because of their molecular diversity that can be modified to fine-tune their physicochemical properties [22, 23]. These polycations are able to condense large gene into compact structures and to mask the negative DNA charges, necessities for transfecting most types of cells [24, 25]. Generally, since plasmid DNA is a large, negatively charged molecule up to 1 μm in length [26], it is impossible to allow the plasmid DNA itself to be internalized by cells *in vitro* due to the inherent negative charge of the cell membrane. When mixed with a positively charged nonviral vector, it is well known that the molecular size of the plasmid DNA decreases by molecular condensation due to formation of nanoparticle. It is likely that a condensed DNA vector nanoparticle of slightly positive charge can interact electrostatically with the cell membrane and thereby be internalized (Figure 6.1).

Polycations used for the formation of DNA nanoparticle are polyamines that become cationic at physiologic conditions. The most studied polycations used for gene complexation and delivery are the branched/linear polyethyleneimine (PEI) [27, 28], poly(L-lysine) [29], poly(dimethyl aminoethyl methacrylate) (pDMAEMA) [30], poly(trimethyl aminoethyl methacrylate) (pTMAEMA) [31], poly(vinylpyridine) [32], chitosan [33], poly(amidoamine) dendrimers [34], polymer micelles [35], and diethylaminoethyl dextran (DEAE-dextran) [36]. Most polycations are toxic to cells and are nonbiodegradable, while the polymers based on amino acids such as poly(Lysine) are immunogenic [29]. More advanced polymeric gene delivery systems employ macromolecules with a very high cationic charge density that act as endosomal buffering systems, thus suppressing the endosomal enzyme activities and protecting the DNA from degradation. The high cationic charge mediates both DNA condensing and buffering capacity that diminish the requirement for the addition of endosomolytic agents [37, 38]. Among the cationic polymers, it is known that the amine residues of PEI could function as an endosomal buffering system, which may suppress the action of endosomal enzymes and thereby protect the plasmid DNA from degradation. The resultant enhanced transfection efficiency [39] is called the "buffering effect." Lysosomal membranes contain V-ATPases, which pump H^+ at the expense of ATP. A charge gradient would theoretically build because of protons being pumped into the lysosomes, but there is also an influx of Cl^-, which relieves this gradient. It has been hypothesized that the increased Cl^- concentration, in turn, increases lysosomal osmolarity, with water rushing in to relieve the gradient, resulting in lysosomal swelling and bursting [38]. Even though lysosomes eventually burst, they are still quite successful at degrading endocytosed particles beforehand. Transgene delivery is often interrupted within endolysosomes because of nucleases that reside within lysosomes. PEI is thought to be a good gene delivery vehicle because of its ability to accept the protons pumped into endolysosomes. By accepting protons, PEI could raise endolysosomal pH, which would alter protein folding within the endolysosome and perhaps inactivate degradative enzymes. The enzymes might be inactivated long enough for endolysosomal bursting to occur before plasmid digestion has taken place. Modified PEI/DNA nanoparticles have been shown to increase transfection efficiency, which lend support to the previously mentioned hypothesis.

6.2.2 siRNA Delivery

Gene silencing efforts in advanced clinical stages have yet to discriminate between the individual alleles of a gene. These approaches assume that either a uniform reduction of both copies of a target gene will not lead to adverse consequences or that a specific window of reduced expression can be tolerated while simultaneously improving disease outcome. There are, however, specific dominant genetic disorders in which a different approach, aimed at silencing only the mutant copy of a gene, may be preferable.

Current methods of gene silencing employ the RNAi pathway, as well as antisense oligonucleotides (ASOs) that trigger RNase H-mediated cleavage of a complementary RNA target. In RNAi-based methods, synthetic miRNA and small hairpin RNA (shRNA) encoded by viral vectors are transcribed and processed into cytoplasmic siRNA, where they trigger RNA-induced silencing complex (RISC)-mediated degradation or translational repression of target RNA. In addition, siRNA and chemically modified single-stranded siRNA (ss-siRNA) can be directly administered to target cells and enter the miRNA pathway further downstream. ASOs are not dependent on the delivery vehicle and can be freely taken up by targets cells to induce RNA degradation, without the risk of saturating the miRNA pathway. The dsRNAs are processed by a combination of RNAse III Dicer, TAR RNA-binding protein (TRBP), and interferon-induced protein kinase R activator (PACT), into siRNAs about 20–24 nucleotides (nt) in length, with phosphate groups at both 5′ ends, and hydroxyl groups and two nucleotide overhangs at both 3′ ends. The siRNAs thus processed consist of a sense strand and a complementary antisense strand. The siRNAs unwind due to the helicase activity of the preRISC domain and with the help of an argonaute protein (AGO) that later cleaves the sense strand. Then, the mature RISC containing the antisense strand along with the AGO protein attaches and cleaves the complementary mRNA. Designing siRNAs that are fully complementary to their target mRNA is of utmost importance. Initially, investigators followed a few empirical guidelines like GC content and the region of the mRNA to be targeted for effective gene silencing.

6.3 Target Drug Delivery

Drug delivery to specific sites within the body is one of the biggest concerns in the field of pharmaceutical and medical sciences. This is due to the fact that chemotherapy drugs, if not targeted properly, can kill not only tumor cells but also healthy tissues. Some researchers are investigating the possibility to develop a new nanocarrier, such as liposome and micelle. Others are studying super-paramagnetic nanoparticles such as iron oxide (SPION), which are widely used as a nanocarrier, thanks to their nontoxicity, biocompatibility, large surface area, and suitable magnetic properties, all of which are favorable for the drug-delivery mechanism. However, due to low colloidal stability in the liquid phase, these particles aggregate very quickly and easily in the human body and thus undergo rapid clearance by macrophages of the spleen, liver, and bone marrow; a phenomenon commonly known as reticuloendothelial system (RES). For this, it is necessary to pass these particles through RES organs and toward the targeted sites. Generally, SPION will deploy appropriate surface modifications to prevent the aggregation at RES. Many studies have focused on improving SPION properties in order to optimize various disease therapies. Other studies are investigating magnetic nanoparticles properties to track and record the progression of the disorder.

6.3.1 Target Therapy in Cells

Cell receptor-mediated internalization and endosomal escape is a very promising approach. The basic research approach of nonviral vectors is to neutralize the negative charge of plasmid DNA and to condense the DNA size by polyion complexation with various polymers and liposomes of positive charge. It is likely that the condensed plasmid DNA nanoparticles with a positive charge effectively interact with cells to internalize, enhancing the efficiency of gene transfection. However, since this complex–cell interaction is based on the simple and nonspecific electrostatic force, the more enhanced and cell-specific cell transfection cannot be always expected. One of practically possible ways to improve this situation is to take advantage of cell receptor systems that play an important role in the cellular uptake of several substances. It cannot always be expected that the cellular uptake of large amounts of genetic material by way of the conventional pinocytosis process will occur. Therefore, the ligands of cell receptors, such as asialoorosomucoid and transferrin, are used to enhance the internalization of plasmid DNA into cells, resulting in enhanced gene expression. The plasmid DNA internalization into liver cells through an asialoglycoprotein receptor is well documented in the literature. A galactose residue is coupled to the PEI and poly(L-lysine) vector of plasmid DNA and enhances the gene expression by way of the asialoglycoprotein receptor. Although it is difficult to artificially regulate the intracellular fate of internalized plasmid DNA, the promoted endosomal escape of plasmid DNA will result in an enhanced level of gene expression. However, it has been shown that intraarterial delivery of plasmid DNA transfected skeletal myofibers quite efficiently in the limb skeletal muscles of rhesus monkey. One research trial was conducted in which a complex of plasmid DNA and poly(L-lysine) was rapidly released from the endocytotic pathway to accelerate DNA entry into the cytoplasm when a peptide derived from virus coats was used in conjunction with the complex. Figure 6.2 indicates how targeting of cell receptors that are located on the surface of cells and extracellular matrix (ECM) could be used for new generation of drug-delivery technology.

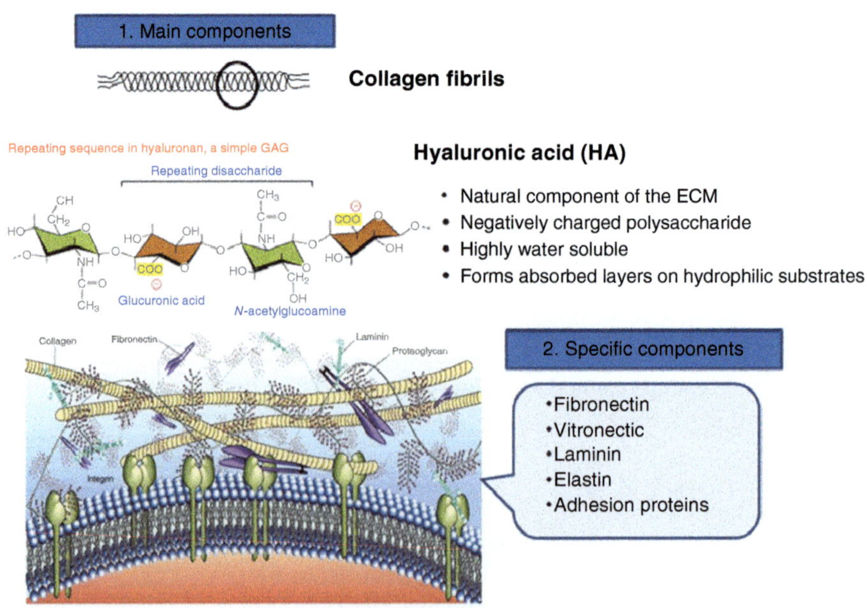

Figure 6.2 ECM components and cell receptor proteins in targeted drug-delivery systems.

6.4 Drug Delivery in Imaging Technology

6.4.1 MRI Technology

Bioimaging is a relatively recent development in the medical field that makes use of digital technology. It relates to methods that noninvasively visualize biological processes in real time. Bioimaging aims to interfere as little as possible with life processes. This new imaging technique produces information about anatomical structure that is linked to functional data, as described by electric and magnetic fields, mechanical motion, and metabolism. It provides comprehensive views of the human body in progressively greater depth and detail. This technology not only provides well identification in research but also exhibits great potential in clinical development.

Bioimaging is thus a powerful tool for viewing the internal workings of the body and its diseases. Moreover, it is often used to gain information on the 3D structure of the observed specimen from the outside, i.e. without physical interference. Some examples of bioimaging as applied to the medical field include X-ray and ultrasound (US) images, MRI, 3D and 4D body images using CT scans, etc. The first multi-lens microscopes were built about 400 years ago and have been used for medical research since ages. However, it was only with the advent of digital photography that scientists could harness their full power using computational tools for image analysis. With progress in digital technology, bioimaging is expected to be cheaper and faster and promises to be the mainstay of medical research.

6.4.2 Ultrasound Technology

Physical stimuli have been demonstrated to enhance the internalization of plasmid DNA into cells. It has been demonstrated that electroporation is effective in enhancing the transfection efficiency of plasmid DNA *in vitro* as well as *in vivo*. However, since the applied electric power is relatively high, cytotoxicity and tissue damage often occurred. As another stimulus, hydrodynamic and hydrostatic pressures have been demonstrated to enhance the expression of plasmid DNA. A large volume of plasmid DNA solution is injected intravenously for a short time period, which results in the generation of increased hydraulic pressure in the blood vessel. It is thought that this hydrodynamic pressure enables the plasmid DNA to physically enhance the internalization into cells present within the target tissue, thereby improving the efficiency of gene expression. For example, in one experiment, a large volume of plasmid DNA solution was injected into the femoral artery of rats for 10 seconds, while all the blood vessels leading to and out of the hindlimb were occluded. This injection increased the level of gene expression within the muscle by up to 40-fold as compared with the intramuscular injection of plasmid DNA. A rapid injection of plasmid DNA solution at a volume almost equivalent to that of total blood into mice via the tail vein induced efficient gene transfection in the lung, spleen, heart, kidney, and liver, although the highest level was observed in the liver. It has been proposed that the injected DNA solution accumulates in the inferior vena cava and flows back to tissues directly linked to this part of vascular system, such as the liver. It is possible that this accumulation results in an enhanced gene expression. The hydrostatic pressure would thereby be effective in pushing DNA into the liver cells. After a large volume of hypertonic aqueous solution of plasmid DNA was injected into the portal vein, the hepatic vein was occluded to increase the osmotic and hydrostatic pressure. The pressure generated was thought to enlarge the pore size of sinusoid fenestrae, thereby resulting in the promotion of plasmid DNA extravasation and consequently enhanced DNA uptake by hepatocytes.

Shooting plasmid DNA into a target tissue is carried out with the use of a "gene gun." In this technique, plasmid DNA coated on gold microspheres is atomized, and the atomized solution is shot at a high pressure into the target tissue. The skin, liver, and muscle tissue have been successfully transfected by this technique, although the transfection efficiency varies from tissue to tissue. The major limitation of this method is related to the shallow penetration of plasmid DNA into the tissue. The level of gene expression is also low, and the duration is very short. This approach allows the plasmid DNA to penetrate directly through the cell membrane into the cytoplasm or even nucleus and to bypass the endosome/lysosome, thus avoiding plasmid DNA degradation.

Recently, US irradiation has been noted and widely applied to therapeutic, operative, and diagnostic procedures in clinical medicine. Among them, this physical therapy is popular since US acts on cells and tissues by way of various thermal or nonthermal mechanisms, sometimes promoting wound healing. The main objective of US therapy is to speed up the healing process of hard or soft tissue injuries. In addition, US has practical advantages in therapeutic usage, such as accessibility, low cost, and safety. US is composed of a propagating pressure wave or sound wave capable of transferring mechanical energy into various body tissues. The energy is absorbed, propagated, or reflected depending on the frequency and tissue type. It has been reported that US-induced mechanical perturbations serve as extracellular stimuli to osteogenic cells. The positive effects of US on bone fracture healing have also been reported. The release profile of drugs from polymeric matrices can be regulated by making use of US, while several reports note the US-enhanced transdermal permeation of drugs. The transfection efficiency of plasmid DNA *in vitro* and *in vivo* was also found to be enhanced by US irradiation. US irradiation in the presence of plasmid DNA enables cells cultured *in vitro* to show enhanced gene expression irrespective of the cell type, such as plant and animal cells. Other reports have revealed that the US-mediated DNA delivery is effective for intact plant cells, leaf segments, and plant embryos.

6.5 Summary and Challenges

The promise of gene therapy has recently been enhanced by major advances in the area of genomics with identification of the structure and function of thousands of genes in the human genome as well as advances in controlling the delivery and expression of plasmid-based gene expression systems *in vivo*. Recent advances in the design and *in vitro* as well as *in vivo* use of gene delivery systems and DNA nanoparticles for gene delivery to cells and tissue have been reviewed in this chapter. While improvements to transfection efficiencies have been made and research toward understanding the underlying mechanism behind several gene delivery vehicles has been carried out, the issues of toxicity, transfection efficiency, and host response still remain as problems for some gene delivery applications. There are many thoughts still to be clarified regarding the application of nonviral gene delivery vehicles to human treatment.

References

1 Southwood, L.L., Frisbie, D.D., Kawcak, C.E., and McIlwraith, C.W. (2004). Delivery of growth factors using gene therapy to enhance bone healing. *Vet. Surg.* 33: 565–578.

2 Mansouri, S., Lavigne, P., Corsi, K. et al. (2004). Chitosan-DNA nanoparticles as non-viral vectors in gene therapy: strategies to improve transfection efficacy. *Eur. J. Pharm. Biopharm.* 57: 1–8.

3 Brokx, R. and Gariepy, J. (2004). Peptide- and polymer-based gene delivery vehicles. *Methods Mol. Med.* 90: 139–160.

4 Anwer, K., Rhee, B.G., and Mendiratta, S.K. (2003). Recent progress in polymeric gene delivery systems. *Crit. Rev. Ther. Drug Carrier Syst.* 20: 249–293.

5 Mackay, J.A., Deen, D.F., and Szoka, F.C. Jr., (2005). Distribution in brain of liposomes after convection enhanced delivery; modulation by particle charge, particle diameter, and presence of steric coating. *Brain Res.* 1035: 139–153.

6 Oupicky, D. and Diwadkar, V. (2003). Stimuli-responsive gene delivery vectors. *Curr. Opin. Mol. Ther.* 5: 345–350.

7 Hodges, B.L. and Scheule, R.K. (2003). Hydrodynamic delivery of DNA. *Expert Opin. Biol. Ther.* 3: 911–918.

8 Cho, Y.W., Kim, J.D., and Park, K. (2003). Polycation gene delivery systems: escape from endosomes to cytosol. *J. Pharm. Pharmacol.* 55: 721–734.

9 Emerich, D.F. and Thanos, C.G. (2003). Nanotechnology and medicine. *Expert Opin. Biol. Ther.* 3: 655–633.

10 Fechheimer, M., Boylan, J.F., Parker, S. et al. (1987). Transfection of mammalian cells with plasmid DNA by scrape loading and sonication loading. *Proc. Natl. Acad. Sci. USA* 84: 8463–8467.

11 Lee, M. and Kim, S.W. (2005). Polyethylene glycol-conjugated copolymers for plasmid DNA delivery. *Pharm. Res.* 22: 1–10.

12 Kabanov, A.V., Felgner, P.L., and Seymour, L.W. (ed.) (1998)). *Self-assembling Complexes for Gene Delivery. From Laboratory to Clinical Trial.* New York: Wiley.

13 Schreier, H. (1994). The new frontier: gene and oligonucleotide therapy. *Pharm. Acta Helv.* 68: 145–159.

14 Duguid, J.G., Li, C., Shi, M. et al. (1998). A physicochemical approach for predicting the effectiveness of peptide-based gene delivery systems for use in plasmid-based gene therapy. *Biophys. J.* 74: 2802–2814.

15 Ledley, F.D. (1996). 'Pharmaceutical approach to somatic gene therapy '. *Pharm. Res.* 13: 1595–1604.

16 Joung, I., Harber, G., Gerecke, K.M. et al. (2005). Improved gene delivery into neuroglial cells using a fiber-modified adenovirus vector. *Biochem. Biophys. Res. Commun.* 328: 1182–1187.

17 Zhan, J., Gao, Y., Wang, W. et al. (2005). Tumor-specific intravenous gene delivery using oncolytic adenoviruses. *Cancer Gene Ther.* 12: 19–25.

18 Dodart, J.C., Marr, R.A., Koistinaho, M. et al. (2005). Gene delivery of human apolipoprotein E alters brain Abeta burden in a mouse model of Alzheimer's disease. *Proc. Natl. Acad. Sci. USA* 102: 1211–1217.

19 Xu, D., McCarty, D., Fernandes, A. et al. (2005). Delivery of MDR1 small interfering RNA by self-complementary recombinant adeno-associated virus vector. *Mol. Ther.* 11: 523–528.

20 Oehmig, A., Fraefel, C., Breakefield, X.O., and Ackermann, M. (2004). Herpes simplex virus type 1 amplicons and their hybrid virus partners, EBV, AAV, and retrovirus. *Curr. Gene Ther.* 4: 385–393.

21 Donahue, R.E., Kessler, S.W., and Bodine, D. (1992). Helper virus induced T cell lymphoma in non-human primates after retroviral gene transfer. *J. Exp. Biol.* 176: 1125–1129.

22 Putnam, D., Gentry, C.A., Pack, D.W., and Langer, R. (2001). Polymer-based gene delivery with low cytotoxicity by a unique balance of side-chain termini. *Proc. Natl. Acad. Sci. USA* 98: 1200–1209.

23 Kabanov, A.V. (1999). Taking polycation gene delivery systems from *in vitro* to *in vivo*. *Pharm. Sci. Technol. Today* 2: 365–372.

24 Haynes, M., Garrett, R.A., and Gratzer, W.B. (1970). Structure of nucleic acid-poly base complexes. *Biochemistry* 9: 4410–4418.

25 De Smedt, S.C., Demeester, J., and Hennink, W.E. (2000). Cationic polymer based gene delivery systems. *Pharm. Res.* 17: 113–119.

26 Abdallah, B., Sachs, L., and Demeneix, B.A. (1995). Non-viral gene transfer: applications in developmental biology and gene therapy. *Biol. Cell* 85: 1–9.

27 Ferrari, S., Moro, E., Pettenazzo, A. et al. (1997). ExGen 500 is an efficient vector for gene delivery to lung epithelial cells *in vitro* and *in vivo*. *Gene Ther.* 4: 1100–1109.

28 Abdallah, B., Hassan, A., Benoist, C. et al. (1996). A powerful nonviral vector for *in vivo* gene transfer into the adult mammalian brain: polyethylenimine. *Human Gene Ther.* 7: 1947–1954.

29 Kimura, T., Yamaoka, T., Iwase, R., and Murakami, A. (2001). Structure/function relationship in the polyplexes containing cationic polypeptides for gene delivery. *Nucleic Acids Res. Suppl.* 1: 203–209.

30 van de Wetering, P., Cherng, J.Y., Talsma, H. et al. (1998). 2-(dimethylamino) ethyl methacrylate based (co)polymers as gene transfer agents. *J. Controlled Release* 53: 145–156.

31 Wolfert, M.A., Schacht, E.H., Toncheva, V. et al. (1996). Characterization of vectors for gene therapy formed by self- assembly of DNA with synthetic block co-polymers. *Human Gene Ther.* 7: 2123–2129.

32 Yaroslavov, A.A., Sukhishvili, S.A., Obolsky, O.L. et al. (1996). DNA affinity to biological membranes is enhanced due to complexation with hydrophobized polycation. *FEBS Lett.* 384: 177–183.

33 MacLaughlin, F.C., Mumper, R.J., Wang, J. et al. (1998). Chitosan and depolymerized chitosan oligomers as condensing carriers for *in vivo* plasmid delivery. *J. Controlled Release* 56: 259–265.

34 Tang, M.X., Redemann, C.T., and Szoka, F.C. Jr., (1996). *In vitro* gene delivery by degraded polyamidoamine dendrimers. *Bioconjugate Chem.* 7: 703–709.

35 Harada-Shiba, M., Yamauchi, K., Harada, A. et al. (2002). Polyion complex micelles as vectors in gene therapy – pharmacokinetics and *in vivo* gene transfer. *Gene Ther.* 9: 407–415.

36 Takai, T. and Ohmori, H. (1990). DNA transfection of mouse lymphoid-cell by the combination of DEAE-dextran-mediated DNA uptake and osmotic shock procedure. *Biochim. Biophys. Acta* 1048: 105–112.

37 Behr, J.P. (1997). The proton sponge: a trick to enter cells the viruses did not exploit. *Chimia* 51: 34–39.

38 Boussif, O., Lezoualc'h, F., Zanta, M.A. et al. (1995). A versatile vector for gene and oligonucleotide transfer into cells in culture and *in vivo*: polyethylenimine. *Proc. Natl. Acad. Sci. USA* 92: 7297–7305.

39 Bieber, T., Meissner, W., Kostin, S. et al. (2002). Intracellular route and transcriptional competence of polyethylenimine-DNA complexes. *J. Controlled Release* 82: 441–449.

7

Early Stage Detection Technology

7.1 Introduction

Many three-dimensional (3D) models currently in practice require expensive equipment, large sample volumes, long incubation times, and/or extensive expertise, and the greatest disadvantage is that they are too far from the nature of human organs. Because of these problems, research and development on drug discovery, regenerative medicine, biotech, and pharmaceutical industries is very costly and takes several years to bring a single drug/product to the market. 3D *in vitro* systems that can mimic organ and tissue structure and can function *in vivo* will be of great benefit for a variety of biological applications from basic biology to toxicity testing and drug discovery. Figure 7.1 shows the differences between two-dimensional (2D) and 3D systems.

There have been several attempts to generate 3D tissue models, but most of these models require costly equipment, and the most serious disadvantage is that they are too far from the mature of human organs *in vivo*. Because of these problems, research and development in drug discovery, toxicity testing, and biotech industries is highly expensive, involves sacrifice of countless animals, and takes several years to bring a single drug/product to the market or to find the toxicity of chemical entities. Several alternative models have been developed by merging biomaterials science, nanotechnology, and biological principles to generate 3D *in vitro* living organs, called "Human Organs-on-Chip," to mimic natural organ/tissues, thus reducing animal testing and clinical trials. Novel-type mechanically and biologically bio-mimicking collagen-based hydrogel that would provide interconnected mini-wells in which 3D cell/organ culture of human samples in a manner similar to human organs with extracellular matrix (ECM) molecules would be possible. These products mimic the physical, chemical, and biological properties of natural organs and tissues at different scales. This chapter will review the outcome of several experimental results so far in this direction and the future perspectives.

Biomedical Engineering: Materials, Technology, and Applications, First Edition. Hossein Hosseinkhani.

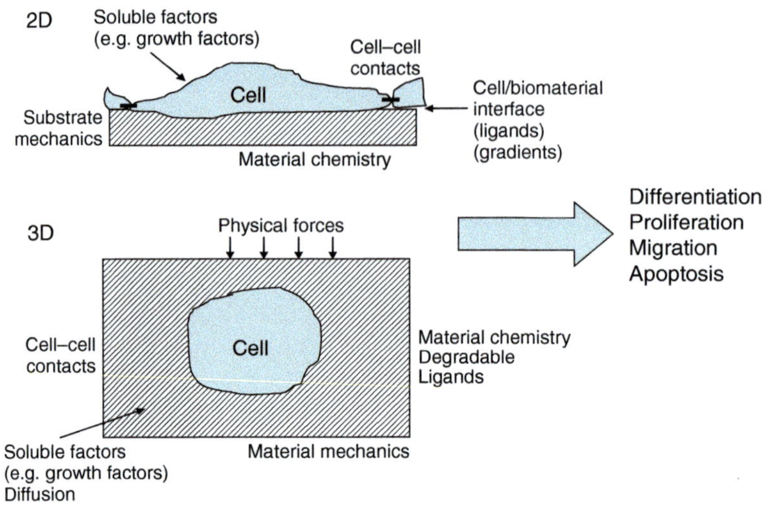

Figure 7.1 Comparison between two-dimensional (2D) and three-dimensional (3D) systems.

7.2 Three-Dimensional (3D) Materials for Biological Application

3D scaffolding materials is an interdisciplinary field that combines engineering and live sciences in order to develop techniques that enables the restoration, maintenance, or enhancement of living tissues and organs on biomaterials. Its fundamental aim is to create natural tissue with the ability to restore missing organ or tissue function, which the organism has not been able to regenerate in physiological conditions. With that, it aspires to improve the health and quality of life for millions of people worldwide and to give solution to the present limitations: rejections and low quantity of donors [1]. Tissue engineering needs scaffolds to serve as a substrate for seeding cells and as a physical support in order to guide the formation of the new tissue [1–6]. The majority of the used techniques utilize 3D polymeric scaffolds, which are composed of natural or synthetic polymers. Synthetic materials are attractive because their chemical and physical properties, such as porosity and mechanical strength, can be specifically optimized for a particular application.

The polymeric scaffold structures are endowed with a complex internal architecture, channels, and porosity that provide sites for cell attachment and maintenance of differentiated function without hindering proliferation [1]. Ideally, a polymeric scaffold for tissue engineering should have the following characteristics: (i) appropriate surface properties promoting cell adhesion, proliferation, and differentiation; (ii) biocompatible; (iii) highly porous, with a high surface area/volume ratio, with an interconnected pore network for cell growth and flow transport of nutrients and metabolic waste; and (iv) mechanical properties sufficient to withstand any *in vivo* stresses [1, 2, 6–10]. The last requisite is difficult to combine with the high porosity in volume of the material. That is why it is necessary to use polymeric matrices with special or reinforced properties, especially if the polymer is a hydrogel. The polymeric scaffold design depends on the regarded applications, but in any case it must achieve structures with the aforementioned characteristics, which are necessary to their correct function. To achieve it, two factors must be considered: (i) materials used, both the porogen and the reticulate polymer, which are infiltrated in the porogen to become a scaffold, and (ii) the structural architecture, both external and internal, basically shown by its porosity (high surface area/volume ratio), geometry, and pore size, as well as keeping in mind that the structures must be easily processed into 3D. On the basis of the extensive range of polymeric materials, different processing techniques have been developed to design and fabricate 3D scaffolds for tissue engineering implants [1, 2, 6, 9–17]. They include (i) phase separation, (ii) gas foaming, (iii) fiber bonding, (iv) photolithography, (v) solid free-form (SFF), (vi) solvent casting in combination with particle leaching, among others. However, none of these techniques have achieved a suitable model of 3D architecture so the scaffolds must be fulfilled using equipment with high cost even for the reasons that will be discussed next. Therefore, using phase separation, a porous structure can be easily obtained by adjusting thermodynamic and kinetic parameters. However, because of the complexity of the processing variables involved in phase-separation technique, the pore structure cannot be easily controlled. Moreover, it is difficult to obtain large pores and may exhibit a lack of interconnectivity [2, 3, 14]. Gas foaming has the advantage of room temperature processing but produces a largely nonporous outer skin layer and a mixture of open and closed pores within the center, leaving incomplete interconnectivity. The main disadvantage of the gas foaming method is that it often result in a non-connected cellular structure within the scaffold [3, 14]. Fiber bonding provides a large surface area for cell attachment and a rapid diffusion of nutrients in favor of cell survival and growth. However, these scaffolds, as the ones used to construct a network of bonded polyglycolic acid (PGA), lacked the structural stability necessary for *in vivo* use. In addition, the technique does not lend itself to easy and independent control of porosity and pore size [1–3]. Photolithography has also been employed for patterning, obtaining structures with high resolution, although this resolution may be unnecessary for many applications of patterning in cell biology. In any case, the disadvantage of this technique is the high cost of the equipment that limits their applicability [18, 19]. SFF scaffold manufacturing methods provide excellent control over scaffold external shape and internal pore interconnectivity and geometry but offer limited microscale resolution. Moreover, the minimum size of global pores is 100 μm. Additionally, SFF requires complex correction of scaffold design for anisotropic shrinkage during fabrication. Moreover, it needs high cost equipment [8]. Finally, solvent casting in combination with particulate leaching method, which involves the casting of a mixture of monomers and initiator solution and a porogen in a mold, polymerization, and followed by leaching out of the porogen with the proper solvent to generate the pores, is inexpensive but still has to overcome some disadvantages in order to find engineering applications, namely, the problem of residual porogen remains, irregular-shaped pores, and insufficient interconnectivity [6, 20].

A new method for producing highly ordered porous scaffolds will be used. This novel, simple, and inexpensive method of scaffold production permits to obtain highly regular pore shapes and arrangements in the form of orthogonal cylindrical channels, which are parallelly aligned with full and controlled interconnectivity, which gives rise to mechanically stable structures without porogen remains. The proposed scaffolds may find applications as structures that facilitate either tissue regeneration or repair during reconstructive operations [1, 21–23]. One intended application of these structures is for nerve surgery. The new structure could also find applications in other areas in which the pore morphology may play an essential role, such as membranes [24] and filters [25].

7.3 Current Approaches of Nanofabrication by Using Lithography

In the past few years, many high-throughput screening (HTS) technologies together with significant efforts on developing diagnostics tools such as biosensors have revolutionized the process of drug discovery to screen hundreds of thousands of compounds against disease targets to find novel drugs. Typically HTS studies are performed through a combination of modern robotics, data processing and control software, liquid handling devices, and sensitive detectors. However, robots and other equipment required for HTS studies are often expensive. Large volumes of invaluable chemicals are required to perform such tests. Another limitation is the time it requires to assay individual samples. At the end, even these sets of facilities are fixed, the experimental data are not trustable to apply for clinical application; therefore animal testing and clinical trials are required in order to get the approval of the product (drug) after a very long time processing. The above problems are some of the many limitations of current technologies of using materials in many aspects of biological applications. Despite much effort at developing methods of enabling testing of many conditions, it still takes a highly specialized and expensive screening lab to run a HTS operation, and these studies are often limited to large-to-moderately sized research institutions. A combinational technology of microengineering, biomaterial sciences, and biological principles will be a promising way to develop 3D *in vitro* living organs with controlled microvasculature and tissue architecture. These 3D models will be useful for *in vitro* diagnostics and drug screening applications as well as for transplantation. Furthermore, 3D *in vitro* living organs significantly contribute in developing microscale biosensors that will be more portable and scalable for point-of-care sample analysis and real-time diagnosis. Modern biosensors based on micro- and nanoscale techniques have the potential to greatly enhance methods of detecting foreign and potentially dangerous toxins and may result in cheaper, faster, and easier-to-use analytical tools.

In this aspect, microscale technologies are potentially powerful tools for addressing some of the challenges in 3D *in vitro* models. Microelectromechanical systems (MEMS), which are an extension of techniques used in the semiconductor and microelectronics industries, can be used to control features at length scales $<1\,\mu m$

to >1 m^3. In the past few years, microfabrication has been increasingly used in biomedical and biological applications. This is partially due to the emergence of techniques such as soft lithography to fabricate microscale devices without the use of expensive "clean rooms" and photolithographic equipment.

When it comes to patterning PEG hydrogels, we can typically use two techniques, photolithography and CFL. In photolithography, we use a mask to selectively cross-link an underlying UV cross-linkable hydrogel. The unreacted part is easily washed with water, and finally we can get successful patterns after developing. On the other hand, in CFL, a prepatterned poly(dimethylsiloxane) (PDMS) stamp is placed on top of polymer surface directly after spin coating or spin casting. To provide mobility to the polymer, we can increase the temperature above glass transition temperature, or if the polymer is mobile itself, it can spontaneously fill into the void space of the stamp. In usual organic polymers, we use organic solvents such as toluene or ethanol so that the contact angle is normally less than 90°. In this case, the polymer takes up the void space according to the well-known Young's equation. In contrast, in most biological systems, the polymer is only soluble in water or PBS or sometimes in methanol or ethanol such that the contact angle is probably larger than 90°. In this case, the polymer cannot rise into the void and recedes toward the bottom, which can be termed capillary depression.

Although PEG–DMA works fine with protein or cell patterning, it has a shortcoming in terms of long-term stability. Typically, water attacks the edge of the pattern so the film gradually came off. Patterning with a polymer that has a random copolymer of PEG methyl ether methacrylate and 3-(trimethoxysilyl)propyl methacrylate has been done. Since this polymer contains methoxysilyl group, it can spontaneously react with silanol groups on hydrophilic substrates, leading to robust interface even without cross-linking step. If you treat the surface at high temperature, the interface is more strengthened. One advantage here is that we can control the height of microstructures with a minor alteration.

This needs very simple lithography, which is nanomolding. Its pattern size is 150 nm and its aspect ratio is about 3. One is patterned area and the other is non-patterned area. A broad range of size regime can be realized, ranging from a few microns to a few hundred microns. Also we were able to obtain aggregated cells to individual cell arrays using different feature sizes.

These techniques are compatible with cells and are now being integrated with biomaterials to facilitate fabrication of cell–material composites that can be used for biomedical engineering [19]. In addition, microscale technologies allow for an unprecedented ability to control the cellular microenvironment in culture and miniaturize assays for high-throughput applications. Initial experiments used micromachining technologies on silicon surfaces to generate vascularized systems. Subsequent work on the replica molding of biocompatible polymers such as PDMS from patterned silicon wafers has resulted in the fabrication of biocompatible scaffolds. More recently, microfabricated capillary networks have been fabricated out of biodegradable elastomers such as poly($_{\mathrm{DL}}$-lactide-*co*-glycolide) (PLGA), polyurethane, and poly(glycerol sebacate) (PGS). However, there are potential disadvantages with the use of these polymers, which include its rigid mechanical properties and bulk degradation kinetics. Alternative methods of fabricating scaffolds with micro- and nanoscale resolution include 3D printing, microsyringe deposition, tissue spin casting, and electrospinning of nanofibers. However, these approaches are difficult to perform and scale-up for fabrication of large 3D organs. A potentially powerful approach to engineering the microvasculature is to use cell-laden hydrogels. It has been demonstrated that biomaterials made from hydrogels can be molded to fabricate microchannels [18]. Although this technology has not yet been used for 3D living models, the ability of this approach to mimic organ and tissues is useful in engineering complex, vascularized organs.

Microfabrication approaches have been used to engineer the desired microvasculature directly into the tissue engineering scaffolds. Initial experiments used micromachining technologies on silicon surfaces to generate vascularized systems. Subsequent work on the replica molding of biocompatible polymers from patterned silicon wafers has resulted in the fabrication of biocompatible scaffolds. Originally, such scaffolds were fabricated from PDMS, a transparent, biocompatible elastomer that is widely used in replica molding. PLGA has been used to generate biodegradable microfluidic systems and microfabricated scaffolds by superpositioning and stacking multiple layers of fabricated structures. These artificial capillary networks were coated with fibronectin and seeded with endothelial cells, which grew to confluence within a few days. Thus, to improve the mechanical properties and to allow for surface erosion instead of bulk erosion of biodegradable scaffolds, other polymers such as PGS have been used. PGS films were patterned with the desired microvasculature and bonded to a flat film to create a capillary network. A potential limitation of this approach is that although endothelial cells can be seeded on the microchannels, the functional cells are restricted to regions between two microfabricated plates containing the microvasculature. Therefore, the ability to encapsulate cells within the bulk material of the scaffold is a potentially important contribution to this research. To overcome the limitations of the hydrophobic biodegradable polymers, layer-by-layer microfluidic patterning has been used to generate scaffolds. In this method, cells and matrix biopolymers were sequentially deposited on particular regions within the microchannels to generate 3D structures with controlled location of the deposited cells. Alternative methods of fabricating scaffolds with micro- and nanoscale resolution include 3D printing, microsyringe deposition, tissue spin casting, and electrospinning of nanofibers. In 3D printing, polymer particles and salt are printed using a bonding agent, which forms a porous scaffold once the salt is dissolved. More recently, rapid prototyping methods such as soft lithography and microsyringe deposition have been used to fabricate PLGA and polyurethane scaffolds that comprised multilayer structures with controlled resolution. However, this approach is difficult to scale-up for fabrication of large 3D tissues. A detection system for released proteins from cells in the microwells. Antibodies are printed on the array and binding of the proteins to the antibodies can then be detected by ELISA and a microarray scanner.

Materials technology aim to develop a technique that draws from microscale engineering, novel biomaterials, and biological principles to overcome the limitations of the current approaches to generate 3D *in vitro* living organs. Fabrication technology

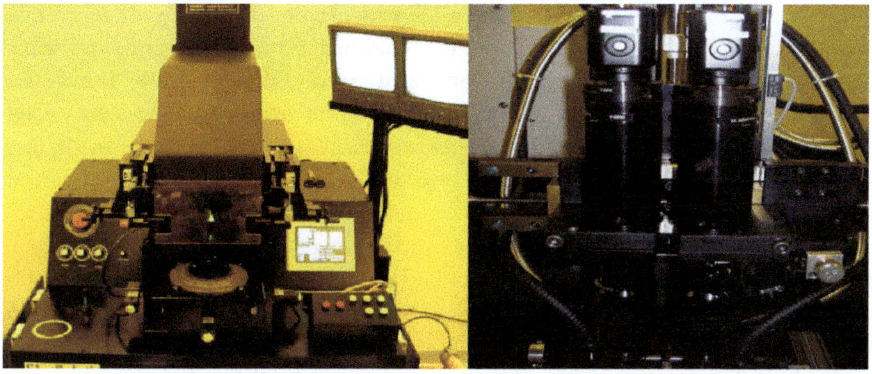

Figure 7.2 Microfabrication lithography devices for the fabrication of silicon template.

such as photolithography is attractive to produce biodevices in a variety of shapes and sizes. For fabrication of 3D *in vitro* living systems, an advanced laboratory of microfabrication facilities based on lithography technology through collaboration with semiconductor industry on the future development of biochip technology is necessary. In addition, using microfluidic channels or micromolding techniques to deposit cells and materials on specific regions of a substrate, which may also be used for microfabrication, is important. To engineer 3D living organ model, soft lithography and photolithography techniques are followed. In this approach, the first layer of cells is cultured on hydrogels following photolithography. Next step is culturing the second layer of the cells inside the empty microwell. This process will be duplicated and continued to culture the third, fourth, and other layers of the cells. By which we are able to create an *in vitro* living organs to be used as intelligent diagnostic tools. Figure 7.2 shows a complete setup microfabrication of devices for the fabrication of template on the surface of silicon to be used for soft lithography. HTS technique offers great flexibility in testing of chemical libraries. For example, various doses of the same chemical can be easily tested from the hydrogel microarrays to assess the minimum toxic dosage. This can be used in arriving at toxic dose levels without expensive and ethically challenging animal experimentation. Also, the system can be modified to conduct other high-throughput experimentation. For example, various differentiation factors at different concentrations can be printed in the chip slide, and its effects on the stem cell differentiation can be studied in a multiplexed manner. One of the significant advantages of performing HTS studies in low volumes is the ability to highly concentrate the molecules that are secreted by the cells.

7.4 Future Approaches

The elucidation of using 3D systems will open many doors and lead to significant improvements in biological tools, drug discovery process, lead identification, and therapeutic approaches.

The miniaturization of this approach allows one to perform many more experiments than previously possible in a simpler manner. 3D culture technology aims to develop set of tools that are simple, inexpensive, portable, and robust that could be commercialized and used in various fields of biomedical sciences such as drug discovery, diagnostic tools, and therapeutic approaches in regenerative medicine. The core of 3D technology is to build up interdisciplinary critical and experimental research media aimed at overcoming the fundamental limits to information processing. Significantly, it will enhance graduate education in academia and produce high-quality researchers and technicians and encourage them to pursue research careers by creating new knowledge in selected areas of focus. It will consist of a top academic and high-tech objective to fabricate, investigate, and implement novel advanced microstructures, nanostructures, and superstructures based on ordered highly functionalized materials to meet the demands of maximum efficient, active hydrogel materials of high and sustained reactivity as well as long-term stability. It will also enable new fundamental research and development for the next generation of biomedical materials as well as exploit such novel structures to develop novel biomedical devices, and transfer the knowledge to academia and industry for a future implementation of novel knowledge and technology. In turn, it should increase the competitiveness among other field of research based on fabrication, investigation, and implementation.

Combinational technology of scaffolding materials and nanotechnology is promising in the development of tissue engineering. Many researchers including our group have widely used tissue-engineered nanoscaffolds for tissue regeneration. Biochip technology in combination with nanomedicine is rapidly under development as one of the great discoveries in the history of biological sciences to be used as platform technology for application in fundamental biology, improved tools and assays, drug discovery, regenerative medicine, drug delivery, and diagnostics tools. Recently, many researchers have considered applying nanomedicine technology as a powerful method in stem cells technology to overcome many limitations in clinical application of stem cells, such as migration of stem cells after *in vivo* transplantation in the body [26]. Combinational technology of biology and engineering is emerging as an integral aspect of today's advanced medicine to provide technology for imaging, cancer treatment, medical tools, bone treatment, drug delivery, diagnostic tests, drug development, and angiogenesis that are all in the field of nanomedicine. The aim is to exploit the improved and often novel physical, chemical, and biological properties of materials at the nanometer scale. Nanomedicine has the potential to enable early detection and prevention and to essentially improve diagnosis, treatment, and follow-up of diseases. Diagnostics, targeted delivery, and regenerative medicine constitute the core disciplines of nanomedicine [27, 28].

Over the years much has been attempted in generating tissue-engineered products. One strategy for engineering 3D-engineered tissues is to cultivate cells within biodegradable scaffolds made from either natural or synthetic materials. A major challenge in 3D tissue engineering is that cells quickly lose their differentiated function. This is in contrast to the behaviors of cells in the body that have the capability to regenerate. Thus, it is desirable to formulate alternative approaches to more precisely control the organization of cells and vascularization of engineered tissues.

Traditional 3D scaffolding approaches are not suitable for generating such complex structures due to lack of control of the tissue architecture and cell–cell interactions. In particular, cells in 2D culture as well as within traditional 3D scaffolds simply do not organize as they do in normal tissue; their metabolic properties are therefore unsuitable for tissue engineering applications. Several strategies aim to make an advance in 3D models by developing the basis for fabricating tissues made from cell-laden hydrogels with engineered microvasculature. Although engineering microscale features into tissue engineering scaffolds has been attempted before, new critical plans are necessary to design 3D system to eliminate the difficulties associated with other microfabricated tissue engineering scaffolds such as uniform cell seeding and use it as a platform technology for drug discovery. Developing techniques that draws from microscale engineering, novel biomaterials, and biological principles to overcome the limitations of the current approaches to generate 3D *in vitro* living organs are critically necessary. These include the inability to generate 3D constructs that mimic the complexity of native tissue structure as well as to generate vascularized structures within 3D tissue culture system. The elucidation of the preceding goals will open many doors and lead to significant improvements in biological tools, drug discovery process, lead identification, and therapeutic approaches. The miniaturization of this approach allows one to perform many more experiments than previously possible in a simpler manner. The 3D *in vitro* technology aims to develop set of tools that are simple, inexpensive, portable, and robust that could be commercialized and used in various fields of biomedical sciences such as drug discovery, diagnostic tools, and therapeutic approaches in regenerative medicine. Also, it will consist of a top academic and high-tech objective to fabricate, investigate, and implement novel advanced microstructures, nanostructures, and superstructures based on ordered highly functionalized materials to meet the demands of maximum efficient, active hydrogel materials of high and sustained reactivity as well as long-term stability. It will also enable new fundamental research and development for the next generation of biomedical materials as well as exploit such novel structures to develop novel biomedical devices, and transfer the knowledge to academia and industry for a future implementation of novel knowledge and technology to the world. In turn, it should increase the international competitiveness of the world into a knowledge-intensive micro- and bioengineering organ-based biosensors by focusing on high-impact research to generate new breakthroughs aimed at solving significant practical problems of biomedical sciences while seeking to extend the boundaries of understanding. It aims to raise research profile of bioengineering as a vibrant center for medical and technological applications through a bottom-up approach that embraces both elements of basic and applied research to enhance the competencies in existing technologies, to seek out promising new areas, and to develop an integrated, cutting-edge research program by growing a pool of top research talent and developing the platforms on which local universities could create research breakthroughs of importance to the world. Since advancement in device development technologies is a significant indicator of the developed societies, rapidly growing market for biomedical devices provides competitive advantage in R&D and commercialization for this field.

References

1 Thomson, R.C., Wake, M.C., Yaszemski, M.J., and Mikos, A.G. (1995). Biodegradable polymer scaffolds to regenerate organs. *Adv. Polym. Sci. (Biopolymers II)* 122: 245–273.

2 Papas, K.K., Long, R.C., Sambanis, A., and Constantinidis, I. (1999). Development of a bioartificial pancreas: I. Long-term propagation and basal and induced secretion from entrapped TC3 cell cultures. *Biotechnol. Bioeng.* 66: 219.

3 Pankajakshan, D., Krishnan, V.K., and Krishnan, L.K. (2007). Vascular tissue generation in response to signaling molecules integrated with a novel poly(epsilon-caprolactone)-fibrin hybrid scaffold. *J. Tissue Eng. Regener. Med.* 1 (5): 389–397.

4 Wintermantel, E., Mayer, J., Blum, J. et al. (1996). Tissue engineering scaffolds using superstructures. *Biomaterials* 17: 83–91.

5 Edelman, E.R., Mathiowitz, E., Langer, R., and Klagsburn, M. (1991). Controlled and modulated release of basic fibroblast growth factor. *Biomaterials* 12: 619–626.

6 Iwakura, A., Tabata, Y., Tamura, N. et al. (2001). Gelatin sheets incorporating basic fibroblast growth factor enhances healing of devascularized sternum in diabetic rats. *Circulation* 104 (Suppl 1): 1325–1329.

7 Mohageri, S., Hosseinkhani, H., Ebrahimi, N.G. et al. (2010). Proliferation and differentiation of mesenchymal stem cell on collagen sponge reinforced with polypropylene/polyethylene terephathalate blend fibers. *Tissue Eng. A* 16: 3821–3830.

8 Hosseinkhani, H., Hosseinkhani, M., Hattori, S. et al. (2010). Micro and nanoscale *in vitro* 3D culture system for cardiac stem cells. *J. Biomed. Mater. Res. Part A* 94: 1–8.

9 Cai, S., Liu, Y., Shu, X.Z., and Prestwich, G.D. (2005). Injectable glycosaminoglycan hydrogels for controlled release of human basic fibroblast growth factor. *Biomaterials* 26: 6054–6067.

10 Ma, T., Li, Y., Yang, S.T., and Kniss, D.A. (2000). Effects of pore size in 3-D fibrous matrix on human trophoblast tissue development. *Biotechnol. Bioeng.* 70: 606–618.

11 Whang, K., Thomas, C.H., Healy, K.E., and Nuber, G. (1995). A novel method to fabricate bioabsorbable scaffolds. *Polymer* 36: 837–842.

12 Furong, T., Hosseinkhani, H., Estrada, G., and Kobayashi, H. (2007). Quantitative method for the analysis of cell attachment using the aligned scaffold structure. *J. Phys.* 61: 587–590.

13 Taboas, J.M., Maddox, R.D., Krebsbach, P.H., and Hollister, S.J. (2003). Indirect solid free form fabrication of local and global porous, biomimetic and composite 3D polymer-ceramic scaffolds. *Biomaterials* 24: 181–194.

14 Cai, Q., Yang, J., Bei, J., and Wang, S. (2002). A novel porous cells scaffold made of polylactic-dextran blend by combining phase-separation and particle-leaching techniques. *Biomaterials* 23: 4483–4492.

15 Hosseinkhani, H., Azzam, T., Kobayashi, H. et al. (2006). Combination of 3-D tissue engineered scaffold and non-viral gene enhance *in vitro* DNA expression of mesenchymal stem cells. *Biomaterials* 27: 4269–4278.

16 Oxley, H.R., Corkhill, P.H., Fitton, J.H., and Tighe, B.J. (1993). Macroporous hidrogels for biomedical applications: methodology and morphology. *Biomaterials* 14: 1064–1072.

17 Smith, L.A. and Ma, P.X. (2004). Nano-fibrous scaffolds for tissue engineering. *Colloids Surf., B* 10: 125–131.

18 Shugens, C., Maquet, V., Grandfils, C. et al. (1996). Biodegradable and macroporous polylactide implants for cell transplantation: 1. Preparation of macroporous polylactide supports by solid–liquid phase separation. *Polymer* 37 (6): 1027–1038.

19 Lindstrom, S., Iles, A., Persson, J. et al. (2010). Nanoporous titania coating of microwell chips for stem cell culture and analysis. *J. Biomech. Sci. Eng.* 5: 272–279.

20 Mikos, A.G., Sakarinos, G., Vacanti, J.P., et al. (1996). Biocompatible polymer membranes and methods of preparation of three dimensional membrane structures. Patent number: 5514378, issued 7 May 1996.

21 Tian, F., Hosseinkhani, F., Hosseinkhani, M. et al. (2008). A quantitative analytical method to measure cell adhesion on aligned micro- and nanofibers. *J. Biomed. Mater. Res. A* 84: 291–299.

22 Hosseinkhani, H., Hosseinkhani, M., Khademhosseini, A. et al. (2008). DNA nanoparticles encapsulated in 3-D tissue engineered scaffold enhance osteogenic differentiation of mesenchymal stem cells. *J. Biomed. Mater. Res. A* 85: 47–60.

23 Hosseinkhani, H. and Hosseinkhani, M. (2009). Tissue engineered scaffolds for stem cells and regenerative medicine. In: *Trend in Stem Cells Biology and Technology*, 367–387. USA: HUMANANA Press Inc.

24 Nikpour, M., Chaouk, H., Mau, A. et al. (1999). Porous conducting membranes based on polypyrrole-PMMA composites. *Synth. Met.* 99: 121–126.

25 Cavallini, A., Notarnicola, M., Berloco, P. et al. (2000). Use of macroporous polypropylene filter to allow identification of bacteria by PCR in human fecal simples. *J. Microbiol. Methods* 39: 265–270.

26 Mahmoudi, M., Hosseinkhani, H., Hosseinkhani, M. et al. (2011). Magnetic resonance imaging tracking of stem cells *in vivo* using iron oxide nanoparticles as a tool for the advancement of clinical regenerative medicine. *Chem. Rev.* 111: 253–280.

27 Hosseinkhani, H. (2011). Editorial: on Nanomedicine. *Int. J. Nanotechnol.* 8: 615–617.

28 Sarabi, R.S., Sadeghi, E., Hosseinkhani, H. et al. (2011). Polyrotaxane capped quantum dots as new candidates for cancer diagnosis and therapy. *J. Nanostruct. Polym. Nanocompos.* 7: 18–31.

8

Regenerative Medicine

8.1 Introduction

Regenerative medicine is an interdisciplinary field that combines engineering and live sciences in order to develop techniques that enables the restoration, maintenance, or enhancement of living tissues and organs. Its fundamental aim is to create natural tissue with the ability to restore missing organ or tissue function, which the organism has not been able to regenerate in physiological conditions. With that, it aspires to improve the health and quality of life for millions of people worldwide and to give solution to the present limitations: rejections and low quantity of donors [1]. The term regenerative medicine is often used synonymously with tissue engineering, although those involved in regenerative medicine place more emphasis on the use of stem cells to produce tissues. Clinical trials on humans could begin in as little as four years, said the company, which hopes to help supply an estimated US$6 billion market for organs and a similar market for cellular therapies to treat diseases such as diabetes. Everyday thousands of people of all ages are admitted to hospitals because of the malfunction of some vital organ. Because of a dearth of transplantable organs, many of these people will die. In perhaps the most dramatic example, the American Heart Association reports only 2300 of the 40 000 Americans who needed a new heart in 1997 got one. Lifesaving livers and kidneys likewise are scarce, as is skin for burn victims and others with wounds that fail to heal. The ever-growing demand for donor organs to meet the needs of individuals on waiting lists will likely never be met. While roughly 100 000 people have transplants in the United States, more than 10 million have implants. There are 20 000 transplants annually, but 2 million implants. In short, the need for organs cannot be met by traditional methods of transplantation. Regenerative medicine and tissue engineering may change that.

Tissue engineering is an interdisciplinary field that applies principles and methods of engineering toward the development of biological substitutes to improve the function of damaged tissue and organs [2]. The motivation of using of tissue engineering in regenerative medicine is due to the following: (i) since 1970s, organ transplantation has become a common therapeutic approach for end-stage organ failure patients; (ii) demand $\gg$ supply (UNOS National Patient Waiting List), for example, 19 095 patients (1989) and 80 766 patients (December 2002); and (iii) cost of organ replacement therapy: US$305 billion (US, 2000). The interdisciplinary approach of

Biomedical Engineering: Materials, Technology, and Applications, First Edition. Hossein Hosseinkhani.

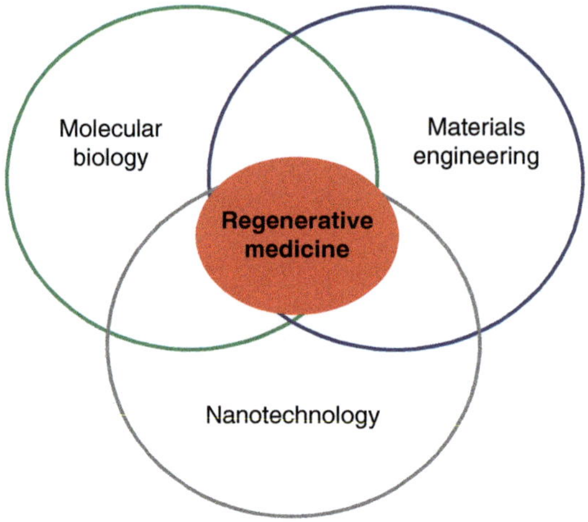

Figure 8.1 Interdisciplinary approach of tissue engineering.

tissue engineering is combinational technology of using molecular biology, materials engineering, and reconstructive surgery [3]. Tissue engineering needs scaffolds to serve as a substrate for seeding cells and as a physical support in order to guide the formation of the new tissue and is designed to regenerate natural tissues or to create biological substitutes for defective or lost organs by using cells [4–8]. There is no doubt that a sufficient supply of nutrients and oxygen to the transplanted cells is vital for their survival and functional maintenance [9]. Mesenchymal stem cells (MSCs) have known as one of the most cells used in tissue engineering and rapidly improved by several researches to evaluate their therapeutic applications [1]. Figure 8.1 illustrates the interdisciplinary approach of tissue engineering.

8.2 Tissue-Engineered Scaffolds

Material design of scaffold for cell proliferation and differentiation is one of the key technologies for tissue engineering. In conventional cell culture such as static tissue culture dish (two-dimensional [2D]), the initial rate of cell growth is higher, but the proliferation stop once the cells reached confluence. Porous materials with three-dimensional (3D) structures have been investigated for the cell scaffold because they have larger surface for cell attachment and proliferation than 2D materials and are preferable to assist the formation of 3D cell constructs that may resemble the structure and function of body tissues. In addition, the 3D scaffold also plays an important role in the substrate for *in vitro* cell culture to increase the number of cells as high as clinically applicable. 3D scaffolds able of regenerating or restoring tissue and/or organs have begun to revolutionize medicine and biomedical science. Scaffolds have been used to support and promote the regeneration of tissues. Different processing techniques have been developed to design and fabricate 3D scaffolds for tissue engineering implants. However, there is not a simple, inexpensive method that fulfills the main characteristics that a scaffold should have for application in tissue engineering.

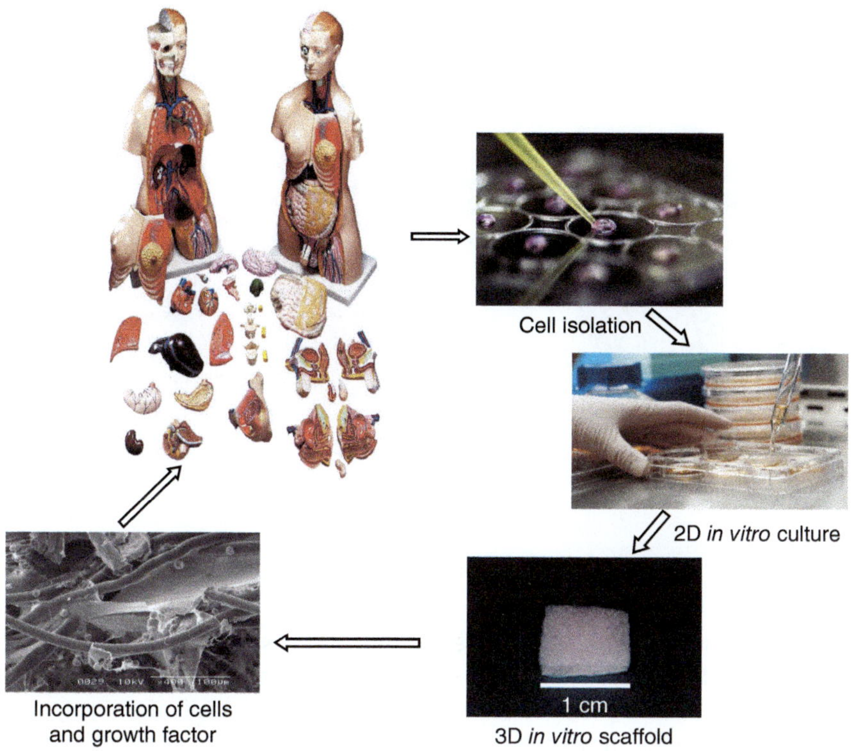

Figure 8.2 General concepts of tissue engineering approaches toward tissue/organ regeneration.

The proliferation of cells in the 3D scaffold needs oxygen and nutrition supply. In this circumstance, the 3D scaffold materials should provide such an environment for cells living in distance. Diffusion of nutrients, bioactive factors, and oxygen through 3D scaffolds are sufficient for survival of large numbers of cells for extended periods of time. A major constraint of biodegradable polymer scaffolds for vascular tissue engineering is poor cell adhesion and lack of signals for new tissue generation. The presence of ECM within the scaffold is desirable for growth of stem cells and *in vitro* formation of remodeled vascular conduit [10]. Tissue regeneration can be achieved by the following three key steps: cell proliferation, cell seeding in a suitable scaffold, and the maintenance of the differentiation phenotype of the engineered tissues [11].

Figure 8.2 shows the general concept of tissue engineering approaches toward tissue/organ regeneration.

The property of scaffold material for cell attachment is one of the major factors contributing their morphology, proliferation, functions, and the subsequent tissue organization [12–19]. Ideally, a polymeric scaffold for tissue engineering should have the following characteristics: (i) appropriate surface properties promoting cell adhesion, proliferation, and differentiation; (ii) biocompatible; (iii) highly porous, with a high surface area/volume ratio, with an interconnected pore network for cell growth and flow transport of nutrients and metabolic waste; and (iv) mechanical properties

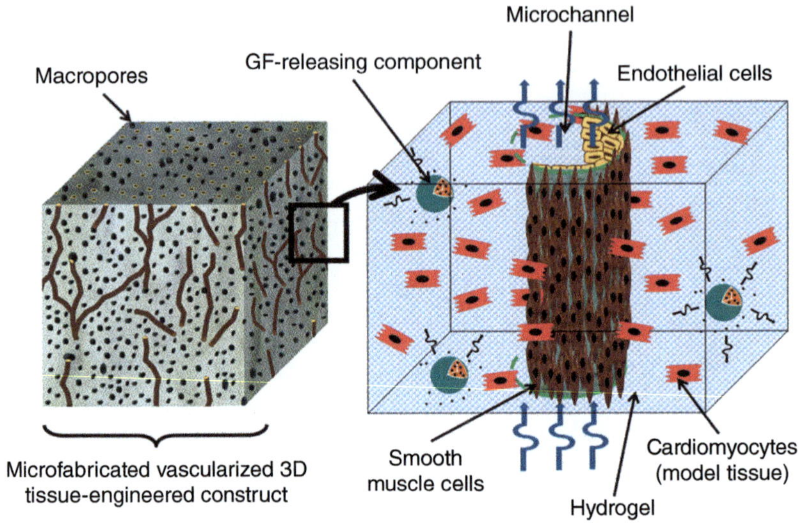

Figure 8.3 3D hydrogel system in tissue-engineered constructs.

sufficient to withstand any *in vivo* stresses [20–24]. The last requisite is difficult to combine with the high porosity in volume of the material. Thus, it is necessary to use polymeric matrices with special or reinforced properties, especially if the polymer is a hydrogel. Figure 8.3 shows the ideal 3D hydrogel system in tissue-engineered constructs.

On basis of the extensive range of polymeric materials, different processing techniques have been developed to design and fabricate 3D scaffolds for tissue engineering implants [25–33]. These include (i) phase separation, (ii) gas foaming, (iii) fiber bonding, (iv) photolithography, (v) solid free-form (SFF), (vi) and solvent casting in combination with particle leaching [29, 34–46]. However, none of these techniques have achieved a suitable model of 3D architecture, so the scaffolds must be fulfilled using equipment with high cost, even for the reasons that will be discussed next. So, using phase separation, a porous structure can be easily obtained by adjusting thermodynamic and kinetic parameters. However, because of the complexity of the processing variables involved in phase-separation technique, the pore structure cannot be easily controlled. Moreover, it is difficult to obtain large pores due to a lack of interconnectivity. Gas foaming has the advantage of room temperature processing but produces a largely nonporous outer skin layer and a mixture of open and closed pores within the center, leaving incomplete interconnectivity. The main disadvantage of the gas foaming method is that it often results in a non-connected cellular structure within the scaffold [34]. Fiber bonding provides a large surface area for cell attachment and a rapid diffusion of nutrients in favor of cell survival and growth. However, these scaffolds, as the ones used to construct a network of bonded polyglycolic acid (PGA), lacked the structural stability necessary for *in vivo* use. In addition, the technique does not lend itself to easy and independent control of porosity and pore size [32, 33].

Photolithography has also been employed for patterning, obtaining structures with high resolution, although this resolution may be unnecessary for many applications of patterning in cell biology. In any case, the disadvantage of this technique is the high cost of the equipment that limits their applicability [36]. SFF scaffold manufacturing methods provide excellent control over scaffold external shape and internal pore interconnectivity and geometry but offer limited microscale resolution. Moreover, the minimum size of global pores is 100 μm. Additionally, SFF requires complex correction of scaffold design for anisotropic shrinkage during fabrication and needs high cost equipment. Finally, solvent casting in combination with particulate leaching method, which involves the casting of a mixture of monomers and initiator solution and a porogen in a mold, polymerization, and followed by leaching out of the porogen with the proper solvent to generate the pores, is inexpensive but still has to overcome some disadvantages in order to find engineering applications, namely, the problem of residual porogen remains, irregular-shaped pores, and insufficient interconnectivity [36]. The proposed scaffolds may find applications as structures that facilitate either tissue regeneration or repair during reconstructive operations. The new structure could also find applications in other areas in which the pore morphology plays an essential role, such as membranes and filters [35, 40].

8.3 Specific Mediated Biomaterials

The design of materials that can regulate cell behavior such as proliferation and differentiation is a key component for the fabrication of tissue engineering scaffolds. From the viewpoint of immune system response of the body, the implanted biomaterials should mimic the structure and biological function of native ECM, in terms of both chemical composition and physical structure as reported by Ma et al. [47]. Therefore, in order to mimic the biological function of ECM proteins, the scaffold materials used in tissue engineering need to be chemically functionalized to promote tissue regeneration as ECM does. Collagen and elastin as ECM proteins are made from fibers in dimension smaller than micrometers [47]. It seems that artificial nanoscaled fibers have great potential application in the field of biomaterials and tissue engineering.

The initial report showed that nanoscaled features influenced cell behaviors [48]. Nanoscaled surface topography has been found to promote osteoblast adhesions [49]. It has been demonstrated that osteoblast adhesion, proliferation, alkaline phosphatase (ALP) activity, and ECM secretion on carbon nanofibers increased by decreasing fiber diameter in the range of 60–200 nm, whereas the adhesion of other kinds of cells such as chondrocytes, fibroblasts, and smooth muscle cells was not influenced [44, 50]. It has been supposed that the nanoscaled surface affects the conformation of adsorbed adhesion proteins such as vitronectin, thus affecting the cell behaviors [51]. In addition, the nanoscaled dimensions of cell membrane receptors such as integrins should also be considered. There are three different approaches toward the formation of nanofibrous materials: phase separation, electrospinning, and self-assembly [52]. Phase separation and self-assembling of biomolecules can generate smaller diameter nanofibers in the same range of natural ECM, while electrospinning generate large diameter nanofibers on the upper end of the range of natural ECM [47]. Electrospinning is a common technique used to fabricate tissue engineering scaffolds. It is an easy technique, is cheap, and can be applied for many different types of polymers. Recent study demonstrated that fabricated PGA/collagen nanofibers through electrospinning significantly enhanced cell adhesion compared with PGA/collagen microfibers [53].

One of the common approaches to produce fibers similar to ECM proteins such as collagen is self-assembly. It has been shown that peptide amphiphile (PA) that contain a carbon alkyl tail and several other functional peptide formed nanofibers through self-assembly by mixing cell suspensions in media with dilute aqueous solutions of the peptide [54]. These self-assembled nanofibers have been used recently to study selective differentiation of neural progenitor cells [55]. Another type of peptide containing 16 alternating hydrophobic and hydrophilic amino acids was fabricated to self-assemble into nanofibers under appropriate pH values [56]. Nanoscaled fibers produced by self-assembly of amphiphilic peptide may be a promising approach in designing the next generation of biomaterials for drug delivery and tissue engineering.

It would be beneficial for biomedical applications if scaffold materials could promote the adhesion and growth of cells on their surfaces. The sequence of arginine–glycine–aspartic acid (RGD) has been discovered as a cell attachment sequence in various adhesive proteins present in the ECM and found in many proteins, such as fibronectin, collagen type 1, vitronectin, fibrin, and von Willebrand factor [57]. It has been recognized that the sequence of RGD interacts with various types of integrin receptors of mammalian cells. Ever since the RGD sequence has been discovered as a cell attachment sequence in adhesive proteins of the ECM, there have been several efforts to synthesize bioactive peptides incorporating RGD for therapeutic purpose [58]. Micro- and nanopatterned scaffolds have been investigated less well in regard to stem cells, although two recent studies highlight their attractiveness [59]. In their study, Silva et al. included a five amino acid, laminin-specific cell-binding domain (which binds to specific integrins on cell surface) at the hydrophilic head of their amphiphilic structures and showed that neural stem cells could be induced to differentiate into neurons when cultured within peptide gel [60]. In contrast, cells grown in control scaffolds without the laminin-specific domain or on 2D tissue culture plastic coated with laminin solution differentiated much less. This was hypothesized to be largely as a result of the density of the cell-binding ligands to which the cells were exposed, indicating clearly the importance of extracellular matrix in influencing cell function. Recent studies have indicated that when the laminin-specific domain in the amphiphilic molecule was replaced with the amino acid sequence, arginine–glycine–aspartate (RGD), a common cell-binding domain in many extracellular matrix proteins, especially collagen, differentiation of MSC to osteoblasts was significantly enhanced compared with amphiphilic nanofibers without this sequence or to 2D controls. This is because the interaction of MSC integrins receptors with RGD of the peptide enhanced cell attachment on peptide nanofibers. The proliferation of cells in the 3D scaffold needs oxygen and nutrition supply. In this circumstance, the 3D scaffold materials should provide such an environment for cells. The artificial scaffolds formed by self-assembling molecules not only provides a suitable support for cells proliferation but also serves as a medium through which diffusion of soluble factors and migration of cells can occur. The result of the cell attachment and proliferation revealed that diffusion of nutrients, bioactive factors, and oxygen through these highly hydrated networks is sufficient for survival of large numbers of cells for extended periods of time.

8.3.1 Self-Assembled Systems in Regenerative Medicine Therapy

For successful tissue regeneration, the cells constituting tissue to be regenerated, such as matured, progenitor precursor, are necessary. Considering the proliferation activity and differentiation potential of cells, stem cells are practically promising. Among them, MSC have been widely investigated to be used by themselves or combining with scaffolds necessary for promotion of cell proliferation and differentiation. It was found that MSC have an inherent nature to differentiate into not only osteogenic linage cells but also chondrogenic, myogenic, adipogenic, and neurogenic lineages [60–65]. MSCs have been experimentally used to demonstrate their *in vivo* potential to induce the regeneration of mesenchymal tissues [66–70]. Since it is reported that the cells are effective in inducing the regeneration of tissues other than mesenchym, their feasibility in the cell source for regenerative medicine is highly expected. They are practically isolated from patient's themselves [71–74].

Another use of self-assembled peptides would be in tissue engineering. Considering the usage of cells in the body, it is no doubt that a sufficient supply of nutrients and oxygen to the transplanted cells is vital for their survival and functional maintenance [72]. Without a sufficient supply, only a small number of cells pre-seeded in the scaffold or migrated into the scaffold from the surrounding tissue would survive. Rapid formation of a vascular network at the transplanted site of cells must be a promising way to provide cells with the vital supply. This process of generating new microvasculature, termed neovascularization, is a process observed physiologically in development and wound healing [75]. Basic fibroblast growth factor (bFGF) has been shown to promote such an angiogenesis process [75, 76]. The growth factors stimulate the appropriate cells (e.g. endothelial cells), already present in the body, to migrate from the surrounding tissue, proliferate, and finally differentiate into blood vessels [75]. However, one cannot always expect the sustained angiogenesis activity when these proteins are only injected in the solution form probably because of their rapid diffusional excretion from the injected site. One possible way for enhancing the *in vivo* efficacy is to achieve its controlled release over an extended time period by incorporating the growth factor in a polymer carrier. If this carrier is biodegraded, harmonized with tissue growth, it will work as a scaffold for tissue regeneration in addition to a carrier matrix for the growth factor release.

8.3.2 Cartilage Regeneration and Biomineralization in Self-Assembled Systems

As understood from the findings, proteins and peptides can self-assemble into various structures like nanotubes, nanovesicles, and 3D peptide matrices with interwoven nanofibers. Macroscopic 3D peptide matrices can be engineered to form various shapes by changing the peptide sequence. Self-assembled peptide materials encouraged cell proliferation and differentiation. These peptide materials were also able to support various types of cell attachments. The ability of the peptides to support attachment of mouse neuronal cells has been fully studied [55]. The primary mouse neuron cells form active connections with the peptide scaffolds that form a valuable area of research for studying about neuron regeneration. In regenerative medicine, these peptide matrices are used to cultivate chondrocyte extracellular matrix that can be used to repair cartilage tissue. Cartilage tissue engineering has been performed by placing the primary chondrocytes and MSCs into these self-assembled peptide hydrogels to produce collagen and glycosaminoglycans. These peptide matrices can also be used in regeneration of bone by incorporating a phosphorylated serine that can attract and organize calcium ions to form hydroxyapatite crystals and functionalizing them with a cell adhesion motif like arginine–glycine–aspartic acid complex. The research studies are not limited only to natural amphiphilic peptides. There are many research trails that have indicated on the synthesizing of complex amphiphilic peptides by joining hydrophilic peptides into long alkyl chains. The peptide end of the molecule was designed to function and regulate biomineralization. For example, bone is produced as a result of deposition of calcium and phosphate ions to form hydroxyapatite crystals. This process is known as mineralization. Serine (nonessential amino acid) is incorporated with the synthetic amphiphilic peptide complex; it serves to attract and organize calcium and phosphate ions to form hydroxyapatite crystals. Furthermore, adding a cell adhesion motif has functionalized the synthetic amphiphilic peptides. RGD has shown to attach to the C-terminal of the peptide. This can be used to study the ability of the bone cells to differentiate, proliferate, and adhere to a biomaterial surface like titanium. Titanium is the most widely used biomaterial surface to produce orthopedic implants, dental implants, and hip replacements. Despite of its excellent biocompatibility, titanium implants still fail. Most orthopedic implants have a lifetime of fifteen years to the maximum. In order to produce a newer version of titanium implants that can stay in the body for a longer period of time, its surface has to be modified with nanosized surface patterns so that bone cells (osteoblasts) differentiate and migrate into these patterns for better bone-implant adhesion. For such a purpose, these synthetic amphiphiles can be used to regulate and control the osteoblasts.

8.3.3 Angiogenesis in Self-Assembled Systems

Tissue engineering is designed to regenerate natural tissues or to create biological substitutes for defective or lost organs by making use of cells. Considering the usage of cells in the body, it is no doubt that a sufficient supply of nutrients and oxygen to the transplanted cells is vital for their survival and functional maintenance. Without a sufficient supply, only a small number of cells pre-seeded in the scaffold or migrated into the scaffold from the surrounding tissue would survive. Rapid formation of a vascular network at the transplanted site of cells must be a promising way to provide cells with the vital supply. This process of generating new microvasculature, termed neovascularization or angiogenesis, is a process observed physiologically in development and wound healing. It is recognized that bFGF functions to promote such an angiogenesis process. The growth factors stimulate the appropriate cells (e.g. endothelial cells), already present in the body, to migrate from the surrounding tissue, proliferate, and finally differentiate into blood vessels. However, one cannot always expect the sustained angiogenesis activity when these proteins are only injected in the solution form probably because of their rapid diffusional excretion from the injected site. One possible way for enhancing the *in vivo* efficacy is to achieve its controlled release over an extended time period by incorporating the growth factor in a polymer carrier. If this carrier is biodegraded, harmonized with tissue growth, it will work as a scaffold for tissue regeneration in addition to a carrier matrix for the growth factor release.

Some studies have demonstrated that bFGF promoted angiogenesis when used in combination with delivery matrices and scaffold [16, 77–82]. A recent study has indicated that a 3D network of self-assembled nanofibers was formed by mixing of bFGF suspension with aqueous solution of peptide amphiphile as an injectable carrier for controlled release of growth factors. The feasibility of prevascularization by the bFGF release from the 3D networks of nanofibers has been tested [83]. Previous works have encapsulated bFGF within alginate, gelatin, agarose/heparin, collagen, and poly(ethylene-*co*-vinyl acetate) carriers [81, 83, 84]. According to the results of these studies, it is conceivable to incorporate the angiogenic factor to a sustained releasing system prior to the implantation. The bFGF-incorporated releasing system requires surgery for implantation, which is not welcomed. In contrary, the bFGF incorporated in self-assembled peptide could be delivered to living tissues by simply injecting peptide amphiphile and bFGF solutions. The injected solutions would form hydrogel at the injected site of tissue and the release bFGF induced significant angiogenesis around the injected site, in marked contrast to bFGF injection, or peptide amphiphile injection alone [83]. *In vitro* release profile indicates that the prolonged release of bFGF is continued for 750 hours.

When the release was halted, approximately 90% of total loaded protein had been released. However, it is possible to increase the cumulative number of released molecules by increasing the concentration of the loading solution because the amount of loaded protein may be increased as the concentration of loading solution increases. This result is attributed to an increasing driving force, i.e. concentration difference for protein diffusion. However, the type of interaction forces acting between bFGF and peptide amphiphile molecules is not clear at present. *In vivo* degradation rate of self-assembled peptide amphiphile nanofibers and *in vivo* release profiles of bFGF were estimated in terms of the radioactivity loss of ^{125}I-labeled peptide amphiphile and ^{125}I-labeled bFGF. Therefore, injection of radioiodinated peptide amphiphile and bFGF into the mouse subcutis was carried out to evaluate the time profile of radioactivity remaining *in vivo*. The peptide amphiphile was degraded with time in the body although the degradation rate was slow. The radioactivity of self-assembled peptide amphiphile nanofibers incorporating ^{125}I-labeled bFGF decreased with time although the bFGF radioactivity was retained for longer time periods than that of free ^{125}I-labeled bFGF injection. The decrement order and pattern of bFGF and peptide amphiphile radioactivity reveals that the *in vivo* release rate of bFGF was faster than the *in vivo* degradation rate of peptide amphiphile. Taken together, the results of *in vitro* and *in vivo* release profile indicate that bFGF was released from self-assembled peptide amphiphile nanofibers in the body as a result of combination of diffusion and degradation mechanism. The bFGF was originally characterized *in vitro* as a growth factor for fibroblasts and capillary endothelial cells and *in vivo* as a potent mitogen and chemoattractant for a wide range of cells. The most important of concern regarding the delivery of proteins is whether or not the protein released in the body actually retains its biological activity. To evaluate protein activity, *in vitro* culture techniques are normally employed because of their simplicity and convenience, compared with *in vivo* animal experiments. However, any *in vitro* non-degradation system cannot be applied to evaluate the biological activity of released bFGF. Thus, to obtain information on the retention of bFGF activity, vascularization is directly assessed after subcutaneously injection of peptide amphiphile with bFGF in animals. Histological examination demonstrated that vascularization is remarkable around the injection site of self-assembled peptide amphiphile nanofibers incorporated bFGF, in contrast to sites injected with an aqueous solution of bFGF. Injection of bFGF in the form of a solution is not effective in inducing vascularization at all and injection of bFGF-free peptide amphiphile alone does not induce any vascularization effect. This is in good accordance with the results of immunofluorescence analysis. The result also demonstrated that the number of smooth muscle cells of capillaries' wall as well as endothelial cells of vessels' wall greatly enhanced when release of bFGF is controlled by peptide amphiphile. The prolonged period of angiogenesis by the bFGF-incorporated self-assembled peptide amphiphile nanofiber is ascribed to the prolonged time period of bFGF release. The amount of tissue hemoglobin, which is measured by bFGF-induced neovascularization, notably increased within one day of injection of peptide amphiphile incorporated with bFGF, and the increased level is retained for several days, followed by a slow return over the time range studied. On the other hand, injection of an aqueous solution containing the same dose of bFGF, as a bFGF incorporated in self-assembled peptide amphiphile nanofibers, did not increase the amount of hemoglobin at the injection site over the time range studied; the level of tissue hemoglobin remained at approximately the same level as that found on injection of bFGF-free peptide amphiphile or in untreated mice. No increase in the amount of hemoglobin is observed even when the amount of bFGF in solution that was injected is increased to $10\,\mu g\,\mu l^{-1}$. This must be due to a rapid elimination of bFGF from the injection site. In contrast, the bFGF incorporated in self-assembled peptide amphiphile nanofibers enabled us to reduce the dose that was effective in inducing significant vascularization to $0.2\,\mu g\,\mu l^{-1}$. This finding strongly suggests that the bFGF-incorporated self-assembled peptide amphiphile nanofibers still maintain its biological activity even though exposed to an *in vivo* environment. It is highly possible that the slow degradation of the bFGF-incorporated self-assembled peptide amphiphilic nanofibers achieves a longer period of bFGF release, resulting a prolonged angiogenesis effect. The *in vivo* degradation profile of self-assembled peptide amphiphile nanofibers indicates that the retention period of the self-assembled peptide amphiphile nanofibers-induced vascularization effect is shorter than the degree of degradation of peptide amphiphile. bFGF seems to be released from self-assembled peptide amphiphile as a result of diffusion mechanism. The enhanced vascularization is due to the sustained release of bFGF.

They hypothesized that self-assembly hydrogel comprising peptide amphiphile and VEGF can be used to fabricate tissue engineering scaffolds for angiogenesis. To test this hypothesis, 3D hydrogel of self-assembled peptide amphiphile were fabricated by mixing VEGF suspension with aqueous solution of peptide amphiphile as an injectable carrier for controlled release of growth factors. We demonstrate the feasibility of this approach to induce angiogenesis by investigating that the release kinetics of VEGFrelease from the 3D hydrogel improves angiogenesis. An injectable hydrogel can be formed by mixing of a peptide aqueous solution with VEGF suspension. *In vitro* release profile of VEGF from peptide gel and *in vivo* degradation of peptide were investigated through radioactivity remaining of radio-labeled VEGF and peptide injected subcutaneously to the back of rat. In addition, angiogenesis induced by the released VEGF was assessed in a rat model using histological and immunological examinations. It was demonstrated that the injection of aqueous solution of peptide together with VEGF into the back of rats resulted the formation of a transparent 3D hydrogel at the injected site and induced significant angiogenesis around the injected site, in marked contrast to VEGF injection alone or peptide injection alone. The comparison of *in vitro* degradation of peptide with *in vivo* release profile of VEGF indicated that the govern mechanism of controlled release of VEGF was through degradation of peptide in the body.

8.3.4 Bone Regeneration in Self-Assembled Systems

Bone defects and fracture non-union are common problems, affecting as many as thousand patients in the world every year, and are difficult to heal using current therapies. Previously, these cases have been treated by surgery, using techniques such as autologous bone grafting or artificial bone grafting. It has been reported that bone morphogenetic proteins (BMPs), transforming growth factor-β (TGF-β), and bFGF can induce bone formation in both ectopic and orthotopic sites *in vivo* [85, 86]. However, the use of these proteins alone requires large amounts of protein because of its short half-life, and one cannot always expect the sustained biological activity when these proteins are only injected in the solution form probably because of their rapid diffusional excretion from the injected site. One possible way for enhancing the *in vivo* efficacy is to achieve its controlled release over an extended time period by incorporating the growth factor in a polymer carrier. It is known that many growth factors in the body have short half-life lives. To overcome this limitation, growth factors have been encapsulated within different types of polymeric carriers. A potential limitation of the previously developed systems is that they require surgery for implantation. Injectable self-assembled peptides incorporated with osteoinductive growth factors is superior compared with implantable scaffolds. BMP-2-incorporated self-assembled peptide amphiphile was found to be useful for growth factor release [85]. It has been hypothesized that self-assembly hydrogels comprising peptide amphiphile and BMP-2 can be used to fabricate tissue engineering scaffolds to induce ectopic bone formation [85]. To test this hypothesis, 3D networks of self-assembled peptide amphiphile nanofibers were fabricated by mixing BMP-2 suspension with aqueous solution of peptide amphiphile as an injectable carrier for controlled release of growth factors. They demonstrated the feasibility of this approach to induce ectopic bone formation by showing that BMP-2released from the 3D networks of nanofibers enhanced ectopic bone formation.

The injected solutions of peptide and TGF-β formed a transplant gel, and the sustained release of TGF-β induced significant ectopic bone compared with TGF-β injection. As a flexible delivery system, these scaffolds can be adapted for sustained release of many different growth factors and biomolecules. The ALP activity increased rapidly and saturated at three weeks, while the temporal changes in the osteocalcin (OCN) content increased steadily with time, which was in good accordance with the course of bone formation in the subcutaneous tissue. TGF-β-incorporated self-assembled PA nanofibers significantly increased both the ALP and OCN levels compared with free TGF-β injection. ALP activity is an ectoenzyme, produced by osteoblasts, which is likely to be involved in the degradation of inorganic pyrophosphate to provide a sufficient local concentration of phosphate or inorganic pyrophosphate for mineralization. Therefore, ALP is a useful marker for osteoblast activity. OCN, also known as bone Gla protein, is a highly conserved noncollagenous protein that contains three γ-carboxyglutamic acid residues that allow it to bind calcium. Although the function of OCN is not quite clear, it is well recognized that only osteoblasts or cells with osteoblastic nature produce OCN. OCN is already known to play an important role in the process of ossification for bone formation. Like ALP, OCN is also selected as a marker of osteogenic differentiation [87–92].

One important approach for bone healing and bone ingrowth is the use of growth factors. Some of these factors, e.g. bone morphogenic proteins (BMPs) [93, 94], TGF-β [95, 96], insulin-like growth factor (IGF), and bFGF, act as local regulators of cellular activity and seem to have osteoinductive and angiogenetic potential [97–103]. The growth factor of bFGF has been previously used as angiogenesis factor. Interestingly, the same group has used bFGF as osteoinductive growth factors to induce bone at skull defect of rabbit [104]. bFGF is a growth factor to trigger proliferation of capillary endothelial cells in addition to osteoblast. It has been reported that bFGF stimulated the cells involved in osteogenesis and angiogenesis in a vascularized bone graft [105]. Although bFGF is not a strong osteoinductive growth factor compared with other growth factors such as BMP-2, the present study indicates that the incorporation of bFGF in self-assembled peptides nanofibers as an injectable tissue-engineered scaffold is promising technique to induce significant bone at the injected site.

8.3.5 Nerve Regeneration in Self-Assembled Systems

Ellis-Behnke et al. used self-assembled peptide as 3D scaffold for neural regeneration medicine [106]. In their animal model, they first made brain wounds defect that was carried out using postnatal day-2 Syrian hamster pups. The experimental groups of animals were then treated by injection into the wound of 10–30 ml of 1% RADA16/99% water. Histological observation showed that only in the peptide scaffold-injected animals, the brain tissue appears to have reconnected itself together in all survival times. In addition, axons labeled from their retinal origin with a tracer molecule were found to have grown beyond the tissue bridge. Also, functional tests proved a significant restoration of visual function in all peptide scaffold-treated animals. In 2011, Liang's group succeeded to speed up the process of central nervous system recovery after injury that is needed for real-time measurement of axon regeneration to assess the extent of injury, as well as the optimal timing and delivery of therapeutics and rehabilitation [107]. In their study, they developed a chronic animal model with an *in vivo* measurement technique to provide a real-time monitoring and feedback system. They showed a successful chronic injury model to measure central nervous system regeneration, combined with an *in vivo* measurement system to provide real-time feedback during every stage of the regeneration process. Their results indicated that a chronic optic tract lesion was able to heal, and axons were able to regenerate, when treated with a self-assembling nanofiber peptide scaffold. In a recent study developed by researchers from Sichuan University in China, Meng et al. [108] showed that earlier studies have demonstrated that peptide RADA16-I (AcN-RADARADARADARADA-CONH$_2$) could repair spinal cord injury and optical pathway and restore visual function. The objective of their research was to investigate the role of RADA16-I in the regeneration of peripheral nerve injury in rats. Sciatic nerve injury was performed at female Sprague–Dawley rats. Two groups were conducted: crush group and sever group. There were RADA16-I treatment and blank control in each group. Sense nerve conduction parameters were evaluated on day 30 in the injured sciatic nerve by means of electrophysiological recording. Their results indicated that sense nerve conduction in test groups showed a faster healing rate than blank control. In another interesting work developed by the department of neurology, Northwestern University in the United States, Tysseling et al. [109] showed that injection into the injured spinal cord of peptide amphiphile (PA)-conjugated isoleucine–lysine–valine–alanine–valine (IKVAV) peptide mimicking laminin structure supports of the neural extracellular matrix improved functional recovery after spinal cord injury in two different species, rat and mouse, and in two different injury models, contusion and compression. The results clearly showed that the improvement required the IKVAV epitope and was not observed with the injection of an amphiphile displaying a nonbioactive sequence. They also showed that the IKVAV PA-injected groups also trended higher both in the total number of neurons adjacent to the lesion and in the number of long propriospinal tract connections from the thoracic to the lumbar cord. IKVAV PA injection did not alter myelin thickness, total axon number caudal to the lesion, axon size distribution, or total axon area.

8.3.6 Cardiac Regeneration in Self-Assembled Systems

Stem cells have shown great interest in cardiac tissue engineering. In a work developed by Lee et al. [110], embryonic stem cells were suspended in RADA16-II peptide scaffold solutions and injected in the myocardium of 10-week old mice.

They injected self-assembling peptides into the myocardium to create a 3D microenvironment. After 7, 14, and 28 days, these microenvironments recruited both endogenous endothelial and smooth muscle cells, and exogenously injected cells survived in the microenvironments. They concluded that self-assembling peptides created injectable microenvironments that promoted vascularization. In another work, they also developed drug delivery system by using a biotinylated version of RADA-II to demonstrate a slow release of IGF-1 in infatuated rat myocardia [111]. The biotin sandwich strategy allowed binding of IGF-1 and did not prevent self-assembly of the peptides into nanofibers within the myocardium. In conjunction with cardiomyocytes transplantation, they showed that cell therapy with IGF-1 delivery by biotinylated nanofibers significantly improved systolic function after experimental myocardial infarction. In an interesting study, Davis's group hypothesized that a novel approach to promote vascularization would be to create injectable microenvironments within the myocardium that recruit endothelial cells and promote their survival and organization [110]. They demonstrated that self-assembling peptides could be injected and that the resulting nanofiber microenvironments were readily detectable within the myocardium. Furthermore, the self-assembling peptide nanofiber microenvironments recruited progenitor cells that express endothelial markers, as determined by staining with isolectin and for the endothelial-specific protein platelet–endothelial cell adhesion molecule-1. Vascular smooth muscle cells were recruited to the microenvironment and appear to form functional vascular structures. In their animal model (adult, male C57BL/6 mice aged 8–10 weeks), after tracheal intubation, the hearts were exposed by separation of the ribs. The peptide gel (10 μl, RAD16-II peptide [AcN-RARADADARARADADA-CNH_2]) was injected into the free wall of the left ventricle (LV) through a 30-gauge needle while the heart was beating. After injection, the chests were closed, and animals were allowed to recover under a heating lamp. After the endothelial cell population, cells that express α-sarcomeric actin and the transcription factor Nkx2.5 infiltrate the peptide microenvironment. When exogenous donor green fluorescent protein–positive neonatal cardiomyocytes were injected with the self-assembling peptides, transplanted cardiomyocytes in the peptide microenvironment survived and also augmented endogenous cell recruitment. They demonstrated that self-assembling peptides created nanofiber microenvironments in the myocardium and that these microenvironments promoted vascular cell recruitment. Because these peptide nanofibers may be modified in a variety of ways, this approach may enable injectable tissue regeneration strategies.

8.3.7 Tooth Regeneration in Self-Assembled Systems

Recent advances in tissue engineering have drawn scientists to test the possibility of tooth engineering and regeneration. Tooth regeneration is normally referred to as the regeneration of the entire tooth or root that can be integrated into the jawbone. This technology is still at its infancy, and when it matures, it may be used to restore missing teeth and replace artificial dental implants when the tooth is damaged but still in a reparable condition; regeneration of parts of the tooth structure can prevent or delay the loss of the whole tooth. To engineer and regenerate a whole tooth, the cell source, tissue engineering strategies, and specific scaffolds needed should be chosen appropriately. Dental pulps as a source of adult stem cells show potential for skeletal tissue engineering. The dental pulp is a highly specialized mesenchymal tissue characterized by the presence of odontoblasts and by the fact that it is surrounded by a rigid mineralized tissue. Dental caries are still one of the most prevalent infectious diseases in the world. Currently, replacement of decayed soft and mineralized tissue together with some materials implantations are the most applicable in dental technology. Galler et al. [112] applied a novel approach to develop engineered dental tissues. In their work, two dental stem cell lines were combined with peptide–amphiphile. These self-assembled peptides formed 3D networks of nanofibers, and living cells encapsulated inside them. Cell–matrix interactions were tailored by incorporation of the cell adhesion sequence RGD and an enzyme-cleavable site. Stem cells from human-exfoliated deciduous teeth and dental pulp stem cells were cultured in self-assembled peptide scaffolds for four weeks using osteogenic medium culture. Stem cells from human-exfoliated deciduous teeth showed a spindle-shaped morphology, high proliferation rates, and collagen production, resulting in soft tissue formation. In contrast, dental pulp stem cells reduced proliferation but exhibit an osteoblast-like phenotype, express osteoblast marker genes, and deposit mineral. They concluded that their technology could be suitable for engineering both soft and mineralized matrices for dental tissue regeneration. In the future, the success of regenerative endodontic therapy will depend on the ability to yield a functional pulp tissue within cleaned and shaped root canal systems to revitalize teeth. This may be achieved by an *in vivo* approach, where pulp tissue is regenerated *in situ* into root canals, or by an *ex vivo* approach, which implies a *de novo* engineered pulp relying on the tissue engineering triad (dental pulp stem cells, scaffold, and growth factors).

8.3.8 Soft Tissue Regeneration in Self-Assembled Systems

The skin functions to provide a physical and chemical interface that protects the host against invasion by toxins and microorganisms and prevents dehydration that can result from loss of barrier function. The loss of skin integrity and function due to wound injury has led to efforts designed to better comprehend the molecular and cellular mechanisms that can optimize wound repair. Schneider et al. [113] used an integrated approach, using nanobiotechnology to augment the rate of wound re-epithelialization by combining self-assembling peptide (RADA16-I, Ac-RADARADARADARADA-$CONH_2$) nanofiber scaffold and epidermal growth factor (EGF). This bioscaffold was tested in a bioengineered human skin equivalent (HSE) tissue model that enabled wound re-epithelialization to be monitored in a tissue that recapitulates molecular and cellular mechanisms of repair known to occur in human skin. They found that self-assembled peptide underwent molecular self-assembly to form unique 3D structures that stably covered the surface of the wound, suggesting that this scaffold may serve as a viable wound dressing. They measured the rates of release of EGF from the SAP scaffold and determined that

EGF was only released when the scaffold was in direct contact with the HSE. By measuring the length of the epithelial tongue during wound re-epithelialization, they found that self-assembled peptides scaffolds containing EGF accelerated the rate of wound coverage by fivefold when compared to controls without scaffolds and by 3.5-fold when compared to the scaffold without EGF. In another work published by Kao et al. [114], a biocompatible peptide hydrogel scaffold as a 3D synthetic skin that does not contain animal-derived materials or pathogens were prepared. They investigated preparation methods, proliferation, and functional expression of fibroblasts in the synthetic dermis and differentiation of keratinocytes in the epidermis. Synthetic dermis was prepared by mixing fibroblasts with peptide hydrogel, and synthetic skin was prepared by forming an epidermal layer using keratinocytes on the synthetic dermis. A fibroblast-rich foamy layer consisting of homogeneous peptide hydrogel subsequently formed in the synthetic dermis, with fibroblasts aggregating in clusters within the septum. The epidermis consisted of three to five keratinocyte layers. Immunohistochemical staining showed human type I collagen, indicating functional expression around fibroblasts in the synthetic dermis, keratinocyte differentiation in the epidermis, and expression of basement membrane proteins. The number of fibroblasts tended to increase until the second week and was maintained until the fourth week but rapidly decreased in the fifth week. In the synthetic dermis medium, the human type I collagen concentration increased after the second week to the fifth week. They suggested that peptide hydrogel acts as a synthetic skin scaffold that offers a platform for the proliferation and functional expression of fibroblasts and keratinocytes.

Traumatic injury or surgery may trigger extensive bleeding. However, conventional hemostatic methods have limited efficacy and may cause surrounding tissue damage. Cheng et al. [115] used self-assembling peptides and specifically extend fragments of functional motifs derived from fibronectin and laminin to evaluate the capability of these functionalized peptides in the effect of hemostasis and liver tissue regeneration. From the results, these peptides could self-assemble into nanofibrous network structure and became gelation into hydrogel with pH adjustment. In animal study, the efficacy of hemostasis was achieved immediately within seconds in rat liver model. The histological analyses by histological staining and immunohistochemistry revealed that self-assembled peptide with these functionalized motifs significantly enhanced liver tissue regeneration. They concluded that self-assembled peptides may have potential as pharmacological tools to extensively advance clinical therapeutic applications in hemostasis and tissue regeneration in the field of regenerative medicine.

8.3.9 Other Forms of Self-Assembled Systems in Regenerative Medicine Therapy

Self-assembling systems are the emerging field of science that utilize biological molecules for biological applications. Molecular self-assembly technology takes advantage of the unique properties of biological molecules like amphiphilic peptides by utilizing their self-assembling property for the bioengineering of molecular templates and supramolecular structures. Amphiphiles are molecules containing a nonpolar hydrophobic region, and a polar hydrophilic region will self-assemble in aqueous solution to form distinct structures such as micelles, vesicles, and tubules. When suspended in an aqueous solution, the nonpolar hydrophobic regions of amphiphilic molecules are attracted toward each other and away from water (hydrophobic effect). The shape and dimensions of supramolecular structures formed from such assemblies will then depend on different factors, such as the structure of the polar head group and the shape of each amphiphile. Several self-assembling amphiphilic peptide and protein systems that self-assemble to form various nanostructures like nanofibers, nanotubes, vesicles, helical ribbons, and fibrous scaffolds have been described extensively for their potential applications in the field of biotechnology. The artificial scaffolds formed by self-assembling molecules not only provides a suitable support for cell proliferation but also serves as a medium through which diffusion of soluble factors and migration of cells can occur. The results of the cell attachment and proliferation revealed that diffusion of nutrients, bioactive factors, and oxygen through these highly hydrated networks is sufficient for survival of large numbers of cells for extended periods of time. As understood from the findings, proteins and peptides can self-assemble into various structures like nanotubes, nanovesicles, and 3D peptide matrices with interwoven nanofibers. Macroscopic 3D peptide matrices can be engineered to form various shapes by changing the peptide sequence.

Self-assembled peptide materials encouraged cell proliferation and differentiation. These peptide materials are also able to support various types of cell attachments. Self-assembling amphiphilic peptides are molecules engineered with hydrophilic heads and hydrophobic tails that, under correct conditions, can self-assemble into a network of nanofibers with the head sticking out into solution and the tails hidden in the core of the thread. These networks have the advantage of porosity, and living cells can be combined with such materials before the scaffold is fabricated. Silva's group included a five amino acid, laminin-specific cell-binding domain (which binds to specific integrins on cell surface) at the hydrophilic head of the peptide amphiphile and indicated that neural stem cells could be induced to differentiate into neurons when cultured within the laminin-specific domain or on 2D tissue culture plate coated with laminin solution that differentiated much less [60]. Self-assembling nanofibers have been used to repair the spinal cord injury in a mouse model. The injection of peptide amphiphile increased the number of oligodendroglia at the site of injury. Furthermore, the nanofibers promoted regeneration of both descending motor fibers and ascending sensory fibers through the lesion site. Ellis-Behnke' group showed that peptide nanofibers enabled the regenerated axons to reconnect to target tissues with sufficient density to promote functional return of vision, as evidenced by visually elicited orienting behavior [106].

8.4 Future Prospects

As stated earlier, there is a large need for novel therapies for the repair and regeneration of a multitude of tissues. Specifically, the primary significance is the development and application of novel technology for the identification of cell/biomaterial/signaling factor systems that promote tissue regeneration with optimal functional properties. There is currently great interest in harnessing these unique properties because, by using stem cells, it may be possible to generate cells in the laboratory that can be used to either replace damaged tissues inside the body or to identify safer and more effective medicines.

Self-assembly of biological molecules forms the basic principle in the formation of complex biological structures. The topic discussed previously on nanostructured biological materials through self-assembly of peptides and proteins give us a wide knowledge about the basic principle underlying molecular self-assembly of proteins and peptides. It gives a general overview of different kinds of self-assembling protein and peptide systems. Throughout the chapters discussed earlier, it clearly guides the reader about the potential applications of these self-assembled structures that can be applied for fabricating wider range of novel biomaterials for the use in medical applications. With appropriate references and examples, it opens up the reader's mind incorporating a wider range of knowledge about the self-assembling proteins and peptide systems. These macroscopic structures have inspired the researchers to use them in various areas of science like electronics, biotechnology, nanotechnology, and medicine.

Discussed in this chapter are a few applications and the advantages of self-assembled proteins and peptides in regenerative medicine therapy. Even though there are certain technological hurdles, these can be overcome by understanding the drawbacks of the individual systems and finding an alternative to overcome this drawbacks. However, it is necessary to create new and other alternative methods if we encounter any problems using the current technology in delivering biomolecules, such as proteins, growth factors, and DNA. Such systems have previously been created as one alternative method to enhance the *in vitro* and *in vivo* localization of biomolecules. Thus, self-assembling proteins and peptides and its versatile applications have kindled the interests of numerous scientists and researchers to find newer ways to apply it in the field of regenerative medicine. Tissue engineering is evolving rapidly with much more potential impacts in treatment of diseases like heart failures and diabetes. Numerous studies are being conducted worldwide to discover newer solutions that can be effectively applied to treat these diseases using self-assembled proteins and peptides. Results obtained in regenerative medicine are inspiring the scientific community to discover new innovative non-invasive tools at the micro- and nanoscale levels for such purposes.

References

1 Privé, G.G. (2009). *Curr. Opin. Struct. Biol.* 19: 379.
2 Yeh, J.I., Du, S., Tortajada, A. et al. (2005). *Biochemistry* 44: 16912.
3 Zhao, X., Nagai, Y., Reeves, P.J. et al. (2006). *Proc. Natl. Acad. Sci. U.S.A.* 103: 17707.
4 Kiley, P., Zhao, X., Vaughn, M. et al. (2005). *PLoS Biol.* 3: 1180.

5 Matsumoto, K., Vaughn, M., Bruce, B.D. et al. (2009). *J. Phys. Chem. B* 113: 75.
6 McGregor, C.L., Chen, L., Pomroy, N.C. et al. (2003). *Nat. Biotechnol.* 21: 171.
7 Dazzi, F. and Horwood, N.J. (2007). *Curr. Opin. Oncol.* 19: 650.
8 Hosseinkhani, H. (2012). *Curr. Drug Safety* 7: 37.
9 Langer, R.S. and Vacanti, J.P. (1999). *Sci. Am.* 280: 86.
10 Langer, R.J. and Vacanti, P. (1993). *Science* 260: 920.
11 Hosseinkhani, M., Hosseinkhani, H., Khademhosseini, A. et al. (2007). *Stem Cells* 25: 571.
12 Hosseinkhani, H., Hosseinkhani, M., Tian, F. et al. (2006). *Biomaterials* 27: 5089.
13 Wintermantel, E., Mayer, J., Blum, J. et al. (1996). *Biomaterials* 17: 83.
14 Hosseinkhani, H., Inatsugu, Y., Hiraoka, Y. et al. (2005). *Tissue Eng.* 11: 1459.
15 Hosseinkhani, H., Inatsugu, Y., Hiraoka, Y. et al. (2005). *Tissue Eng.* 11: 1476.
16 Colton, C.K. (1995). *Cell Trans.* 4: 415.
17 Pankajakshan, D., Krishnan, V.K., and Krishnan, L.K. (2007). *J. Tissue Eng. Regener. Med.* 1: 389.
18 Hosseinkhani, H., Hosseinkhani, M., and Kobayashi, H. (2006). *J. Bioact. Compat. Polym.* 21: 277.
19 Hosseinkhani, H., Hosseinkhani, M., and Kobayashi, H. (2006). *Biomed. Mater.* 1: 8.
20 Xie, Y., Yang, S.T., and Kniss, D.A. (2001). *Tissue Eng.* 7: 585.
21 Hosseinkhani, H., Azzam, T., Kobayashi, H. et al. (2006). *Biomaterials* 27: 4269.
22 Noshi, T., Yoshikawa, T., Ikeuchi, M. et al. (2000). *J. Biomed. Mater. Res.* 52: 621.
23 Hosseinkhani, H., Yamamoto, M., Inatsugu, Y. et al. (2006). *Biomaterials* 27: 1387.
24 Ma, T., Li, Y., Yang, S.T., and Kniss, D.A. (2000). *Biotechnol. Bioeng.* 70: 606.
25 Hosseinkhani, H., Hosseinkhani, M., and Khademhosseini, A. (2006). *Yakhte Med. J.* 8: 204.
26 Thomson, R.C., Wake, M.C., Yaszemski, M.J., and Mikos, A.G. (1995). *Adv. Polym. Sci.* 122: 245.
27 Hutmatcher, D.W. (2000). *Biomaterials* 21: 2529.
28 Whang, K., Thomas, C.H., Healy, K.E., and Nuber, G. (1995). *Polymer* 36: 837.
29 Taboas, J.M., Maddox, R.D., Krebsbach, P.H., and Hollister, S.J. (2003). *Biomaterials* 24: 181.
30 Gomes, M.E., Ribeiro, A.S., Malafaya, P.B. et al. (2001). *Biomaterials* 22: 883.
31 Hutmatcher, D.W., Schantz, T., Zein, I. et al. (2001). *J. Biomed. Mater. Res.* 5: 203.
32 Cai, Z. and Cheng, G. (2003). *J. Mater. Sci. Lett.* 22: 153.
33 Cai, Q., Yang, J., Bei, J., and Wang, S. (2002). *Biomaterials* 23: 4483.
34 Zein, I., Hutmacher, D.W., Tan, K.C., and Teoh, S.H. (2002). *Biomaterials* 23: 1169.
35 Lin, A.S.P., Borrows, T.H., Cartmell, S.H., and Guldberg, R.E.M. (2003). *Biomaterials* 24: 481.

36 Oxley, H.R., Corkhill, P.H., Fitton, J.H., and Tighe, B.J. (1993). *Biomaterials* 14: 1064.
37 Chirila, T.V., Constable, I.J., Crawford, G.J. et al. (1993). *Biomaterials* 14: 26.
38 Zhang, X., Jiang, X.N., and Sun, C. (1997). *Sens. Actuators, A* 77: 149.
39 Chen, G., Ushida, T., and Tateishi, T. (2001). *Mater. Sci. Eng., C* 17: 63.
40 Chen, G., Ushida, T., and Tateishi, T. (2001). *Biomaterials* 22: 2563.
41 Kane, R.S., Takayama, S., Ostuni, E. et al. (1999). *Biomaterials* 20: 2363.
42 Hosseinkhani, H., Hong, P.D., Yu, D.S. et al. (2012). *Int. J. Nanomed.* 27: 3035.
43 Hutmacher, D.W., Ng, K.W., Kaps, C. et al. (2003). *Biomaterials* 24: 4445.
44 Blacher, S., Maquet, V., Schils, F. et al. (2003). *Biomaterials* 24: 1033.
45 Nikpour, M., Chaouk, H., Mau, A. et al. (1999). *Synth. Met.* 99: 121.
46 Cavallini, A., Notarnicola, M., Berloco, P. et al. (2000). *J. Microbiol. Methods* 39: 265.
47 Ma, Z., Kotaki, M., Inai, R., and Ramakrishna, S. (2005). *Tissue Eng.* 11: 101.
48 Schmidt, C.E. and Leach, J.B. (2003). *Biomed. Eng.* 5: 293.
49 Kang, H.W., Tabata, Y., and Ikada, Y. (1999). *Biomaterials* 20: 1339.
50 Plant, G.W., Harvey, A.R., and Chirila, T.V. (1995). *Brain Res.* 671: 119.
51 Shugens, C.H., Maquet, V., Grandfils, C. et al. (1996). *Polymer* 37: 1027.
52 Maquet, V., Martin, D., Scholtes, F. et al. (2001). *Biomaterials* 22: 1137.
53 Rosenberg, M.D. (1963). *Science* 139: 411.
54 Webster, T.J., Siegel, R.W., and Bizios, R. (1999). *Biomaterials* 20: 1221.
55 Pricea, R.L., Waidb, M.C., Haberstroha, K.M., and Webster, T.J. (1877). *Biomaterials* 2003: 24.
56 Elias, K.L., Price, R.L., and Webster, T.J. (2002). *Biomaterials* 23: 3279.
57 Webster, T.J., Schalder, L.S., Siegel, R.W., and Bizios, R. (2001). *Tissue Eng.* 7: 291.
58 Smith, L.A. and Ma, P.X. (2004). *Biointerfaces* 10: 125.
59 Kameoka, J., Verbridge, S.S., Liu, H. et al. (2004). *Nano Lett.* 4: 2105.
60 Silva, G.A., Czeisler, C., Niece, K.L. et al. (2004). *Science* 303: 1352.
61 Tian, F., Hosseinkhani, H., Hosseinkhani, M. et al. (2008). *Biomed. Mater. Res. Part A* 84: 291.
62 Hartgerink, J.D., Beniash, E., and Stupp, S.I. (2001). *Science* 294: 1684.
63 Hong, Y., Legge, R.L., Zhang, S., and Chen, P. (2003). *Biomacromolecules* 4: 1433.
64 Ruoslahti, E. and Pierschbacher, M.D. (1987). *Science* 238: 491.
65 Williams, J.A. (1992). *Pathol. Biol.* 40: 813.
66 Evans, N.D., Gentleman, E., and Polak, J.M. (2006). *Mater. Today* 12: 26.
67 Hosseinkhani, H., Hosseinkhani, M., Tian, F. et al. (2006). *Biomaterials* 27: 4079.
68 Vandenburgh, H., Del Tatto, M., Shansky, J. et al. (1996). *Hum. Gene Ther.* 7: 2195.
69 Petite, H., Viateau, V., and Bensaid, W. (2000). *Biotechnolgy* 18: 959.
70 Peretti, G.M., Randolph, M.A., Villa, M.T. et al. (2000). *Tissue Eng.* 6: 567.
71 Woerly, S., Plant, G.W., and Harvey, A.R. (1996). *Biomaterials* 17: 301.

72 Ishaug, S.L., Crane, G.M., Miller, M.J. et al. (1997). *J. Biomed. Mater. Res.* 36: 17.

73 Freed, L.E. and Vunjak-Novakovic, G. (1998). *Adv. Drug Delivery Rev.* 33: 15.

74 Freed, L.E., Vunjak-Novakovic, G., and Langer, R.J. (1993). *Cell Biochem.* 51: 257.

75 Niklason, L.E., Gao, J., Abbott, W.M. et al. (1999). *Science* 284: 489.

76 Ma, P.X. and Langer, R. (1999). *J. Biomed. Mater. Res.* 44: 217.

77 Schneider, A.I., Maier-Reif, K., and Graeve, T. (1999). *In Vitro Cell Dev. Biol. Anim.* 35: 515.

78 Papas, K.K., Long, R.C., Sambanis, A., and Constantinidis, I. (1999). *Biotechnol. Bioeng.* 66: 219.

79 Griffith, M., Osborne, R., Munger, R. et al. (1999). *Science* 286: 2169.

80 Kloth, S., Ebenbeck, C., and Kubitza, M.S. (1995). *FASEB J.* 9: 963.

81 Polverini, P.J. (1995). *Crit. Rev. Oral Biol. Med.* 6: 230.

82 Ware, J.A. and Simons, M. (1997). *Nat. Med.* 3: 158.

83 Thompson, J.A., Anderson, K.D., Dipietro, J.M. et al. (1988). *Science* 241: 1349.

84 Tabata, Y., Nagano, A., and Ikada, Y. (1999). *Tissue Eng.* 5: 127.

85 Tabata, Y. and Ikada, Y. (1999). *Biomaterials* 20: 2169.

86 Cai, S., Liu, Y., Shu, X.Z., and Prestwich, G.D. (2005). *Biomaterials* 26: 6054.

87 Hosseinkhani, H., Hosseinkhani, M., Khademhosseini, A., and Kobayashi, H. (2007). *J. Controlled Release* 117: 380.

88 Hosseinkhani, H., Hosseinkhani, M., Tian, F. et al. (2007). *Tissue Eng.* 13: 1.

89 Shin, H., Zygourakis, K., Farach-Carson, M.C. et al. (2004). *J. Biomed. Mater. Res.* 69: 535.

90 Yoshikawa, T., Ohgushi, H., and Tamai, S. (1996). *J. Biomed. Mater. Res.* 32: 481.

91 Lee, Y.M., Seol, Y.J., Lim, Y.T. et al. (2001). *J. Biomed. Mater. Res.* 54: 216.

92 Wang, Y., Uemura, T., Dong, J. et al. (2003). *Tissue Eng.* 9: 1205.

93 Dohi, Y., Ohgushi, H., and Tabata, S. (1992). *J. Bone Miner. Res.* 7: 1173.

94 Subramani, K., Pathak, S., and Hosseinkhani, H. (2012). *Digit. J. Nanomater. Biosci.* 7: 85.

95 Urist, M.R. (1965). *Science* 150: 853.

96 Urist, M.R., Lietze, A., and Dawsen, E. (1984). *Clin. Orthop.* 187: 277.

97 Yu, Y., Yang, J.L., Chapman-Sheath, P.J., and Walsh, W.R. (2002). *J. Biomed. Mater. Res.* 60: 392.

98 Blumenfeld, I., Srouji, S., Lanir, Y. et al. (2002). *Exp. Gerontol.* 37: 553.

99 Rifkin, D.B. and Moscatlli, D. (1989). *J. Cell Biol.* 109: 1.

100 Abraham, J.A., Wlang, J.L., Tunolo, A. et al. (1986). *EMBO J.* 5: 2523.

101 Wans, J.S. and Aspenberg, P. (1996). *J. Orthop. Res.* 14: 316.

102 Globus, R.K., Petterson-Bückendahl, P., and Gospodarowicz, D. (1988). *Endocrinology* 123: 98.

103 Gospodarowicz, D., Ferrara, N., Schweigerer, G., and Neufeld, G. (1987). *Endocrinol. Rev.* 8: 95.

104 Jingnski, S., Hedemann, A., Kana, S.K. et al. (1990). *J. Orthop. Res.* 8: 364.

105 Tabata, Y., Yamada, K., Miyamoto, S. et al. (1998). *Biomaterials* 19: 807.

106 Ellis-Behnke, R.G., Liang, Y.X., You, S.W. et al. (2006). *Proc. Natl. Acad. Sci. U.S.A.* 103: 5054.
107 Liang, Y.X., Cheung, S.W., Chan, K.C. et al. (2011). *Nanomedicine* 7: 351.
108 Meng, H., Chen, R., Xu, L. et al. (2012). *Life Sci. J.* 9: 42.
109 Tysseling, V.M., Sahni, V., Pashuck, E.T. et al. (2010). *J. Neurosci. Res.* 88: 3161.
110 Davis, M.E., Motion, J.P., Narmoneva, D.A. et al. (2005). *Circulation* 111: 442.
111 Davis, M.E., Hsieh, P.C., Takahashi, T. et al. (2006). *Proc. Natl. Acad. Sci. U.S.A.* 103: 8155.
112 Galler, K.M., Cavender, A., Yuwono, V. et al. (2008). *Tissue Eng. Part A* 14: 2051.
113 Schneider, A., Garlick, J.A., and Egles, C. (2008). *PLoS One* 3: e1410.
114 Kao, B., Kadomatsu, K., and Hosaka, Y. (2009). *Tissue Eng. Part A* 15: 2385.
115 Cheng, T.Y., Wu, H.C., Chang, W.H. et al. (2012). *Technical Proceedings of the 2012 NSTI Nanotechnology Conference and Expo*, vol. 3. Taiwan: Taipei.

9

Nanobiotechnology

9.1 Introduction

Nanotechnology has shown great achievements of applications in medicine, photonics, and electronics [1–9]. In general any structure, material, or formulation with at least one dimension ranging between 1 and 100 nm is regarded as a nanostructure, nanomaterial, or nanoparticle (NP) [10]. In contrast to their bulk counterparts, NPs present excellent properties including high surface-to-volume ratio and superb electrical, magnetic, mechanical, optical, and thermal behavior, as well as high surface energy [11, 12]. These nanomaterials exist in diverse structures in which each has its own advantages and applications [13–18]. Their dispensability and reactivity in different solvents can be adjusted by tailoring surface chemistry [19, 20]. NPs have been employed in biomedical studies for various purposes such as sensing, imaging, diagnosis, and treatment [21–26]. The application of NPs or nanoscale formulations in medicine and health is generally referred to as nanomedicine, involving biology, chemistry, physics, medical sciences, and engineering [27]. Although this advanced field of medicine has emerged for only several decades, it has led to several challenges in traditional medicine such as systemic and organ toxicity, impaired target specificity, and poor bioavailability [28–30]. Thus, a broad range of medical products (more than 200) including diagnostic agents, medical devices, and therapeutics have been clinically approved or are being evaluated due to nanomedicine research [31]. There are, however, several hurdles that eliminate the clinical translocation of NPs [32–34]. (i) It is difficult to synthesize NPs capable of encapsulating effective therapeutic molecules with activated release. (ii) Difficulty exists to efficiently deliver NPs to the target site for *in vivo* studies. (iii) The efficacy and toxicity of NPs are significant obstacles. (iv) The clinical translocation and industrial production of NPs in a large scale with reasonable cost are challenging. Hence, these barriers must be completely overcome to expand the efficacy and application of NPs for clinical purposes [35]. Rapid development of molecular nanotechnology and nanovaccinology that exploit the improved and often novel physical, chemical, and biological properties of materials at the nanometer scale must be considered before clinical application due to side effects. Nanotechnology has the potential to enable early detection and prevention and to essentially improve diagnosis, treatment, and follow-up of diseases. It is committed to supporting activities such as theranostics, as

Biomedical Engineering: Materials, Technology, and Applications, First Edition. Hossein Hosseinkhani.

well as enabling diagnostic devices and therapeutics to be combined for real benefit to patients. Growing advances in nanotechnology have stimulated researchers to create innovative technologies to enhance and modify the quality of new devices. On the other hand, little is known about the potential risks of the nanoscale materials. Also, control of adverse effects is critical in the development of new technology. The study of side effects of nanomaterials on the ecosystems and living organisms, along with the prevention and improvements of these adverse effects, is termed nanotoxicology. Nanomedicine, nanotechnology, and nanotoxicology are primarily aimed to improve human life by reducing risks [36]. Nanomaterials have been widely developed and applied in commercial products such as electronic sensors, energy equipment, sun creams, and biomedical devices.

There has not been sufficient evaluation of nanomaterial interfaces with tissue and organs, although several investigations have raised questions if nanoscale materials should be treated as carcinogens or asbestos. Since nanotechnology has high impact on consumer and biomedical products, it is important to perform toxicity research on the basic interaction of nanomaterials with natural tissues. It is extremely important to understand the exact role of NPs *in vivo*. When NPs are administrated within a human body, the effect of NPs-loaded therapeutics followed by their distribution into tissues (i.e. kidneys, lung, and liver) should be precisely elucidated. Clearance of NPs can potentially occur through pulmonary, hepatic, and renal clearance. They may not degrade or eliminate, which makes predicting long-term behavior particularly difficult. NPs may accumulate in various organs and interact with off-target cells. They may degrade and eliminate from the body due to their very small size [37–39]. There are few studies addressing whether nanomaterials are (bio)compatible with natural organ, tissues, or cells. This chapter discusses the chemical, physical, and biological properties of various nanomaterials (both natural and synthetic) with specific focus on recent technologies related to the fabrication and process of NPs with different systems. More importantly, we discuss the *in vivo* fate of NPs including routes of administration, biodistribution, metabolism, routes of clearance, and blood biocompatibility.

9.2 Classification of Nanomaterials

Physical, chemical, and biological properties of many conventional materials change when interacting with NPs. NPs have a greater surface area per weight than larger particles, causing them to be more reactive to certain other molecules. A broad range of NP types exists, and the NPs have enjoyed extensive applications. Since there is no standardized nomenclature, terms such as nanomedicals/nanomedicines, engineered nanomaterials (ENMs), NPs, and nonbiological complex drugs have been applied in the literature. Most nanomaterials can reproduce the biological properties of macromolecules [40]. Figure 9.1 shows the direct interactions of biological cells under nanomaterials interactions.

Examples are inorganic NPs such as upconverting NPs [41], superparamagnetic iron oxide NPs (SPIONs) [42], gold nanoclusters (AuNCs) [43], mesoporous

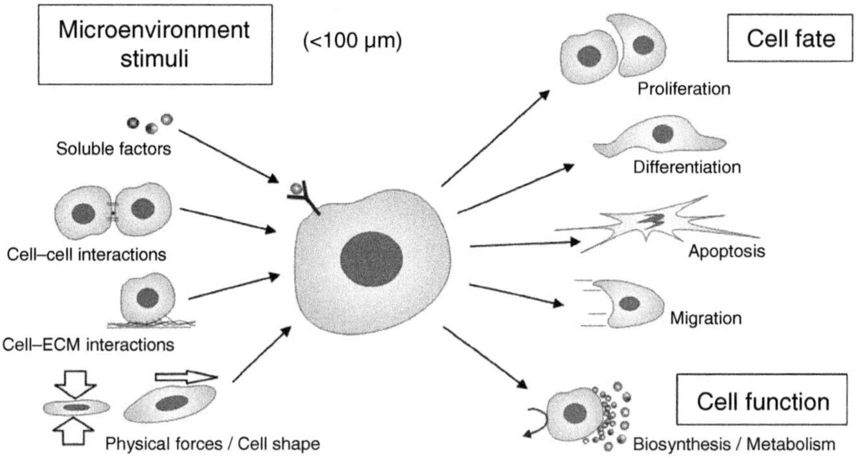

Figure 9.1 Interactions of biological cells with nanomaterials.

silica NPs (MSNPs), and quantum dots (QDs) [44], as well as polymeric nanostructures [45], nanotubes [46], ribonucleic acid (RNA) NPs [47], graphene [48], nanodiamonds (NDs) [49], and protein constructs [50]. Most of these NPs possess excellent properties, including shape- and size-dependent magnetic, optical, and electronic characteristics. The quality of these properties is affected by their fabrication methods, characterizations, and modifications [51–54]. However, transfection efficiency is not high, and further material improvement is needed.

9.2.1 Polymeric NPs

Polymeric nanocarriers (PNCs) are promising delivery systems in nanomedicine due to their high stability, ease of fabrication, enhanced accumulation in tumor, higher drug circulation in blood, and increasing solubility of hydrophobic drugs. They can be divided into polymeric micelles (PMs), dendrimers, nanogels, and nanocapsules. The design of polymeric NPs relies on the route of administration, therapeutic application, and target/specific site (tissue, cell, and organ). These NPs are constructed from biodegradable and/or nonbiodegradable polymers. Biodegradable polymers, extensively applied for sustained release of biomolecules [55–59], can be formed by natural (polysaccharide, protein, and peptide) and/or synthetic polymers [poly(ε-carpolactone) (PCL) and poly(lactic acid) (PLA)]. However, the main synthetic nonbiodegradable polymers showing low cytotoxicity are polyethylene glycol (PEG) [60], *N*-(2-hydroxypropyl)methacrylamide (HPMA) copolymer [61], and poly(vinylpyrrolidone) (PVP) [62]. Administration routes for PNC are intravenous injection, mucosal, oral, and dermal/transdermal delivery. Intravenous injection is the most common route of delivery; however the latter methods are less invasive. For successful delivery in all routes, PNCs need to overcome biological obstacles including blood, skin, extracellular matrix, and mucus. They, therefore, must be used with advanced constituents. Safe clinical applications of polymeric NPs, their payloads, and their chemistry play significant roles, since they affect biocompatibility, stability, biodistribution, cellular fate, and biodegradability. Hence, these features enable polymeric NPs to enhance the diagnosis and treatment of diseases such as urinary tract, pulmonary, and viral infections, as well as cancer and cardiovascular disease. Biopolymeric NPs such as sugars and protein with their natural biodegradability and biocompatibility are desirable for drug delivery. Protein NPs have received considerable attention, since their amphiphilicity increases their interaction with both physiological fluids and hydrophobic drugs [63]. These NPs can be formed through natural proteins by several methods, including chemical cross-linking, spontaneous self-assembly, salt-induced coacervation, and emulsification [64, 65]. Furthermore, a wide range of functional groups such as carboxyl, amino, and thiol facilitates chemical interactions with therapeutic agents [66]. Basic examples of natural proteins applied for drug delivery are casein, elastin, albumin, gliadin, and gelatin [67–72]. Albumins are more promising, as they can serve as a physiological transporter. They bind naturally to various therapeutic agents such as endogenous molecules and drugs [67].

9.2.1.1 Dendrimers

Another important class of polymers is dendrimer, i.e. synthetic, hydrobranched three-dimensional polymeric NPs first discovered in 1978 [73]. Their structures consist of three major components: a core (which alters the shape of dendrimer), an interior (which alters the host–guest characteristics of dendrimer), and the shell (which can be functionalized). The most commonly used dendrimers are polyamidoamine (PAMAM), polypropyleneimine (PPI), poly(L-lysine), poly(glycerol-succinic acid) (PGLSA-OH), and poly(2,2-bis(hydroxymethyl)propionic acid (bis-MPA). Dendrimers have been widely applied in diverse applications including cancer due to properties such as low toxicity, bioattachment capability, tunable solubility, monodispersity, efficient membrane transport and controllable structure [74, 75], drug/gene delivery [76–78], and siRNA delivery [79]. PAMAM is known to induce a hemolytic and cytotoxic effect due to the presence of a cationic surface; thus it hinders the clinical applications of dendrimers [80]. This can be addressed by modification of PAMAM with functional groups such as PEGylation, amino acid conjugation, acetylation, and hydroxyls that further provide a facile way for *in vivo* applications of dendrimers [81, 82].

9.2.1.2 Polymeric Micelles

Micelles are also considered as one of the most effective PNCs particularly for the delivery of poorly water-soluble drugs/genes [83] and hydrophobic agents [84, 85]. These colloidal particles with a usual size of 5–100 nm are formed from a hydrophobic core and hydrophilic shell. The hydrophobic part of micelle acts as a container for hydrophobic drugs, DNA, or protein by chemical or physical binding, while the hydrophilic part protects the hydrophobic part from biological invasions. PMs display properties superior to drug carriers due to their higher stability in the blood stream, loading capacity, larger solubilization power, longevity, and therapeutic potential. They are also able to prolong circulation and reduce the toxicity of anticancer drugs simply by encapsulating them [86, 87]. For example,

PMs embedded with cisplatin decrease nephrotoxicity [86]. The cisplatin-loaded core is protected by a PEG shell that increases blood circulation longevity by reducing acute kidney accumulation of PMs. PMs are the only available nanocarriers that overcome multiple drug resistance (MDR) by absolute targeting using several methods, including passive targeting, temperature- and pH-responsive drug delivery systems (DDs), as well as folate-mediated DDs [88]. Kataoka and coworker developed several PMs based on PEG–poly(amino acid) block copolymers [89]. Of these formulations, which are already in clinical trials in Asia and North America, encapsulated paclitaxel (PTX) with PMs is likely to get approval soon, since it is in Phase III clinical trial.

9.2.1.3 Polycationic NPs

Polycationic NPs have recently been considered as promising candidates for nonviral gene delivery, in part because of their molecular diversity that can be modified to fine-tune their physicochemical properties [90–92]. These polycations are able to condense large genes into compact structures and to mask negative DNA charges, necessary for transfecting most types of cells. Generally, since plasmid DNA is a large, negatively charged molecule up to 1 μm in length, it is not possible to allow the plasmid DNA itself to be internalized by cells *in vitro* due to the inherent negative charge of the cell membrane. When mixed with a positive-charged nonviral vector, the molecular size of the plasmid DNA decreases by molecular condensation due to formation of NPs. Polycations used for the formation of DNA NPs are polyamines that have become cationic under physiologic conditions. The most studied polycations employed for DNA NP and delivery are the branched/linear polyethyleneimine (PEI) [93, 94], poly(L-lysine) [95], poly(dimethyl aminoethyl methacrylate) (pDMAEMA) [96], poly(trimethyl aminoethyl methacrylate (pTMAEMA) [97], poly(vinylpyridine) [98], chitosan [99], poly(amidoamine) dendrimers [100], polymer micelles [101], and diethylaminoethyl dextran (DEAE-dextran) [102]. Most polycations are toxic to cells and nonbiodegradable, while polymers based on amino acids such as poly(lysine) are immunogenic [103].

9.2.2 Lipid NPs

Lipid NPs (LNPs) can be classified into liposomes (lipid bilayer vesicular nanostructures), solid lipid NPs (SLNs), nanostructured lipid carriers, and nonlamellar lipid liquid crystalline interfaces [104, 105]. They are of increasing interest in drug delivery, theranostics, and multimodal imaging mainly due to their lipophilicity to more easily interact with and enter cells, and their high drug loading capacity [106]. For example, Li et al. employed clathrin to modify SLN-loaded salinomycin for the drug delivery applications [107]. The clathrin-modified SLN-SAL (CMSLN-SAL) decreases premature drug release while securing burst intracellular drug release. Reduction modification of premature drug release and ensured burst intracellular drug release was obtained (i) by enhancing the mechanical strength of SLN-SAL and (ii) by lipid membrane fusion after the clathrin shell was depolymerized through HSC70 as a cytoplasm enzyme. Furthermore, higher inhibitive effects and much more intracellular uptake on cancer cells were observed for CMSLN-SAL as opposed to SLN-SAL. LNPs possess a hydrophobic shell or core and hydrophilic surface for ease of interaction within the body. Having hydrophobic/hydrophilic regions allow the loading of both hydrophobic and hydrophilic therapeutic agents for targeted delivery and multimodality imaging [108–110]. In immunotherapy applications, LNPs provide more benefits than other vectors: (i) both composition and components can be easily varied to enhance delivery efficacy and decrease toxicity, (ii) immune cell targeting ligands or adjuvants can be added to adjust the immune response, and (iii) LNPs have been successfully exploited for mRNA vaccine delivery [111–114]. Recent examples of LNPs for mRNA delivery in cancer immunotherapy were reported by Sayour et al. [115] and Oberli et al. [116]. Cationic lipids are other types of chemical vectors composed of positively charged lipids, such as dioctadecylamido-glycylspermine (DOGS, Transfectam™), *N*-[1-(2,3-dioleyloxy)propyl]-*n*,*n*,*n*-trimethylammonium chloride (DOTMA) [117, 118], 1,2-dimyristyloxypropyl-3-dimethy-hydroxyethyl ammonium bromide (DMRIE) [119], and poly-L-lysine-modified silica nanoparticle (PMS-NP). They are novel nonviral vectors for gene delivery [120]. A cationic liposome, Lipofectin™, is commercially available for gene transfection. A coprecipitate of plasmid DNA with calcium phosphate is applied to cells for transfection [121]. Natural polymeric materials have been used to form NPs to transfer proteins, drugs, and genes *in vitro* as well as *in vivo* due to lower immune rejection [122–141]. These NPs have been chosen as safe drug delivery systems partly due to their nontoxicity, biodegradability, low cost, scalability, reproducibility, and high-throughput capability. These features make them potentially beneficial for future clinical trials.

9.2.3 Magnetic NPs

Iron oxide nanoparticles (IONPs) or magnetic NPs are important in various applications, mainly due to their low toxicity, biocompatibility, supermagnetic characteristics, and simple synthesis. The most commonly used IONPs are magnetite (Fe_3O_4) and maghemite (γ-Fe_2O_3), which serve as a platform for drug delivery [142], *in vivo* tracking of labeled cells [143], gene therapy [144], controlled drug delivery [145] against tumors, vascular imaging [146–148], magnetic separation of cells or molecules [149], and overcoming the severe multidrug resistance of cancer cells [150]. Magnetic NPs were first commercially introduced in the 1990s for disease diagnosis. They are still in use (e.g. Gastromark®/Lumirem®) for imaging applications. In contrast to other NPs, the size of magnetic particles (usually from a few to tens of nanometers) can be easily modified. Also, they can be simply altered by an external magnetic force. Additionally, the properties of magnetic NPs improve the signal of close protons, enabling them as contrast agents in magnetic resonance imaging (MRI). These characteristics make them useful in multimodal and non-traditional multimodal imaging techniques [151]. Physical characteristics of NPs are defined by their inorganic core, while surface properties are significant to ensure the specific localization and biocompatibility of NPs with tissues, cells, and proteins [152]. Surface modification of magnetic NPs with peptides [153], aptamers [154], antibodies [14], and other small molecules [155] enable these NPs for site-specific and multimodal imaging. Surface functionalization of magnetic core with hydrophilic polymers such as PEG and dextran have also broadened their biological applications [156, 157].

Among magnetic NPs, SPIONs with an average diameter of 5–10 nm have received significant attention in diverse fields of nanomedicine [158–160], ranging from cancer [161], tissue repair [162, 163], targeted drug delivery [164], and MRI [165], to gene delivery [162]. For instance, Feridex (which enters the reticuloendothelial system, RES) is exploited for the detection of tumors in liver lesions [166, 167], and Combidex is used for the diagnosis of tumors in lymph node (LN) metastases. In particular, SPIONs have been extensively investigated as contrast agents in MRI [168, 169], because of their superparamagnetism properties. Likewise, SPIONs can be modified with a wide range of materials. They induce low toxicity, since they degrade in the body after long-term administration. Nonetheless, the fast blood clearance from the circulation, poor physiological stability, and the lack of target specificity as drug delivery carriers eliminate their clinical applications [170–174]. They can also accumulate in the lungs, liver, and spleen. Addressing these challenges may not be easy. The addition of new functions can lead to more complex and expensive materials that are difficult to control [175]. To solve this, Hou and coworkers developed MTX-PEG-CS-IONPs-Cy5.5 nanocomposites through functionalizing the surface of iron oxide NPs with chitosan (CS-IONPs) in the presence of fluorescent cyanin dye (Cy5.5) and methotrexate PEG (MTX-PEG) for targeted cancer therapy [176]. In contrast to the free MTX and the PEG-CS-IONPs-Cy5.5 injection *in vitro* and *in vivo*, MTX-PEG-CS-IONPs-Cy5.5 exhibits higher cell internalization, greater tumor accumulation, and fewer adverse effects.

On the other hand, ultrasmall superparamagnetic iron oxide nanoparticles (USPIOs) with the smaller size of 50 nm are able to pass phagocytosis, resulting in long blood circulation times [177–179]. Hence, they can reach tumor [180], LN [181], and inflammation sites [182]. In a recent example, Gao and coworkers introduced an easy method for fabrication of ultrasmall size magnetite nanoparticles (uBSPIO) for MRI imaging [183]. uBSPIO was synthesized under mild conditions with bovine serum albumin (BSA) as a scaffold, followed by the combination of albumin with luteinizing hormone-releasing hormone (LHRH) peptide. The resulting uBSPIO (~ 5 nm) exhibits very high transverse relaxivity, superparamagnetism, good stability, and monodispersity. Also, uBSPIO with good biocompatibility shows suitable *in vivo* MRI ability. LHRH–uBSPIO actively target the target tumor. Other types of magnetic NPs such as extremely small iron oxide NPs (ESIONPs) [184], micrometer-sized iron oxide particles (MPIOs) [185], water-soluble superparamagnetic iron oxide NPs (WSIONPs) [186], water-dispersible ferrimagnetic iron oxide NPs (WFIONPs) [187], and ferrimagnetic iron oxide NPs (FIONs) [188] have expanded therapeutic applications of iron oxide NPs.

9.2.4 Gold NPs

Although gold nanoparticles (AuNPs, also GNPs) were used as coloring agents in ancient times, the first scientific research was reported by Michael Faraday in 1852 through the reduction of gold chloride via phosphorous. This is due to their size-, shape-, and color-dependency, and tunable surface plasmon, as well as ease of functionalization and electronic and optical properties, AuNPs have been extensively applied in controlled release [189], cancer [190–194], gene delivery and antibacterial procedure [195], inflammatory diseases [196], genetic materials [197], cell tracking [198], thrombus imaging [199], photothermal treatment [200], tissue regeneration [201], and immunoassay [202]. They exist in the range of 1–120 nm with various shapes such as rods [203], cages [204], spheres [205], and shells [206]. Common methods for the preparation of AuNPs are stabilization of gold NPs using thiol as introduced by Giersig and Mulvaney [207] and the biphasic reduction of Au^{3+} introduced by Brust et al. [208]. The seed-mediated growth method reported by Murphy et al. [209] and by El-Sayed and colleagues [210] has also become popular due to fabrication of AuNPs in various shapes and sizes. The terms "cluster" and "colloid" are frequently used in literature identically, except that the former commonly applies to particles having diameters less than 10 nm, whereas the latter generally is used for larger particles. In recent research, Conde et al. synthesized dendrimer–dextran hydrogels patch doped with drug–NP and siRNA–NP conjugates for local colorectal cancer using gene, drug, and phototherapy in a mouse model [211]. The adhesive hydrogel patch improves stability and allows local delivery of embedded NPs. The drug gold nanorods mediate the conversion of near-infrared radiation into heat. This leads to the release of a thermally and chemotherapeutic effect, causing cell damage. Short interfering RNA (siRNA)–gold nanospheres have been employed to deliver siRNAs against Kras, a key oncogene driver.

Recently the use of AuNPs in nanosensors has attracted attention due to excellent optical and physical properties, namely, a quantized charging effect, nonlinear optical characteristics, surface enhanced Raman scattering (SERS), fluorescence resonance energy transfer (FRET), and localized surface plasmon resonance (LSPR) [212]. These features make AuNPs superior scaffolds for the utilization of advanced biological and chemical sensors [213, 214]. AuNPs possess several advantages compared to other nanomaterials: (i) the highly stable AuNPs can be prepared with a simple fabrication method; (ii) they have important optoelectronic characteristics; (iii) AuNPs display better biocompatibility using suitable ligands [205]; (iv) by altering the shape, size, and surrounding chemical environment, the properties of AuNPs can be easily adjusted; and (v) they can be applied for the selective detection and binding of biological targets and small molecules such as proteins, DNA, and mRNA [215, 216].

Studies on the use of AuNPs into mammalian cells are of considerable interest. The first efforts to examine the biodistribution of colloidal gold were carried out on rabbits [217], mice [218], and rats [219, 220] in the 1960s–1980s. After parenteral administration, colloidal golds were caught by liver cells, expelled via bile, and removed from the body with feces. Gold particles were mostly found in Kupffer (Kupffer–Browicz) cells after injection. Scott et al. were probably the first to observe that AuNPs phagocytosis is associated with their size (40 nm gold particles show clearance and maximal accumulation) [217]. Danscher and coworkers found that Kupffer cells are important for the removal of AuNPs from the body [221]. In this study, AuNPs with the size of 2–40 nm were injected intravenously into mice. Only 10% of AuNPs were captured by the spleen, while the other 90% were captured by Kupffer cells of the liver. The same group presented the data on localization of 40 nm AuNPs in lysosomes/endosomes of Kupffer cells. Results show that they can remain for six months [222].

9.2.5 Silica NPs

Silica NPs (SNPs or SiO_2) are "Generally Recognized As Safe" (GRAS) by the US Food and Drug Administration (FDA), and they have been recently employed in cancer therapy for a first-in-human clinical trial [223]. Their unique properties, such as good biocompatibility, surface functionalization, ease of fabrication, hydrophilic surface, and low cost, make them a suitable platform for cancer [224], tissue engineering [225, 226], targeted delivery [227], neuro-therapeutics [228], and imaging [229]. Recently, Sun and colleagues examined the ability of SiO_2 NPs to disrupt the blood–brain barrier (BBB) *in vitro* and *in vivo* [230]. They demonstrated that SiO_2 NPs induce a cytoskeleton arrangement and tight junction loss, promoting an inflammatory response and vascular endothelial growth factor (VEGF) release in brain microvessel endothelial cells (BMECs). This initiates astrocytes to reinforce the generation of VEGF and enhance the aquaporin-4 (AQP4) expression, therefore, leading to disruption of BBB by a complex immunoregulatory loop between astrocytes and BMECs in the presence of SiO_2. They also found that the BBB dysfunction caused by SiO_2 NPs can be efficiently protected through inhibition Rho-kinase (ROCK)/c-Jun N-terminal kinases (JNKs) and reactive oxygen species (ROS). *In vivo* data in the brains of Sprague Dawley (SD) rats prove that SiO_2 NPs cause astrocyte activation, BBB paracellular opening, and oxidative stress. These results suggest that SiO_2 NPs have great potential in nanodiagnostics and neuro-therapeutics.

SNPs do not interfere with magnetic radiation and are transparent in the near infrared (NIR), ultraviolet, and visible zones of the electromagnetic spectrum. In 2011, fluorescent SNPs (Cornell dots) for targeted molecular imaging of cancer was approved by the FDA for the first time in a human clinical trial [223, 231], acknowledging the excellent potential of SNPs for clinical studies. SNPs can be categorized as mesoporous and nonporous (solid) structures. Nonporous SNPs are normally synthesized by the Stöber method [232], direct micelle-assisted methods [233], and reverse microemulsions (water-in-oil) approach [234]. In contrast, MSNPs are obtained via a surfactant-templated sol–gel method [235, 236], and they are characterized by high surface areas, large pore volumes, tunable pore sizes, and rigid and stable frameworks. The shape, size, pore structure, and size of MSNPs can be controlled by adjusting the concentration and composition of surfactants during fabrication, thus the morphology and size of nonporous SNPs can be highly controlled. Nonporous SNPs are largely used for delivery of drugs/biomolecules through conjugation or encapsulation, whereas MSNPs deliver active payloads via chemical or physical adsorption [237, 238]. Payload release from nonporous SNPs is controlled by the degradation of silica or chemical linkers. By contrast, release from MSNPs is controlled via altering the inner surface of the pores to control the binding affinity with drugs or by using a gatekeeper [237]. Gatekeepers are able to release drugs or biological molecules in response to external or internal stimuli that make them important for smart drug delivery. Examples of external and internal stimuli are magnetic fields, light and temperature, pH, and biomolecules. In response to external or internal stimuli, the gates are opened and release drugs. MSNPs for biomedical applications are classified in three main sections, including MCM-41, MCM-48, and SBA-15, each indicating different structural properties and pore sizes. One example is the use of gated MCM-41 for the treatment of inflammatory diseases [239]. In particular, ε-poly-L-lysine-gated MSNPs loaded with VX-765 (as an inflammatory drug) were developed in macrophages *in vitro* and *in vivo*. When encapsulated into MSNPs, VX-765 was able to reach the inflammation site and was c. fourfold more effective. MSNPs have the capability of releasing drugs selectivity at the inflammatory site.

9.2.6 Quantum Dots

QDs are fluorescent semiconductor nanocrystals first demonstrated as optical probes for cell and molecule labeling [240, 241]. They consist of a core and a shell layer with a size of a few nanometers, showing unique optical properties in comparison to conventional fluorescent proteins or organic dyes. QDs possess size-dependent tunable electronic and optical characteristics from visible to NIR range, broad absorption, narrow emission band, long fluorescent lifetime, and excellent brightness. QDs have a large surface area for biofunctionalization [242]. They also exhibit great resistance against photobleaching, which makes them useful for long-term *in vivo* imaging at the single-molecule level [243, 244]. The main synthesis method for preparation of QDs is solution phase, which provides highly luminescent QDs with high colloidal and optical stability and a narrow emission bandwidth. Toxicity is the major drawback of QDs due to the release of metal ions toward cells [245, 246]. Surface modification is a feasible way to reduce toxicity, rendering QDs suitable in biomedical areas. This can not only improve quantum yield of QDs [247] but also make them water soluble. For example, cadmium-based QDs such as cadmium selenide (CdSe) were applied for surface modification of zinc sulfide (ZnS) polymer shells or SiO_2 layers to prevent possible toxicity of materials [246]. Commonly used surface modifiers are antibodies, proteins, and peptides, which allow QDs to form a single or mixture of several recognition moieties, targeting specific tissues or organs for molecular imaging [248, 249]. In addition, surface modification enables QDs to act as monitoring indicators for disease therapy and targeted drug delivery. Surface-modified QDs have been widely applied in a wide range of biomedical imaging applications, including cancer targeting, imaging oral carcinoma [250, 251], imaging breast cancer, and monitoring treatment efficacy [252], as well as monitoring protease activity [253]. Reasons for such specific applications of QDs in cancer imaging can be attributed to (i) QDs own great quantum yields, high photostability, large effective Stokes shifts, tunable emission wavelengths, and high molar extinction coefficients [242, 254], (ii) QDs' readily functionalization with different molecules [255, 256], and (iii) NIR emission via QDs showing deep tissue penetration [257, 258]. Thus, they have been widely applied in both *in vitro* and *in vivo* imaging [242, 255, 256, 259]. In particular, an indium phosphate core/zinc acetate shell QD (InP/ZnS QD) has received considerable interest due to lower toxicity than other cadmium-based QDs, such as cadmium sulfide (CdS) and CdSe QDs [260–263]. Recently, Wang et al. developed InP/ZnS QD-based micelles in the presence of an anti-epidermal growth factor receptor (EGFR) nanobody for breast cancer therapy [264]. The NIR fluorescent QDs were functionalized with polylactide–PEG (PLA–PEG) to form QD micelles. They were further loaded with aminoflavone (AF) as an anticancer drug. The cellular uptake and cytotoxicity of QD-based micelles in MDA-MB-468 cancer cells were enhanced by an anti-EGFR nanobody 7D12 conjugation. The AF-encapsulated nanobody conjugate better accumulates in a tumor, resulting in higher efficient tumor regression. The resultant NPs exhibited no sign of systemic toxicity in a mouse model.

The hydrophobicity of QDs precludes biological and biomedical applications. Several approaches have been introduced to meet this challenge, such as using ligand exchange [265] and hydrophilic polymer encapsulation [266]. For example, the active moieties-decorated surface coronas, as hydrophilic molecules, enable QDs with long circulation, active tumor-targeting, and superb biocompatibility [267, 268]. On the contrary, it is generally known that the use of a specific concentration of surface modifier can not only increase the particles sizes of QD but also decrease their fluorescence intensity [269]. Preparing QD-based nanodrugs with a regulated and constant repetition quality is sophisticated due to the unknown drug uploading processes [267, 270]. One solution is the employment of ingenious self-assembly methods, such as electrostatic interaction, and host–guest approaches driven by adjustable physical or chemical interactions [271, 272]. For example, Deng and coworkers report the fabrication of H_2O_2 and pH-responsive QDs by ingenious self-assembly strategy for tumor therapy [273]. The combination of sulfobetaine-α-cyclodextrin (Sb-CD), doxorubicin (DOX) as an anticancer drug, and alkyl chain-modified "Host" ZnAgInSe/ZnS formed a QD-based prodrug (Sb-CD-DOX-ZAISe/ZnS) through host–guest strategy; QD drug uploading and water transfer were properly integrated. The resultant NPs display passive tumor-targeting and imaging for malignant tumors in metaphase. The NPs were coated with folic acid (FA) through electric attraction (Sb-CD-DOXZAISe/ZnS@FA) and present pH-adjusted charge reversal ability together with higher curative effect than free DOX.

9.2.7 Carbon Nanomaterials

Carbon nanotubes (CNTs) are fibrous particles formed by graphene sheet. They are classified into two main types: single-walled nanotubes (SWNTs) and multiwalled nanotubes (MWNTs). The former consists of one layer of graphene with a diameter of 0.5–3 nm and a length of 20–1000 nm, while the latter is formed from multiple layers of graphene having dimensions ranging between 1.5–100 nm and 1–50 μm [274, 275]. CNTs can also be classified as fullerenes, carbon nanobuds, carbon nanotorous, carbon nanohorns, and carbon nanopeapods [276, 277]. CNTs are synthesized using three methods: arc-discharge [276, 278], laser-ablation, and chemical vapor deposition (CVD). Due to their superior electrical and mechanical characteristics, CNTs have been applied in the electrodes of lithium batteries [279–281]. They indicate wide light absorption in the NIR photoluminescence, UV–Vis–NIR region, unique photothermal response, excellent Raman signals, and large surface area, which make them important as a wearable health indicator [282], *in vivo* imaging [283], and tissue engineering [284]. Koo et al. developed a new method for synthesis of a wearable electrocardiogram (ECG) cardiac monitor based on voltage-dependent color-tunable organic light-emitting diodes (CTOLEDs) and ultrathin CNT [282]. A p-MOS inverter consisting of four CNT transistors was fabricated to obtain high amplification of ECG signals. In the CTOLEDs, an ultrathin exciton-blocking layer (EBL) was applied between two adjacent emission layers to modify the balance charges. This resulted in the production of different colors in response to applied voltages. The design of CTOLEDs and ultrathin CNT not only are extremely resilient to very high folding and bending of human skin, but it also exhibits real-time ECG signals using adjustable colors, thus having superior reliability and stability after repeated deformations. However, CNTs are highly insoluble in aqueous solution or organic solvent, which hinders their biomedical applications. Hence, modifications of CNTs to improve solubility and expand applications are important. Three major modification methods for CNTs are (i) chemical modification of the π-conjugated skeleton of CNT, (ii) physical adsorption of many functional (bio)molecules, and (iii) filling of the inner empty cavity of CNT [285]. For example, one of the highly synthesized and simple modification methods for soluble CNT is 1,3-dipolar cycloaddition of azomethine, which can be applied for biomedical applications [285].

9.2.7.1 Carbon Dots

Carbon dots (CDs) are another class of CNTs first discovered in 2006. These zero-dimensional materials have characteristics greater than other traditional QDs and organic fluorophores, such as higher photostability against blinking and photobleaching, large two-photon excitation cross sections, and tunable emission [286]. In addition, CDs are hydrophilic and can be produced economically, thus enabling them to be applied in bioimaging [287], electrocatalysis [288, 289], optical sensing [290], and photocatalysis [291, 292]. The hydrophilicity of CDs is due to the presence of various functional groups including carboxyl, carbonyl, epoxy, and hydroxyl on their surfaces. This leads to high modification with biological, organic, or polymeric materials [293]. CDs also possess suitable biocompatibility, cell permeability, and lower toxicity [286, 294]. They are, therefore, attractive for drug delivery [295], intracellular sensing [286, 294], and photodynamic therapy [296]. In a recent study, a tumor extracellular microenvironment-responsive drug delivery based on CDs was developed for imaging-guided delivery [297]. This nanocarrier was synthesized by complexing of PEG-functionalized poly(allylamine) (PAH) and dimethylmaleic acid (DMMA) with cisplatin(IV) prodrug, followed by chemically modified CDs (CDs-Pt(IV)). CDs-Pt(IV)@PEG-(PAH/DMMA) exhibit positive charge at tumor extracellular microenvironment (pH 6.8) and negative charge at normal condition (pH 7.4), which leads to improved internalization by cancer cells, controlled intracellular drug release, prolonged circulation time in blood, multicolor bioimaging, effective accumulation at the tumor site, and facilitated endosome escape.

9.2.7.2 Nanodiamonds

As a member of the nanocarbon family, NDs can be fabricated through detonation, high-temperature–high-pressure methods, or CVD [298]. They are characterized by a large surface area, chemical stability, low costs, octahedral symmetry, and strong structure [299, 300]. NDs produced by detonation are most promising, since they can be easily modified with other (bio)molecules via covalent bonds or non-covalent interactions [301]. They have become indispensable in biomedicines, particularly for the delivery of drugs and biomolecules inside living cells [302, 303], because of the high compatibility between cell structure and ND materials [304]. In a notable example, Chow and colleagues report a highly effective method based on nanodiamond drug delivery to overcome chemoresistance in hepatic cancer stem cells [305]. Since epirubicin is chemotherapeutic, it was physically adsorbed to NDs and formed the stable nanodiamond–epirubicin drug complex (EPND) that improved tumor cell retention and enhanced endocytic uptake. Compared to unmodified standard chemotherapeutic, EPND has better efficacy in killing cancer stem and normal cancer cells. It efficiently inhibits secondary tumor formation. Fluorescent NDs are also effective tools in biomedical imaging due to nitrogen-vacancy centers, also known as auto-fluorescent structural defects, which consist of a neighbor vacancy and a substitutional nitrogen [306, 307]. They not only provide advantages of low toxicity and tunable surface chemistry of NDs, but they also possess semiconductor QDs properties. Thus, they are hold great potential for *in vivo* imaging.

9.2.8 Nanofibers

The properties of nanofibers have caused researchers and companies to consider using them in several fields. Nanofibers are an exciting new class of material used for various value-added applications such as skeletal muscle regeneration [308], vascular regeneration [309], nerve regeneration [310], and drug delivery [311]. An understanding of the mechanical properties of single nanofibers may be the first step in creating ideal materials. Three different approaches toward the formation of nanofibrous materials have emerged: self-assembly, electrospinning, and phase separation [312]. Each of these approaches is very different and has a unique set of characteristics, lending to its development as a scaffolding system. For instance, self-assembly can generate small diameter nanofibers at the lowest end of the range of natural extracellular matrix collagen, while electrospinning only generates large diameter nanofibers at the upper end of the range of natural extracellular matrix collagen. Electrospun fibers have emerged as a culture platform for different types of cells, since their structural parameters (such as pore size and fiber diameter with the typical range of 0.1–10 mm) are similar to single-cell processes – cell proliferation, alignment, migration, etc. [313, 314]. The chemical, mechanical, and geometrical properties of electrospun fiber scaffolds alter protein expression levels and apparent cell behavior [315–319]. Guo and coworkers developed a 3D hybrid scaffold for cardiac tissue engineering using an electrospinning technique [320]. The scaffold was formed from a photocurable methacrylated gelatin (GelMA) hydrogel shell that mimics the native cardiac tissue structure and aligns conductive nanofiber yarn networks (NFYs-NET composed of CNTs, polycaprolactone, and silk fibroin). NFYs-NET not only promotes cardiomyocytes (CMs) maturation but also controls cellular orientation. 3D scaffolds individually control cellular orientation on different layers in a 3D environment and improve aligned CMs maturation on each layer. They synthesize an endothelialized myocardial tissue via the co-culture of endothelial cells (ECs) within the hydrogel shell and CMs on the NFYs-NET layer.

Phase separation, on the other hand, generates nanofibers in the same range as natural extracellular matrix collagen and allows for the design of macropore structures. These attempts at an artificial extracellular matrix have the potential to accommodate cells and guide their growth with subsequent tissue regeneration. Self-assembly (i.e. the autonomous organization of molecules into patterns or substrates without human intervention) is common throughout nature and technology. Self-assembly of natural or synthetic macromolecules produces nanoscaled supramolecular structures and nanofibers. Specifically designed amphiphilic peptides containing a carbon alkyl tail and several other functional peptide regions have been synthesized and self-assembled into nanofibers with a diameter of 7.6 ± 1 nm [321]. These self-assembled nanofibers have been used to study selective differentiation of neural progenitor cells [322]. Another peptide containing 16 alternating hydrophobic and hydrophilic amino acids has been induced to self-assemble into nanofibers under appropriate pH values [323]. Natural polysaccharides such as alginate, gellan gum, hyaluronic acid (HA), and chitosan are another class of materials that can be self-assembled into fibrous systems. Despite their excellent physicochemical properties for fabrication of cell/drug delivery for tissue regeneration [324–326], only few studies have developed polysaccharides fibers [327, 328]. For example, Sant et al. report synthesis of self-assembled fibrous hydrogels via anionic gellan gum and cationic chitosan, which repeat the hierarchy of natural collagen fibers [329]. They functionalize gellan gum with methacrylate through photo-cross-linking to enhance the stability of fibrous structures. The two oppositely charged polysaccharides form polyelectrolyte complexation by electrostatic interaction. A microfluidic method has been applied to direct the preparation of hydrogel fibers and self-assembly. Additionally, these fibers exhibit biochemical mimicry of native collagen and high mechanical properties that make them potentially valuable for drug/gene delivery and tissue engineering.

9.2.9 Nanowires

Nanowires are one-dimensional thin structures having length-to-width aspect ratios typically higher than 10 and diameters less than 300 nm. They can be synthesized using various methods, the most common of which are metal-NP-mediated procedures (e.g. vapor–liquid–solid, solution–liquid–solid), direct deposition (e.g. seed-induced nanowire growth, screw dislocation-driven nanowire growth), oriented attachment, template-directed methods, lithography, and hydrothermal methods [330]. Nanotechnology based on nanowires promises a new generation of devices benefiting from large surface-to-volume ratios, small active volumes, quantum confinement effects, and integration in complex architectures on the nanoscale. Due to their novel electronic, mechanical, optical, and thermal characteristics, nanowires are a powerful platform in the development of photovoltaics [331], lithium-ion batteries [332], electronics [333], photonics [334], and sensors [335, 336]. They have also been applied in cancer [337] and drug/gene delivery [338, 339]. However, the biomedical and biological applications of nanowires remained relatively unexplored before the pioneering research of the Yang Group in 2007. They successfully cultured mouse embryonic stem cells on a nanowire array for gene delivery [340]. Since then, nanowires and variants such as nanostraws, nanoneedles, and nanopillars have become the mainstream for optical/electrical stimulation and intracellular delivery. These are all related to incredible properties of nanowires, including varied small cross section, high aspect ratio, and high surface-to-volume ratio. The cross sections of nanowires are much smaller than the typical cell dimensions, which provides easy penetration into cell membranes [341]. Also, their high aspect ratio, reputed to be their most significant property for bioassay, provides possible electrical and optical signals across cellular membranes [342, 343]. The high surface-to-volume ratio of nanowires also allows suitable interaction with cells for biomedical areas [344, 345]. Generally, interfacing living cells with nanowire arrays for biomedical and biological studies can be categorized into two main groups: cell-penetrating and non-penetrating. The ability of nanowires to penetrate into living cells leads to minimal disruption of cells, thus inducing minor toxicity. This is beneficial in the systematic finding of signaling components that are applied in exploring the role of intracellular signaling pathway, as well as in activation of Toll-like receptors [346, 347]. It is also beneficial for the transfection of immune cells in leukemogenesis [348].

Yang and colleagues introduced a sandwich assay for monitoring endogenous enzyme activity in living cells via two vertical silicon nanowires to immobilize cells and present enzymatic substrates [349]. This method provides measurement of various enzymatic functions when matched with mass spectrometry and fluorescence microscopy. Cui and coworkers utilized iridium oxide nanotubes that improve intracellular access duration and sensitivity compared to solid nanopillar electrodes [340]. Using vertical nanowire arrays, the Park Group developed a scalable intracellular platform for electrical interfacing to neuronal circuits [342]. This platform is able to perform, record, and stimulate neuronal activities along with its compatibility with optogenetic techniques and fluorescence microscopy, allowing interface with numerous neurons.

In contrast to several successful intracellular and biomolecular studies, the exact mechanism of cell–nanowire interface is not fully understood. Melosh and colleagues show that both actin cytoskeleton and plasma membrane act as synergistic obstacles for cell penetration of nanostraws [350]. Three different cases for nanostraw–cell interactions have been reported: (i) nanostraws did not penetrate but had access to an intact lipid bilayer, (ii) they passed through the cell membrane but not into the cytoskeleton, and (iii) they passed through the cell membrane and the cytoskeleton. By using cobalt ion delivery assay, the researchers measured the nanostraw penetration rates into cells (about 7%) [351]. Recently, Lieber and colleagues report a biomimetic method for internalization of nanowires into primary neuronal cells [352]. Semiconductor nanowires were modified with a trans-activating transcriptional (TAT) activator as a cell-penetrating peptide to facilitate the nanowires internalization. Results confirm c. 15% internalization efficiency. It took 30–40 minutes for nanowires to become completely internalized into mouse hippocampal neurons. In a notable example, Melosh and coworkers employed a continuum elastic cell mechanics model via impaling and adhesion-mediated mechanisms to clarify how penetration occurs [353]. Cell penetration via adhesion mechanism was much more impressive, and both cell stiffness and nanowire geometry played significant roles. Stiffer cells were dependent to nanowire geometry, while they exhibited greater penetration. The static models in this research are only able to explain when nanowire penetration occurs, yet the dynamic processes are critical. To address this, a dynamic nanowire–cell interface was reported using two mechanical models, namely, by substrate adhesion force and cell traction forces, combined with experimental cell-penetration measurements [354]. The adhesion model indicates that nanowire penetration occurs over a limited period, whereas the traction model shows higher cell penetration during a longer time. Nanowire penetration values measured from a cobalt ion delivery assay were compared with two models, and the role of cell properties and nanowire geometry was systematically determined to clarify the significant factors for penetration.

There are several studies regarding the use of non-penetrating nanowires for effective cell capture and separation. For example, Wang et al. fabricated a 3D cell capture substrate based on a silicon nanopillar modified with antibody against epithelial cell adhesion molecule (anti-EpCAM) that indicates highly efficient capture of circulating tumor cells and cell viability (84–91%) [355]. Lee and coworkers designed functionalized nanowire arrays capable of separating single cells from the heterogeneous mixture of cell populations [356]. Silicon nanowires with diameters of 100–200 nm were functionalized with streptavidin. They selectively isolated primary $CD4^+$ T lymphocytes from mouse whole splenocytes with high purity (~88%). Using 3D silicon nanowire-functionalized streptavidin further enabled an outstanding tumor cell capture (A549 cell) efficiency on selective areas of the micro-patterned platform [357]. A highly efficient cell capture platform capable of quantitative analysis and multiplexed phenotyping of rare trafficking leukocytes in cerebrospinal fluid samples collected from Alzheimer's disease patients is reported [358]. Several silicon nanowires were modified with capture agents against specific leukocyte phenotypic markers. The nanowire platform shows enhanced rare cell capture efficiency due to improved cell–substrate interactions and high-affinity binding. These 3D nanostructured materials can be applied in diagnosis or measurement of neurodegenerative pathology. Unlike the success of non-penetrating nanowires for enhanced separation efficiency, their mechanism was not examined in detail. Lee and colleagues studied cell surface morphology in contact with nanopillar surfaces and important factors that influence the mechanism for separation efficiency [359]. Quartz nanopillars alter filopodial morphologies (e.g. average number, width of protruded filopodia per cell, and length), which significantly affects the separation efficiency of CD^{4+} T lymphocytes.

Researchers discovered metal-like conductivity along protein filaments [360]. Physicists and biologists have teamed up to build a living transistor, an unusual kind of bacteria that produces long stringy filaments outside its body and conducts electrons better than some metals. Scientists describe these filaments as a "living nanowire" that can be a step forward in merging biological systems and electronic devices, leading to tiny organic super-batteries or biological superconductors made for a fraction of the cost of existing silicon-based chips. Strings of nanowires, called *pili,* allow bacteria to be rid of electrons that are a by-product of its digestive process. Humans and animals rid electrons through breathing, but bacteria living in anaerobic zones do not have oxygen to carry the electrons. Each spaghetti-like strand is 10–20 times longer than the bacteria itself, according to Tuominen. This discovery may lead to a range of new conducting nanomaterials that are living and naturally occurring as well as being nontoxic.

9.2.10 Nanorods

In nanotechnology, nanorods are defined as one-dimensional nanostructures with unique morphology, showing an aspect ratio of about 3–5. They are fabricated through direct chemical procedures using semiconducting or metal materials. Fabrication methods usually contain a mixture of ligands as shape control agents that link to various faces of the nanorod seed with different stabilities. Consequently, various faces of the nanorod seed with different rates cause an elongated shape [361–363]. Nanorod reflectivity can be adjusted by altering orientation with an applied electric field. Nanorods also offer fascinating applications as theranostic agents in the field of biomedicine, particularly in cancer treatment. They are able to absorb photons in the NIR and produce heat in the presence of IR radiation. In view of their small size (between 20 and 300 nm), they tend to accumulate in tumor cells. This is because of the tendency of tumor cells to grow on new blood vessels. As a result, when nanomaterials such as nanorods are administrated into a patient's bloodstream, they prefer to accumulate at tumor sites. In the response to suitable IR radiation wavelengths, IR photons activate nanorods and lead to heat generation, which further removes the cancerous tissue from body [364]. Nanorods can be produced mainly from gold, silica, and zinc oxide nanorods. Nanorods have attracted interest in applications such as pancreatic [365], skin [366], head, and neck cancer [367], multimodal imaging [368], and tissue engineering because of their unique properties [369]. In particular, gold nanorods (AuNRs) with anisotropic shapes possess strong surface plasmon resonance (SPR) in the NIR region [370] and visible region at c. 520 nm [371]. In those regions light is able to penetrate into the deep tissue area [372]. The absorption profile of gold nanorods includes two absorption bands: one due to light absorbed along the short axis (transverse) and the other due to absorption along the long axis (longitudinal). The longitudinal absorbance of AuNR depends on the aspect ratio (length/width), where the higher aspect ratio causes longer wavelengths for longitudinal band, and the transverse remains unchanged. For example, the longitudinal absorbance of AuNR can be tuned by selection of a suitable aspect ratio, which facilitates preparation of ideal networks for specific application such as the first NIR window in the range of ~650–950 nm [373]. Generally, two biological transparency windows are located in the NIR regions, known as the first NIR window (650–950 nm) and the second NIR window (1000–1350 nm). Yeh and colleagues developed a rattle-like AuNR that destroys cancer cells in response to continuous-wave diode laser in the first window (808 nm) and the second window (1064 nm) [373]. AuNRs are also able to physically bind therapeutic molecules such as antibodies [374] and interferons [375] on their surface through electrostatic and hydrophobic interaction [376]. Therefore, these properties of AuNR offer significant applications in optical signal transmission [377], cellular activity [378], biological sensing [379], and photothermal and photodynamic therapies [380]. Hahn and coworkers synthesized a hyaluronate–gold nanorod/death receptor 5 antibody (HA–AuNR/DR5 Ab) for noninvasive transdermal treatment of skin cancer [366]. The HA–AuNR/DR5 Ab significantly decreases tumor growth for up to five days.

Apart from the aforementioned studies that concern the use of AuNR, other types of nanorods including gadolinium oxide, zinc oxide, tellurium, and copper bismuth sulfide nanorods have been applied in different fields [381, 382]. In a notable example, ternary semiconductor nanorods composed of copper bismuth sulfide (Cu3BiS3) with various length/diameter ratios were obtained via a facile organic route [383]. When dispersed in an aqueous solution, PEGylated Cu3BiS3 NR shows strong and wide photoabsorption wavelengths from UV to NIR. Researchers used a 1064 nm-laser-driven photothermal agent and compared results to other reported photothermal therapy (PTT) agents in the second NIR window [384–386], and PEGylated Cu3BiS3 NR exhibit higher photothermal conversion efficiency (40.7%). These nanorods also possess high photothermal transduction efficiency, large absorption cross section, and high tissue penetration depth at 1064 nm, as well as low cytotoxicity, resulting in outstanding tumor photothermal ablation in the second NIR window. The injected nanorods in tumors indicate intense CT and photoacoustic (PA) signals and trigger mild PTT in the second NIR window irradiation. Hence, such excellent nanorods hold profound implications for many aspects of biomedical studies, particularly radiotherapy as an important tumor therapy. In another study, a micropatterned polymer nanorod through templating method was employed. A thin layer of gelatin on a nylon mesh with microscale openings was deposited by electrospray [369]. Then, the gelatin polymer was transferred onto an anodized aluminum oxide (AAO) membrane as a template. Such cost-effective and easy procedures can be adjusted by altering the characteristics of the AAO membrane, such as inter-channel and channel-diameter spacing as well as preparation of nylon mesh dimensions and electrospray time.

9.2.11 Self-assembly of Proteins and Peptides

Self-assembly can be classified in two types: static and dynamic [387]. Most research studies on self-assembly have focused on the static type, while studies on dynamic self-assembly are still poorly understood. Static self-assembly contributes to systems that are at global or local equilibrium and do not dissipate energy [387]. In static self-assembly, formation of the ordered structure may require energy, but it is stable once formed. A few examples of static self-assembly phenomenon tailored by nature are lipid molecules forming oil droplets in water, four hemoglobin polypeptides forming a functional hemoglobin protein, and the combination of RNA and ribosomal proteins to form a functional ribosome. Dynamic self-assembly occurs when the formation of an ordered state of equilibrium requires dissipation of energy. In other words, the interactions responsible for the formation of structures or patterns between components occur only if the system dissipates energy [387]. Self-assembly takes place at molecular, mesoscopic, and macroscopic scales. Based on these criteria, self-assembly has been classified as molecular and nanoscale self-assemblies (classic form of self-assembly in chemistry involving atoms, molecules, and crystal formation) and meso- and macroscopic self-assemblies (e.g. engineered microparts). Molecular and nanoscale self-assembly can be further classified as intramolecular and intermolecular [388]. In intramolecular self-assembly, molecules are often complex polymers that can assemble from the random coil into a well-defined stable structure (secondary and tertiary structure). Protein folding to form secondary and tertiary amine structures is a good example of intramolecular self-assembly. Intermolecular self-assembly is the ability of molecules to form supramolecular assemblies (quaternary structure). Supramolecular micelle formation by surfactant molecules in solution and self-assembled monolayers (SAMs) on a substrate are classic examples of intermolecular self-assembly. Magnetic, capillary, electrostatic, and gravitational forces play a vital role in self-assembly of structures on the meso and macro scale levels [388].

The self-assembly of proteins and peptides is characterized by physical interactions, including van der Waals interactions (London dispersion, dipole–dipole forces), hydrophobic interactions, electrostatic forces (halogen bonding, hydrogen, and ionic bonding), and π-interactions (π–π stacking, π-anion, π-polar, and π-cation) [25, 389–391]. Such interactions can be regulated by altering environmental factors and manipulating sequences of amino acids [392, 393]. The supramolecular nanostructure of proteins and peptides plays key role for biological, physical, and chemical properties, which significantly affect applications. Self-assembled protein-based nanomaterials are not only able to transport phototherapeutic agents through the nanostructure core and hydrophobic domain, but they also deliver (bio)molecules/therapeutic agents naturally, thus offering significant potential for applications in PTT and drug delivery. Kim and colleagues developed self-assembled protein NPs with a targeting moiety and a tumor-selective oncolytic protein for cancer therapy [394]. Apoptin (Apo) was employed as a tumor-selective apoptotic protein, whereas an EGFR-specific repebody (Rb) was used to function as a tumor-targeting moiety and a fusion partner for generating apoptin NPs in

Escherichia coli. The repebody genetically fused to apoptin. The fusion protein obtained self-assembled supramolecular repebody-apoptin (Rb-Apo) NPs showing great stability and homogeneity as a soluble form in *E. coli*. Rb-Apo NPs indicated a degree of tumor localization higher than maltose binding protein-fused apoptin (MBP-Apo) NPs. Then, the Rb-Apo NPs with 5 and 10 mg kg^{-1} were intravenously injected each three days into xenograft mice. An excellent antitumor activity and low side effects in xenograft mice were detected for Rb-Apo NPs by combined action of tumor-selective apoptosis and targeted delivery by the apoptin and repebody, respectively. Thus, the tumor size decreased in a dose-response manner.

Due to their bioavailability and biocompatibility, self-assembled peptide nanostructures are of interest in biomedical studies, particularly in cancer therapy and drug delivery [395–397]. Therefore, several peptides have been explored for the preparation of supramolecular nanostructures, including aromatic dipeptides [398], polypeptides [399], and amphiphilic peptides [400]. Various peptides and proteins that self-assemble to form different nanostructures like nanotubes, vesicles, helical ribbons, and fibrous scaffolds have been discussed in detail [401–403]. These structures are analyzed to design and fabricate new materials with potential applications in biomedical nanotechnology. The findings in research papers have been discussed and grouped under different headings like self-assembling peptide systems, amphiphilic and surfactant peptides, peptides self-assembly into 3D matrix scaffolds, and designed peptide hydrogels for 3D cell culture and regenerative biology [404, 405]. Wang et al. have designed artificial proteins that self-assemble to form hydrogels [406]. The artificial proteins designed by the researchers consist of an ionic self-complementary peptide group that has an alternating polar and nonpolar arrangement of peptide molecules. These peptides form a stable β-strand and β-sheet structures that self-assemble to form nanofibers. The nanofibers form interwoven matrices that further give a scaffold hydrogel with high water content. This hydrogel structure contains water as its dispersion medium and responds to changes in pH and other environmental factors. Such protein hydrogels can be employed for advanced wound closure and tissue repair in regenerative medicine and tissue engineering. Biodegradable protein hydrogels can be utilized in drug delivery systems, delivering pharmaceutical protein complexes in the treatment of diseases like cancer. Adams et al. designed peptide nanotubes using surfactant-like peptides [407]. These nanotubes can be used as templates for growing metal nanocrystals and to fabricate nanowires. They can also serve as ion channels when incorporated into the phospholipid bilayer of the cell membrane. Recent studies report biomimetic protein structures and peptide systems capable of forming complexes with metals and semiconducting elements. Surface-binding peptides bind covalently with metal surfaces like gold [408]. These peptides can also form complexes with DNA that in turn bind to a metal surface. Such property may be exploited to design and fabricate nanobiosensors. Certain peptides with strong dipoles undergo severe conformational changes between the α-helical and the β-sheet structures. They are called molecular switch peptides. One report relates that AuNPs can be attached to these dipolar peptides to fabricate tiny molecular switches [409]. Hydrogels constructed through self-assembly are receiving increasing attention for a variety of biomedical and biotechnological applications, including scaffolds for regenerative medicine [410, 411], the controlled release of therapeutics [412–415], and defined cell culture matrices [416]. Despite the success of self-assembled proteins and peptides nanomaterials for biomedical applications, their preparation by adjustable procedures remains challenging, because the corresponding nanostructures and detailed alternation of physical interactions of proteins and peptides have not been fully researched. Consequently, complex structural characteristics, as well as the self-assembling mechanisms of protein and peptide nanomaterials, need to be studied extensively to expand their applications.

References

1 Pumera, M. (2010). Graphene-based nanomaterials and their electrochemistry. *Chem. Soc. Rev.* 39: 4146–4157.

2 Wang, J. (2005). Nanomaterial-based electrochemical biosensors. *Analyst* 130: 421–426.

3 Chen, X., Li, C., Grätzel, M. et al. (2012). Nanomaterials for renewable energy production and storage. *Chem. Soc. Rev.* 41: 7909–7937.

4 Chen, G., Roy, I., Yang, C., and Prasad, P.N. (2016). Nanochemistry and nanomedicine for nanoparticle-based diagnostics and therapy. *Chem. Rev.* 116: 2826–2885.

5 Min, Y., Caster, J.M., Eblan, M.J., and Wang, A.Z. (2015). Clinical translation of nanomedicine. *Chem. Rev.* 115: 11147–11190.

6 Li, Y., Wang, Y., Huang, G., and Gao, J. (2018). Cooperativity principles in self-assembled nanomedicine. *Chem. Rev.* 118: 5359–5391.

7 Xu, G., Zeng, S., Zhang, B. et al. (2016). New generation cadmium free quantum dots for biophotonics and nanomedicine. *Chem. Rev.* 116: 12234–12327.

8 Behzadi, S., Serpooshan, V., Tao, W. et al. (2017). Cellular uptake of nanoparticles: journey inside the cell. *Chem. Soc. Rev.* 46: 4218–4244.

9 Du, B., Yu, M., and Zheng, J. (2018). Transport and interactions of nanoparticles in the kidneys. *Nat. Rev. Mater.* 3: 358–374.

10 D'Mello, S.R., Cruz, C.N., Chen, M.-L. et al. (2017). The evolving landscape of drug products containing nanomaterials in the United States. *Nat. Nanotechnol.* 12: 523–529.

11 Prasad, P.N. (2004). *Nanophotonics*. Hoboken, NJ: Wiley-Interscience.

12 Burda, C., Chen, X.B., Narayanan, R., and El-Sayed, M.A. (2005). Chemistry and properties of nanocrystals of different shapes. *Chem. Rev.* 105: 1025–1102.

13 Cui, Y., Wei, Q.Q., Park, H.K., and Lieber, C.M. (2001). Nanowire nanosensors for highly sensitive and selective detection of biological and chemical species. *Science* 293: 1289–1292.

14 Gao, X.H., Cui, Y.Y., Levenson, R.M. et al. (2004). *In vivo* cancer targeting and imaging with semiconductor quantum dots. *Nat. Biotechnol.* 22: 969–976.

15 Kan, S., Mokari, T., Rothenberg, E., and Banin, U. (2003). Synthesis and size-dependent properties of zinc-blende semiconductor quantum rods. *Nat. Mater.* 2: 155–158.

16 Coleman, J.N., Lotya, M., O'Neill, A. et al. (2011). Two-dimensional nanosheets produced by liquid exfoliation of layered materials. *Science* 331: 568–571.

17 Kinnear, C., Moore, T.L., Rodriguez-Lorenzo, L. et al. (2017). Form follows function: nanoparticle shape and its implications for nanomedicine. *Chem. Rev.* 117: 11476–11521.

18 Sharifzadeh, G. and Hosseinkhani, H. (2017). Biomolecule-responsive hydrogels in medicine. *Adv. Healthc. Mater.* 6: 1700801.

19 Kievit, F.M. and Zhang, M.Q. (2011). Surface engineering of iron oxide nanoparticles for targeted cancer therapy. *Acc. Chem. Res.* 44: 853–862.

20 Nam, J., Won, N., Bang, J. et al. (2013). Surface engineering of inorganic nanoparticles for imaging and therapy. *Adv. Drug Deliv. Rev.* 65: 622–648.

21 Langer, R. and Tirrell, D.A. (2004). Designing materials for biology and medicine. *Nature* 428: 487–492.

22 Gatenby, R.A. (2009). A change of strategy in the war on cancer. *Nature* 459: 508–509.

23 Brannon-Peppas, L. and Blanchette, J.O. (2004). Nanoparticle and targeted systems for cancer therapy. *Adv. Drug Deliv. Rev.* 56: 1649–1659.

24 Chauhan, V.P. and Jain, R.K. (2013). Strategies for advancing cancer nanomedicine. *Nat. Mater.* 12: 958–962.

25 Webber, M.J., Appel, E.A., Meijer, E., and Langer, R. (2016). Supramolecular biomaterials. *Nat. Mater.* 15: 13–26.

26 Chabner, B.A. and Roberts, T.G. (2005). Chemotherapy and the war on cancer. *Nat. Rev. Cancer* 5: 65–72.

27 Wagner, V., Dullaart, A., Bock, A.K., and Zweck, A. (2006). The emerging nanomedicine landscape. *Nat. Biotechnol.* 24: 1211–1217.

28 Riehemann, K., Schneider, S.W., Luger, T.A. et al. (2009). Nanomedicine-challenge and perspectives. *Angew. Chem. Int. Ed.* 48: 872–897.

29 Boisselier, E. and Astruc, D. (2009). Gold nanoparticles in nanomedicine: preparations, imaging, diagnostics, therapies and toxicity. *Chem. Soc. Rev.* 38: 1759–1782.

30 Allen, T.M. and Cullis, P.R. (2004). Drug delivery systems: entering the mainstream. *Science* 303: 1818–1822.

31 Etheridge, M.L., Campbell, S.A., Erdman, A.G. et al. (2013). The big picture on nanomedicine: the state of investigational and approved nanomedicine products. *Nanomedicine* 9: 1–14.

32 Jain, R.K. and Stylianopoulos, T. (2010). Delivering nanomedicine to solid tumors. *Nat. Rev. Clin. Oncol.* 7: 653–664.

33 Sanhai, W.R., Sakamoto, J.H., Canady, R., and Ferrari, M. (2008). Seven challenges for nanomedicine. *Nat. Nanotechnol.* 3: 242–244.

34 Ruoslahti, E., Bhatia, S.N., and Sailor, M.J. (2010). Targeting of drugs and nanoparticles to tumors. *J. Cell Biol.* 188: 759–768.

35 Tang, F., Li, L., and Chen, D. (2012). Mesoporous silica nanoparticles: synthesis, biocompatibility and drug delivery. *Adv. Mater.* 24: 1504–1534.

36 Oberdörster, G. (2010). Safety assessment for nanotechnology and nanomedicine: concepts of nanotoxicology. *J. Intern. Med.* 267: 89–105.

37 Kolosnjaj-Tabi, J., Javed, Y., Lartigue, L. et al. (2015). The one year fate of iron oxide coated gold nanoparticles in mice. *ACS Nano* 9: 7925–7939.

38 Chou, L.Y.T., Zagorovsky, K., and Chan, W.C.W. (2014). DNA assembly of nanoparticle superstructures for controlled biological delivery and elimination. *Nat. Nanotechnol.* 9: 148–155.

39 Kreyling, W.G., Abdelmonem, A.M., Ali, Z. et al. (2015). *In vivo* integrity of polymer-coated gold nanoparticles. *Nat. Nanotechnol.* 10: 619–623.

40 Kotov, N.A. (2010). Inorganic nanoparticles as protein mimics. *Science* 330: 188–189.

41 Wilhelm, S., Kaiser, M., Wuerth, C. et al. (2015). Water dispersible upconverting nanoparticles: effects of surface modification on their luminescence and colloidal stability. *Nanoscale* 7: 1403–1410.

42 Colombo, M., Carregal-Romero, S., Casula, M.F. et al. (2012). Biological applications of magnetic nanoparticles. *Chem. Soc. Rev.* 41: 4306–4334.

43 Lin, C.-A.J., Yang, T.-Y., Lee, C.-H. et al. (2009). Synthesis, characterization, and bioconjugation of fluorescent gold nanoclusters toward biological labeling applications. *ACS Nano* 3: 395–401.

44 Parak, W.J., Pellegrino, T., and Plank, C. (2005). Labelling of cells with quantum dots. *Nanotechnology* 16: R9–R25.

45 Zhang, R., Yang, J., Sima, M. et al. (2014). Sequential combination therapy of ovarian cancer with degradable N-(2-hydroxypropyl)methacrylamide copolymer paclitaxel and gemcitabine conjugates. *Proc. Natl. Acad. Sci. U.S.A* 111: 12181–12186.

46 Sharifzadeh, G., Wahit, M.U., Soheilmoghaddam, M. et al. (2016). Kappa-carrageenan/halloysite nanocomposite hydrogels as potential drug delivery systems. *J. Taiwan Inst. Chem. Eng.* 67: 426–434.

47 Guo, P. (2010). The emerging field of RNA nanotechnology. *Nat. Nanotechnol.* 5: 833–842.

48 Nurunnabi, M., Khatun, Z., Reeck, G.R. et al. (2014). Photoluminescent graphene nanoparticles for cancer phototherapy and imaging. *ACS Appl. Mater. Interfaces* 6: 12413–12421.

49 Yu, S.J., Kang, M.W., Chang, H.C. et al. (2005). Bright fluorescent nanodiamonds: no photobleaching and low cytotoxicity. *J. Am. Chem. Soc.* 127: 17604–17605.

50 Lee, Y., Fukushima, S., Bae, Y. et al. (2007). A protein nanocarrier from charge-conversion polymer in response to endosomal pH. *J. Am. Chem. Soc.* 129: 5362–5363.

51 Sapsford, K.E., Tyner, K.M., Dair, B.J. et al. (2011). Analyzing nanomaterial bioconjugates: a review of current and emerging purification and characterization techniques. *Anal. Chem.* 83: 4453–4488.

52 Schaak, R.E. and Williams, M.E. (2012). Full disclosure: the practical side of nanoscale total synthesis. *ACS Nano* 6: 8492–8497.

53 Wolf, L.K. (2012). Sweating the small stuff. *Chem. Eng. News* 90: 48–50.

54 Kovalenko, M.V., Manna, L., Cabot, A. et al. (2015). Prospects of nanoscience with nanocrystals. *ACS Nano* 9: 1012–1057.

55 Chou, L.Y.T., Ming, K., and Chan, W.C.W. (2011). Strategies for the intracellular delivery of nanoparticles. *Chem. Soc. Rev.* 40: 233–245.

56 Yildiz, I., Shukla, S., and Steinmetz, N.F. (2011). Applications of viral nanoparticles in medicine. *Curr. Opin. Biotechnol.* 22: 901–908.

57 Pokorski, J.K. and Steinmetz, N.F. (2011). The art of engineering viral nanoparticles. *Mol. Pharm.* 8: 29–43.

58 Voigt, N., Henrich-Noack, P., Kockentiedt, S. et al. (2014). Surfactants, not size or zeta-potential influence blood–brain barrier passage of polymeric nanoparticles. *Eur. J. Pharm. Biopharm.* 87: 19–29.

59 Rosenholm, J.M., Meinander, A., Peuhu, E. et al. (2009). Targeting of porous hybrid silica nanoparticles to cancer cells. *ACS Nano* 3: 197–206.

60 Rawat, M., Singh, D., and Saraf, S. (2006). Nanocarriers: promising vehicle for bioactive drugs. *Biol. Pharm. Bull.* 29: 1790–1798.

61 Naahidi, S., Jafari, M., Edalat, F. et al. (2013). Biocompatibility of engineered nanoparticles for drug delivery. *J. Control. Release* 166: 182–194.

62 Duncan, R. (2003). The dawning era of polymer therapeutics. *Nat. Rev. Drug Discov.* 2: 347–360.

63 Hawkins, M.J., Soon-Shiong, P., and Desai, N. (2008). Protein nanoparticles as drug carriers in clinical medicine. *Adv. Drug Deliv. Rev.* 60: 876–885.

64 Jahanshahi, M. and Babaei, Z. (2008). Protein nanoparticle: a unique system as drug delivery vehicles. *Afr. J. Biotechnol.* 7: 4926–4934.

65 Coester, C.J., Langer, K., Von Briesen, H., and Kreuter, J. (2000). Gelatin nanoparticles by two step desolvation - a new preparation method, surface modifications and cell uptake. *J. Microencapsul.* 17: 187–193.

66 Weber, C., Coester, C., Kreuter, J., and Langer, K. (2000). Desolvation process and surface characterisation of protein nanoparticles. *Int. J. Pharm.* 194: 91–102.

67 Elzoghby, A.O., Samy, W.M., and Elgindy, N.A. (2012). Albumin-based nanoparticles as potential controlled release drug delivery systems. *J. Control. Release* 157: 168–182.

68 Joachim, E., Kim, I.-D., Jin, Y. et al. (2014). Gelatin nanoparticles enhance the neuroprotective effects of Intranasally administered osteopontin in rat ischemic stroke model, drug delivery. *Transl. Res.* 4: 395–399.

69 Shapira, A., Davidson, I., Avni, N. et al. (2012). β-Casein nanoparticle-based oral drug delivery system for potential treatment of gastric carcinoma: stability, target-activated release and cytotoxicity. *Eur. J. Pharm. Biopharm.* 80: 298–305.

70 Zhen, Z.P., Tang, W., Guo, C.L. et al. (2013). Ferritin nanocages to encapsulate and deliver photosensitizers for efficient photodynamic therapy against cancer. *ACS Nano* 7: 6988–6996.

71 Kim, J.Y., O'Malley, S., Mulchandani, A., and Chen, W. (2005). Genetically engineered elastin-protein a fusion as a universal platform for homogeneous, phase-separation immunoassay. *Anal. Chem.* 77: 2318–2322.

72 Duclairoir, C., Orecchioni, A.M., Depraetere, P. et al. (2003). Evaluation of gliadins nanoparticles as drug delivery systems: a study of three different drugs. *Int. J. Pharm.* 253: 133–144.

73 Buhleier, E., Wehner, W., and Vögtle, F. (1978). "Cascade"- and "nonskid-chain-like" synthesis of molecular cavity topologies. *Synthesis* 9: 155–158.

74 Zhang, C., Pan, D., Luo, K. et al. (2014). Dendrimer–doxorubicin conjugate as enzyme-sensitive and polymeric nanoscale drug delivery vehicle for ovarian cancer therapy. *Polym. Chem.* 5: 5227–5235.

75 Cheng, Y., Zhao, L., Li, Y., and Xu, T. (2011). Design of biocompatible dendrimers for cancer diagnosis and therapy: current status and future perspectives. *Chem. Soc. Rev.* 40: 2673–2703.

76 Soiberman, U., Kambhampati, S.P., Wu, T. et al. (2017). Subconjunctival injectable dendrimer-dexamethasone gel for the treatment of corneal inflammation. *Biomaterials* 125: 38–53.

77 Nandy, B. and Maiti, P.K. (2011). DNA compaction by a dendrimer. *J. Phys. Chem. B* 115: 217–230.

78 Pavan, G.M., Kostiainen, M.A., and Danani, A. (2010). Computational approach for understanding the interactions of UV-degradable dendrons with DNA and siRNA. *J. Phys. Chem. B* 114: 5686–5693.

79 Vasumathi, V. and Maiti, P.K. (2010). Complexation of siRNA with dendrimer: a molecular modeling approach. *Macromolecules* 43: 8264–8274.

80 Malik, N., Wiwattanapatapee, R., Klopsch, R. et al. (2000). Dendrimers: relationship between structure and biocompatibility *in vitro*, and preliminary studies on the biodistribution of 125I-labelled polyamidoamine dendrimers *in vivo*. *J. Control. Release* 65: 133–148.

81 Zhang, F., Magruder, J.T., Lin, Y.A. et al. (2017). Generation-6 hydroxyl PAMAM dendrimers improve CNS penetration from intravenous administration in a large animal brain injury model. *J. Control. Release* 249: 173–182.

82 Hu, Q., Ding, B., Yan, X. et al. (2017). Polyethylene glycol modified PAMAM dendrimer delivery of kartogenin to induce chondrogenic differentiation of mesenchymal stem cells. *Nanomedicine* 13: 2189–2198.

83 Shen, Y. and Jiasheng, T. (2009). Synthesis and characterization of low molecular weight hyaluronic acid-based cationic micelles for efficient siRNA delivery. *Carbohydr. Polym.* 77: 95–104.

84 Van Hasselt, P.M., Janssens, G.E.P., Slot, T.K. et al. (2009). The influence of bile acids on the oral bioavailability of vitamin K encapsulated in polymeric micelles. *J. Control. Release* 133: 161–168.

85 Bromberg, L. (2008). Polymeric micelles in oral chemotherapy. *J. Control. Release* 128: 99–112.

86 Nishiyama, N., Okazaki, S., Cabral, H. et al. (2003). Novel cisplatin-incorporated polymeric micelles can eradicate solid tumors in mice. *Cancer Res.* 63: 8977–8983.

87 Jones, M.-C. and Leroux, J.-C. (1999). Polymeric micelles–a new generation of colloidal drug carriers. *Eur. J. Pharm. Biopharm.* 48: 101–111.

88 Kedar, U., Phutane, P., Shidhaye, S., and Kadam, V. (2010). Advances in polymeric micelles for drug delivery and tumor targeting. *Nanomedicine* 6: 714–729.

89 Cabral, H. and Kataoka, K. 22. The impact of "Development of the polymer micelles carrier system for doxorubicin" on the nanomedicine realm: Original research article: Development of the polymer micelle carrier system for doxorubicin, 2001 references. *J. Control. Release* 190 (2014): 70–72.

90 Hosseinkhani, H., Abedini, F., Ou, K.L., and Domb, A.J. (2015). Polymers in gene therapy technology. *Polym. Adv. Technol.* 26: 198–211.

91 Hosseinkhani, H., Aoyama, T., Ogawa, O., and Tabata, Y. (2002). *In vitro* transfection of plasmid DNA by amine derivatives of gelatin accompanied with ultrasound irradiation. *Pharm. Res.* 19: 1469–1477.

92 Hosseinkhani, H., Aoyama, T., Ogawa, O., and Tabata, Y. (2002). Ultrasound enhancement of *in vitro* transfection of plasmid DNA by a cationized gelatin. *J. Drug Target.* 10: 193–204.

93 Ferrari, S., Moro, E., Pettenazzo, A. et al. (1997). ExGen 500 is an efficient vector for gene delivery to lung epithelial cells *in vitro* and *in vivo*. *Gene Ther.* 4: 1100–1109.

94 Abdallah, B., Hassan, A., Benoist, C. et al. (1996). A powerful nonviral vector for *in vivo* gene transfer into the adult mammalian brain: polyethylenimine, human. *Gene Ther.* 7: 1947–1954.

95 Kimura, T., Yamaoka, T., Iwase, R., and Murakami, A. (2001). Structure/function relationship in the polyplexes containing cationic polypeptides for gene delivery. *Nucleic Acids Res. Suppl.* 1: 203–209.

96 van de Wetering, P., Cherng, J.Y., Talsma, H. et al. (1998). 2-(dimethylamino) ethyl methacrylate based (co)polymers as gene transfer agents. *J. Control. Release* 53: 145–156.

97 Wolfert, M.A., Schacht, E.H., Toncheva, V. et al. (1996). Characterization of vectors for gene therapy formed by self-assembly of DNA with synthetic block co-polymers, human. *Gene Ther.* 7: 2123–2129.

98 Yaroslavov, A.A., Sukhishvili, S.A., Obolsky, O.L. et al. (1996). DNA affinity to biological membranes is enhanced due to complexation with hydrophobized polycation. *FEBS Lett.* 384: 177–183.

99 MacLaughlin, F.C., Mumper, R.J., Wang, J. et al. (1998). Chitosan and depolymerized chitosan oligomers as condensing carriers for *in vivo* plasmid delivery. *J. Control. Release* 56: 259–265.

100 Tang, M.X., Redemann, C.T., and Szoka, F.C. Jr., (1996). *In vitro* gene delivery by degraded polyamidoamine dendrimers. *Bioconjug. Chem.* 7: 703–709.

101 Harada-Shiba, M., Yamauchi, K., Harada, A. et al. (2002). Polyion complex micelles as vectors in gene therapy pharmacokinetics and *in vivo* gene transfer. *Gene Ther.* 9: 407–415.

102 Takai, T. and Ohmori, H. (1990). DNA transfection of mouse lymphoid-cell by the combination of DEAE-dextran-mediated DNA uptake and osmotic shock procedure. *Biochim. Biophys. Acta* 1048: 105–112.

103 Hosseinkhani, H. (2012). 3D *in vitro* technology for drug discovery. *Curr. Drug Saf.* 7: 37–43.

104 Chang, D.P., Barauskas, J., Dabkowska, A.P. et al. (2015). Non-lamellar lipid liquid crystalline structures at interfaces. *Adv. Colloid Interface Sci.* 222: 135–147.

105 Mulet, X., Boyd, B.J., and Drummond, C.J. (2013). Advances in drug delivery and medical imaging using colloidal lyotropic liquid crystalline dispersions. *J. Colloid Interface Sci.* 393: 1–20.

106 Chen, H., Sulejmanovic, D., Moore, T. et al. (2014). Iron-loaded magnetic nanocapsules for pH-triggered drug release and MRI imaging. *Chem. Mater.* 26: 2105–2112.

107 Li, J., Sun, L., Liu, Y. et al. (2019). To reduce premature drug release while ensure burst intracellular drug release of solid lipid nanoparticle-based drug delivery system with clathrin modification. *Nanomedicine* 15: 108–118.

108 Louie, A. (2010). Multimodality imaging probes: design and challenges. *Chem. Rev.* 110: 3146–3195.

109 Fattahi, H., Laurent, S., Liu, F. et al. (2011). Magnetoliposomes as multimodal contrast agents for molecular imaging and cancer nanotheragnostics. *Nanomedicine (London, U.K.)* 6: 529–544.

110 Xing, Y., Zhao, J., Conti, P.S., and Chen, K. (2014). Radiolabeled nanoparticles for multimodality tumor imaging. *Theranostics* 4: 290–306.

111 Midoux, P. and Pichon, C. (2015). Lipid-based mRNA vaccine delivery systems. *Expert Rev. Vaccines* 14: 221–234.

112 Reichmuth, A.M., Oberli, M.A., Jaklenec, A. et al. (2016). mRNA vaccine delivery using lipid nanoparticles. *Ther. Delivery* 7: 319–334.

113 Kranz, L.M., Diken, M., Haas, H. et al. (2016). Systemic RNA delivery to dendritic cells exploits antiviral defence for cancer immunotherapy. *Nature* 534: 396–401.

114 Broos, K., Van der Jeught, K., Puttemans, J. et al. (2016). Particle-mediated intravenous delivery of antigen mRNA results in strong antigen-specific T-cell responses despite the induction of type I interferon. *Mol. Ther.–Nucleic Acids* 5: e326.

115 Sayour, E.J., Grippin, A., De Leon, G. et al. (2018). Personalized tumor RNA loaded lipid-nanoparticles prime the systemic and intratumoral milieu for response to cancer immunotherapy. *Nano Lett.* 18: 6195–6206.

116 Oberli, M.A., Reichmuth, A.M., Dorkin, J.R. et al. (2017). Lipid nanoparticle assisted mRNA delivery for potent cancer immunotherapy. *Nano Lett.* 17: 1326–1335.

117 Felgner, P.L., Gadek, T.R., Holm, M. et al. (1987). Lipofection: a highly efficient, lipid-mediated DNA-transfection procedure. *Proc. Natl. Acad. Sci. U.S.A.* 84: 7413–7418.

118 Malone, R.W., Felgner, P.L., and Verma, I.M. (1989). Cationic liposome-mediated RNA transfection. *Proc. Natl. Acad. Sci. U.S.A.* 86: 6077–6083.

119 Felgner, J.H., Kumar, R., Sridhar, C.N. et al. (1994). Enhanced gene delivery and mechanism studies with a novel series of cationic lipid formulations. *J. Biol. Chem.* 269: 2550–2559.

120 Zhu, S.G., Xiang, J.J., Li, X.L. et al. (2004). Poly(L-lysine)-modified silica nanoparticles for the delivery of antisense oligonucleotides. *Biotechnol. Appl. Biochem.* 39: 179–187.

121 Lee, J.H. and Welsh, M.J. (1999). Enhancement of calcium phosphate-mediated transfection by inclusion of adenovirus in coprecipitates. *Gene Ther.* 6: 676–682.

122 Abedini, F., Ismail, M., Hosseinkhani, H. et al. (2010). Toxicity evaluation of dextran-spermine polycation as a tool for gene therapy *in vitro*. *J. Cell Anim. Biol.* 4: 170–176.

123 Hosseinkhani, H., Chen, Y.R., He, W. et al. (2013). Engineering of magnetic DNA nanoparticles for tumor-targeted therapy. *J. Nanopart. Res.* 15: 1–10.

124 Abdullah, S., Yeo, W.Y., Hosseinkhani, H. et al. (2010). Gene transfer into the lung by nanoparticle dextran-spermine/plasmid DNA complexes. *J. Biomed. Biotechnol.* 2010: 1–10.

125 Hosseinkhani, H., Aoyama, T., Ogawa, O., and Tabata, T. (2002). Liver targeting of plasmid DNA by pullulan conjugation based on metal coordination. *J. Control. Release* 83: 286–301.

126 Amini, R., Hosseinkhani, H., Jalilian, A. et al. (2012). Engineered smart biomaterials for gene delivery. *Gene Ther. Mol. Biol.* 14: 72–86.

127 Abedini, F., Ismail, M., Hosseinkhani, H. et al. (2011). Effects of CXCR4 siRNA/dextran-spermine nanoparticles on CXCR4 expression and serum LDH levels in a mouse model of colorectal cancer metastasis to liver. *Cancer Manag. Res.* 3: 301–309.

128 Abedini, F., Hosseinkhani, H., Ismail, M. et al. (2011). *In vitro* intracellular trafficking of biodegradable nanoparticles dextran-spermine in cancer cell lines. *Int. J. Nanotechnol.* 8: 712–723.

129 Hosseinkhani, H., Aoyama, T., Ogawa, O., and Tabata, T. (2001). *In vitro* transfection of plasmid DNA by different-cationized gelatin with or without ultrasound irradiation. *Proc. Jpn. Acad. Ser. B* 77: 161–166.

130 Hosseinkhani, H., Aoyama, T., Ogawa, O., and Tabata, Y. (2003). Tumor targeting of gene expression by dextran conjugation based on metal coordination. *J. Control. Release* 88: 297–312.

131 Hosseinkhani, H., Aoyama, T., Ogawa, O., and Tabata, T. (2003). Ultrasound enhances the transfection of plasmid DNA by non-viral vector. *Curr. Pharm. Biotechnol.* 4: 109–122.

132 Hosseinkhani, H. and Tabata, Y. (2003). RETRACTED: *In vitro* gene expression by cationized derivatives of an artificial protein with repeated RGD sequences, Pronectin®. *J. Control. Release* 86: 169–182.

133 Hosseinkhani, H. and Tabata, Y. (2004). RETRACTED: PEGylation enhances tumor targeting of plasmid DNA by an artificial cationized protein with repeated RGD sequences, Pronectin®. *J. Control. Release* 97: 157–171.

134 Hosseinkhani, H., Azzam, T., Tabata, Y., and Domb, A.J. (2004). Dextran-spermine polycation: an efficient non-viral vector for *in vitro* and *in vivo* gene transfection. *Gene Ther.* 11: 194–203.

135 Hosseinkhani, H. and Tabata, Y. (2005). Ultrasound enhances *in vivo* tumor expression of plasmid DNA by PEG-introduced cationized dextran. *J. Control. Release* 108: 540–556.

136 Konishi, M., Tabata, Y., Kariya, M. et al. (2005). *In vivo* anti-tumor effect of dual release of cisplatin and adriamycin from biodegradable gelatin hydrogel. *J. Control. Release* 103: 7–19.

137 Hosseinkhani, H., Kushibiki, T., Matsumoto, K. et al. (2006). Enhanced suppression of tumor growth using a combination of NK4 plasmid DNA-PEG engrafted cationized dextran complex and ultrasound. *Cancer Gene Ther.* 13: 479–489.

138 Hosseinkhani, H. and Hosseinkhani, M. (2009). Biodegradable polymer-metal complexes for gene and drug delivery. *Curr. Drug Saf.* 4: 79–83.

139 Hosseinkhani, H. and Tabata, Y. (2006). Self assembly of DNA nanoparticles with polycations for the delivery of genetic materials into cells. *J. Nanosci. Nanotechnol.* 6: 2320–2328.

140 Hosseinkhani, H. (2006). DNA nanoparticles for gene delivery to cells and tissue. *J. Nanosci. Nanotechnol.* 3: 416–461.

141 Hosseinkhani, H., Hosseinkhani, M., Khademhosseini, A. et al. (2008). DNA nanoparticles encapsulated in 3-D tissue engineered scaffold enhance osteogenic differentiation of mesenchymal stem cells. *J. Biomed. Mater. Res. Part A* 85: 47–60.

142 Arami, H., Stephen, Z., Veiseh, O., and Zhang, M. (2011). Chitosan-coated iron oxide nanoparticles for molecular imaging and drug delivery. *Adv. Polym. Sci.* 243: 163–184.

143 Berman, S.M.C., Walczak, P., and Bulte, J.W.M. (2011). Tracking stem cells using magnetic nanoparticles. *Wiley Interdiscip. Rev. Nanomed. Nanobiotechnol.* 3: 343–355.

144 Kievit, F.M., Veiseh, O., Bhattarai, N. et al. (2009). PEI-PEG-chitosan copolymer coated iron oxide nanoparticles for safe gene delivery: synthesis, complexation, and transfection. *Adv. Funct. Mater.* 19: 2244–2251.

145 Qiu, Y., Tong, S., Zhang, L. et al. (2017). Magnetic forces enable controlled drug delivery by disrupting endothelial cell-cell junctions. *Nat. Commun.* 8: 15594.

146 Krishnan, K.M. (2010). Biomedical nanomagnetics: a spin through possibilities in imaging, diagnostics, and therapy. *IEEE Trans. Magn.* 46: 2523–2558.

147 Rosen, J.E., Chan, L., Shieh, D.B., and Gu, F.X. (2012). Iron oxide nanoparticles for targeted cancer imaging and diagnostics. *Nanomedicine* 8: 275–290.

148 McAteer, M.A., Akhtar, A.M., von zur Muhlen, C., and Choudhury, R.P. (2010). An approach to molecular imaging of atherosclerosis, thrombosis, and vascular inflammation using microparticles of iron oxide. *Atherosclerosis* 209: 18–27.

149 Xu, H.Y., Aguilar, Z.P., Yang, L. et al. (2011). Antibody conjugated magnetic iron oxide nanoparticles for cancer cell separation in fresh whole blood. *Biomaterials* 32: 9758–9765.
150 Chen, Y., Chen, H., and Shi, J. (2014). Inorganic nanoparticles-based drug co-delivery nanosystems to overcome the multidrug resistance of cancer cells. *Mol. Pharm.* 11: 2495–2510.
151 Cheon, J. and Lee, J.-H. (2008). Synergistically integrated nanoparticles as multimodal probes for nanobiotechnology. *Acc. Chem. Res.* 41: 1630–1640.
152 Berry, C.C., Wells, S., Charles, S. et al. (2004). Cell response to dextran-derivatised iron oxide nanoparticles post internalisation. *Biomaterials* 25: 5405–5413.
153 von Maltzahn, G., Park, J.H., Lin, K.Y. et al. (2011). Nanoparticles that communicate *in vivo* to amplify tumour targeting. *Nat. Mater.* 10: 545–552.
154 Kim, D., Jeong, Y.Y., and Jon, S.A. (2010). A drug-loaded aptamer–gold nanoparticle bioconjugate for combined CT imaging and therapy of prostate cancer. *ACS Nano* 4: 3689–3696.
155 Low, P.S., Henne, W.A., and Doorneweerd, D.D. (2008). Discovery and development of folic-acid-based receptor targeting for imaging and therapy of cancer and inflammatory diseases. *Acc. Chem. Res.* 41: 120–129.
156 Walkey, C.D., Olsen, J.B., Guo, H. et al. (2012). Nanoparticle size and surface chemistry determine serum protein adsorption and macrophage uptake. *J. Am. Chem. Soc.* 134: 2139–2147.
157 Veiseh, O., Gunn, J.W., and Zhang, M.Q. (2010). Design and fabrication of magnetic nanoparticles for targeted drug delivery and imaging. *Adv. Drug Deliv. Rev.* 62: 284–304.
158 Jun, Y.-W., Lee, J.-H., and Cheon, J. (2008). Chemical design of nanoparticle probes for high-performance magnetic resonance imaging. *Angew. Chem. Int. Ed.* 47: 5122–5135.
159 Yoo, D., Lee, J.-H., Shin, T.-H., and Cheon, J. (2011). Theranostic magnetic nanoparticles. *Acc. Chem. Res.* 44: 863–874.
160 Ling, D. and Hyeon, T. (2012). Chemical design of biocompatible iron oxide nanoparticles for medical applications. *Small* 9: 1450–1466.
161 Zhou, Y., Wang, R., Teng, Z. et al. (2016). Magnetic nanoparticle-promoted droplet vaporization for *in vivo* stimuli-responsive cancer theranostics. *NPG Asia Mater.* 8: e313–e320.
162 Laurent, S., Forge, D., Port, M. et al. (2008). Magnetic iron oxide nanoparticles: synthesis, stabilization, vectorization, physicochemical characterizations and biological applications. *Chem. Rev.* 108: 2064–2110.
163 Gupta, A.K., Naregalkar, R.R., Vaidya, V.D., and Gupta, M. (2007). Recent advances on surface engineering of magnetic iron oxide nanoparticles and their biomedical applications. *Nanomedicine* 2: 23–39.
164 Mahmoudi, M., Sant, S., Wang, B. et al. (2011). Superparamagnetic iron oxide nanoparticles (SPIONs): development, surface modification and applications in chemotherapy. *Adv. Drug Deliv. Rev.* 63: 24–46.

165 Wang, J.T., Cabana, L., Bourgognon, M. et al. (2014). Magnetically decorated multiwalled carbon nanotubes as dual MRI and SPECT contrast agents. *Adv. Funct. Mater.* 24: 1880–1894.

166 Hagspiel, K.D., Neidl, K.F., Eichenberger, A.C. et al. (1995). Detection of liver metastases: comparison of superparamagnetic iron oxide-enhanced and unenhanced MR imaging at 1.5 T with dynamic CT, intraoperative US, and percutaneous US. *Radiology (Oak Brook, IL, U.S.)* 196: 471–478.

167 Reimer, P. and Tombach, B. (1998). Hepatic MRI with SPIO: detection and characterization of focal liver lesions. *Eur. Radiol.* 8: 1198–1204.

168 Lee, N. and Hyeon, T. (2012). Designed synthesis of uniformly sized iron oxide nanoparticles for efficient magnetic resonance imaging contrast agents. *Chem. Soc. Rev.* 41: 2575–2589.

169 Casula, M.F., Floris, P., Innocenti, C. et al. (2010). Magnetic resonance imaging contrast agents based on iron oxide superparamagnetic ferrofluids. *Chem. Mater.* 22: 1739–1748.

170 Arias, J.L., Reddy, L.H., and Couvreur, P. (2012). Fe_3O_4/chitosan nanocomposite for magnetic drug targeting to cancer. *J. Mater. Chem.* 22: 7622–7632.

171 Kong, S.D., Sartor, M., Hu, C.M. et al. (2013). Magnetic field activated lipid-polymer hybrid nanoparticles for stimuli-responsive drug release. *Acta Biomater.* 9: 5447–5452.

172 Ma, Y., Tong, S., Bao, G. et al. (2013). Indocyanine green loaded SPIO nanoparticles with phospholipid-PEG coating for dual-modal imaging and photothermal therapy. *Biomaterials* 34: 7706–7714.

173 Tian, Y., Jiang, X., Chen, X. et al. (2014). Doxorubicin-loaded magnetic silk fibroin nanoparticles for targeted therapy of multidrug-resistant cancer. *Adv. Mater.* 26: 7393–7398.

174 Lin, L.S., Cong, Z.X., Cao, J.B. et al. (2014). Multifunctional Fe_3O_4@polydopamine core–shell nanocomposites for intracellular MRNA detection and imaging-guided photothermal therapy. *ACS Nano* 8: 3876–3883.

175 Huang, P., Song, H., Wang, W. et al. (2014). Integrin-targeted zwitterionic polymeric nanoparticles with acid-induced disassembly property for enhanced drug accumulation and release in tumor. *Biomacromolecules* 15: 3128–3138.

176 Lin, J., Li, Y., Li, Y. et al. (2015). Drug/dye-loaded, multifunctional PEG–chitosan–iron oxide nanocomposites for methotraxate synergistically self-targeted cancer therapy and dual model imaging. *ACS Appl. Mater. Interfaces* 7: 11908–11920.

177 Shen, T., Weissleder, R., Papisov, M. et al. (1993). Monocrystalline iron oxide nanocompounds (MION): physicochemical properties. *Magn. Reson. Med.* 29: 599–604.

178 Wang, Y.-X.J., Hussain, S., and Krestin, G. (2001). Superparamagnetic iron oxide contrast agents: physicochemical characteristics and applications in MR imaging. *Eur. Radiol.* 11: 2319–2331.

179 Li, W., Tutton, S., Vu, A. et al. (2005). First-pass contrast-enhanced magnetic resonance angiography in humans using ferumoxytol, a novel ultrasmall superparamagnetic iron oxide (USPIO)-based blood pool agent. *J. Magn. Reson. Imaging* 21: 46–52.

180 Huh, Y.-M., Jun, Y., Song, H. et al. (2005). *In vivo* magnetic resonance detection of cancer by using multifunctional magnetic nanocrystals. *J. Am. Chem. Soc.* 127: 12387–12391.

181 Harisinghani, M.G., Barentsz, J., Hahn, P.F. et al. (2003). Noninvasive detection of clinically occult lymph-node metastases in prostate cancer. *N. Engl. J. Med.* 348: 2491–2499.

182 Beckmann, N., Cannet, C., Babin, A.L. et al. (2009). *In vivo* visualization of macrophage infiltration and activity in inflammation using magnetic resonance imaging. *Wiley Interdiscip. Rev. Nanomed. Nanobiotechnol.* 1: 272–298.

183 Wang, Y., Xu, C., Chang, Y. et al. (2017). Ultrasmall superparamagnetic iron oxide nanoparticle for T_2-weighted magnetic resonance imaging. *ACS Appl. Mater. Interfaces* 9: 28959–28966.

184 Lee, S.H., Kim, B.H., Na, H.B., and Hyeon, T. (2013). Paramagnetic inorganic nanoparticles as T_1 MRI contrast agents. *Wiley Interdiscip. Rev. Nanomed. Nanobiotechnol.* 6: 196–209.

185 Shapiro, E.M., Skrtic, S., and Koretsky, A. (2005). Sizing it up: cellular MRI using micron-sized iron oxide particles. *Magn. Reson. Med.* 53: 329–338.

186 Jun, Y., Huh, Y.-M., Choi, J. et al. (2005). Nanoscale size effect of magnetic nanocrystals and their utilization for cancer diagnosis via magnetic resonance imaging. *J. Am. Chem. Soc.* 127: 5732–5733.

187 Lee, N., Choi, Y., Lee, Y. et al. (2012). Water-dispersible ferrimagnetic iron oxide nanocubes with extremely high r_2 relaxivity for highly sensitive *in vivo* MRI of tumors. *Nano Lett.* 12: 3127–3131.

188 Lee, N., Kim, H., Choi, S.H. et al. (2011). Magnetosome-like ferrimagnetic iron oxide nanocubes for highly sensitive MRI of single cells and transplanted pancreatic islets. *Proc. Natl. Acad. Sci. U.S.A.* 108: 2662–2667.

189 Li, Z., Liu, F., Yuan, Y. et al. (2017). Multifunctional gold nanoparticle layers for controllable capture and release of protein. *Nanoscale* 9: 15407–15415.

190 Deng, R., Shen, N., Yang, Y. et al. (2018). Targeting epigenetic pathway with gold nanoparticles for acute myeloid leukemia therapy. *Biomaterials* 167: 80–90.

191 Hou, W., Xia, F., Alfranca, G. et al. (2017). Nanoparticles for multi-modality cancer diagnosis: simple protocol for self-assembly of gold nanoclusters mediated by gadolinium ions. *Biomaterials* 120: 103–114.

192 Tao, Y., Li, M., and Auguste, D. (2017). Pattern-based sensing of triple negative breast cancer cells with dual-ligand cofunctionalized gold nanoclusters. *Biomaterials* 116: 21–33.

193 Meir, R., Shamalov, K., Sadan, T. et al. (2017). Fast image-guided stratification using anti-programmed death ligand 1 gold nanoparticles for cancer immunotherapy. *ACS Nano* 11: 11127–11134.

194 Mangadlao, J.D., Wang, X., McCleese, C. et al. (2018). Prostate specific membrane antigen targeted gold nanoparticles for theranostics of prostate cancer. *ACS Nano* 12: 3714–3725.

195 Peng, L.H., Huang, Y.F., Zhang, C.Z. et al. (2016). Integration of antimicrobial peptides with gold nanoparticles as unique non-viral vectors for gene delivery to mesenchymal stem cells with antibacterial activity. *Biomaterials* 103: 137–149.

196 Yang, H., Kozicky, L., Saferali, A. et al. (2016). Turvey endosomal pH modulation by peptide-gold nanoparticle hybrids enables potent anti-inflammatory activity in phagocytic immune cells. *Biomaterials* 111: 90–102.

197 Pissuwan, D., Niidome, T., and Cortie, M.B. (2011). The forthcoming applications of gold nanoparticles in drug and gene delivery systems. *J. Control. Release* 149: 65–71.

198 Chhour, P., Naha, P.C., O'Neill, S.M. et al. (2016). Labeling monocytes with gold nanoparticles to track their recruitment in atherosclerosis with computed tomography. *Biomaterials* 87: 93–103.

199 Kwon, S.P., Jeon, S., Lee, S.H. et al. (2018). Thrombin-activatable fluorescent peptide incorporated gold nanoparticles for dual optical/computed tomography thrombus imaging. *Biomaterials* 150: 125–136.

200 Yata, T., Takahashi, Y., Tan, M. et al. (2017). DNA nanotechnology-based composite-type gold nanoparticle-immunostimulatory DNA hydrogel for tumor photothermal immunotherapy. *Biomaterials* 146: 136–145.

201 Ge, J., Liu, K., Niu, W. et al. (2018). Gold and gold-silver alloy nanoparticles enhance the myogenic differentiation of myoblasts through p38 MAPK signaling pathway and promote *in vivo* skeletal muscle regeneration. *Biomaterials* 175: 19–29.

202 Huang, Y., Gao, L., and Cui, H. (2018). Assembly of multifunctionalized gold nanoparticles with chemiluminescent, catalytic, and immune activity for label-free immunoassays. *ACS Appl. Mater. Interfaces* 10: 17040–17046.

203 Murphy, C.J., Sau, T.K., Gole, A.M. et al. (2005). Anisotropic metal nanoparticles: synthesis, assembly, and optical applications. *J. Phys. Chem. B* 109: 13857–13870.

204 Skrabalak, S.E., Chen, J., Sun, Y. et al. (2008). Gold nanocages: synthesis, properties, and applications. *Acc. Chem. Res.* 41: 1587–1595.

205 Daniel, M.-C. and Astruc, D. (2004). Gold nanoparticles: assembly, supramolecular chemistry, quantum-size-related properties, and applications toward biology, catalysis, and nanotechnology. *Chem. Rev.* 104: 293–346.

206 Bardhan, R., Lal, S., Joshi, A., and Halas, N.J. (2011). Theranostic nanoshells: from probe design to imaging and treatment of cancer. *Acc. Chem. Res.* 44: 936–946.

207 Giersig, M. and Mulvaney, P. (1993). Preparation of ordered colloid monolayers by electrophoretic deposition. *Langmuir* 9: 3408–3413.

208 Brust, M., Walker, M., Bethell, D. et al. (1994). Synthesis of thiol-derivatised gold nanoparticles in a two-phase liquid–liquid system. *J. Chem. Soc. Chem. Commun.* 7: 801–802.

209 Murphy, J., Gole, A.M., Hunyadi, S.E. et al. (2008). Chemical sensing and imaging with metallic nanorods. *Chem. Commun.* 7: 544–557.

210 Dreaden, E.C., Alkilany, A.M., Huang, X. et al. (2012). The golden age: gold nanoparticles for biomedicine. *Chem. Soc. Rev.* 41: 2740–2779.

211 Conde, J., Oliva, N., Zhang, Y., and Artzi, N. (2016). Local triple-combination therapy results in tumor regression and prevents recurrence in a colon cancer model. *Nat. Mater.* 15: 1128–1138.

212 Saha, K., Agasti, S.S., Kim, C. et al. (2012). Gold nanoparticles in chemical and biological sensing. *Chem. Rev.* 112: 2739–2779.

213 Xia, F., Zuo, X., Yang, R. et al. (2010). Colorimetric detection of DNA, small molecules, proteins, and ions using unmodified gold nanoparticles and conjugated polyelectrolytes. *Proc. Natl. Acad. Sci. U.S.A.* 107: 10837–10841.

214 Wilson, R. (2008). The use of gold nanoparticles in diagnostics and detection. *Chem. Soc. Rev.* 37: 2028–2045.

215 Lee, Y.K., Kim, S., Oh, J.-W., and Nam, J.-M. (2014). Massively parallel and highly quantitative single-particle analysis on interactions between nanoparticles on supported lipid bilayer. *J. Am. Chem. Soc.* 136: 4081–4088.

216 Lee, K., Cui, Y., Lee, L.P., and Irudayaraj, J. (2014). Quantitative imaging of single mRNA splice variants in living cells. *Nat. Nanotechnol.* 9: 474–480.

217 Scott, G.B., Williams, H.S., and Marriott, P.M. (1967). The phagocytosis of colloidal particles of different sizes. *J. Exp. Pathol.* 48: 411–416.

218 Singer, J.M., Adlersberg, L., and Sadek, M. (1972). Long-term observation of intravenously injected colloidal gold in mice. *J. Reticuloendothel. Soc.* 12: 658–671.

219 Hardonk, M.J., Harms, G., and Koudstaal, J. (1985). Zonal heterogeneity of rat hepatocytes in the *in vivo* uptake of 17 nm colloidal gold granules. *Histochemistry* 83: 473–477.

220 Renaud, G., Hamilton, R.L., and Havel, R. (1989). Hepatic metabolism of colloidal gold-low-density lipoprotein complexes in the rat: evidence for bulk excretion of lysosomal contents into bile. *Hepatology* 9: 380–392.

221 Sadauskas, E., Wallin, H., Stoltenberg, M. et al. (2007). Kupffer cells are central in the removal of nanoparticles from the organism. *Part. Fibre Toxicol.* 4: 10.

222 Sadauskas, E., Danscher, G., Stoltenberg, M. et al. (2009). Protracted elimination of gold nanoparticles from mouse liver. *Nanomedicine* 5: 162–169.

223 Benezra, M., Penate-Medina, O., Zanzonico, P.B. et al. (2011). Multimodal silica nanoparticles are effective cancer-targeted probes in a model of human melanoma. *J. Clin. Invest.* 121: 2768–2780.

224 Sun, L., Wang, D., Chen, Y. et al. (2017). Core–shell hierarchical mesostructured silica nanoparticles for gene/chemo-synergetic stepwise therapy of multidrug-resistant cancer. *Biomaterials* 133: 219–228.

225 Shi, M., Xia, L., Chen, Z. et al. (2017). Europium-doped mesoporous silica nanosphere as an immune-modulating osteogenesis/angiogenesis agent. *Biomaterials* 144: 176–187.

226 Kim, J.-H., Kim, H., Choi, Y. et al. (2017). Colloidal mesoporous silica nanoparticles as strong adhesives for hydrogels and biological tissues. *ACS Appl. Mater. Interfaces* 9: 31469–31477.

227 An, M., Li, M., and Liu, H. (2017). Silica nanoparticle as a lymph node targeting platform for vaccine delivery. *ACS Appl. Mater. Interfaces* 9: 23466–23475.

228 Song, S., Du, D., Li, L. et al. (2017). *In vitro* study of receptor-mediated silica nanoparticles delivery across blood–brain barrier. *ACS Appl. Mater. Interfaces* 9: 20410–20416.

229 Zhang, Y., Qiu, Y., Lin, L. et al. (2017). Ultraphotostable mesoporous silica-coated gap-enhanced Raman tags (GERTs) for high-speed bioimaging. *ACS Appl. Mater. Interfaces* 9: 3995–4005.

230 Liu, X., Sui, B., and Sun, J. (2017). Blood–brain barrier dysfunction induced by silica NPs *in vitro* and *in vivo*: involvement of oxidative stress and Rho-kinase/JNK signaling pathways. *Biomaterials* 121: 64–82.

231 Bradbury, M.S., Phillips, E., Montero, P.H. et al. (2013). Clinically-translated silica nanoparticles as dual-modality cancer-targeted probes for image-guided surgery and interventions. *Integr. Biol.* 5: 74–86.

232 Stöber, W., Fink, A., and Bohn, E. (1968). Controlled growth of monodisperse silica spheres in the micron size range. *J. Colloid Interface Sci.* 26: 62–69.

233 Huo, Q., Liu, J., Wang, L.Q. et al. (2006). A new class of silica cross-linked micellar core–shell nanoparticles. *J. Am. Chem. Soc.* 128: 6447–6453.

234 Bagwe, R.P., Yang, C., Hilliard, L.R., and Tan, W. (2004). Optimization of dye-doped silica nanoparticles prepared using a reverse microemulsion method. *Langmuir* 20: 8336–8342.

235 Vivero-Escoto, J.L., Huxford-Phillips, R.C., and Lin, W. (2012). Silica-based nanoprobes for biomedical imaging and theranostic applications. *Chem. Soc. Rev.* 41: 2673–2685.

236 Slowing, I.I., Vivero-Escoto, J.L., Trewyn, B.G., and Lin, V.S.Y. (2010). Mesoporous silica nanoparticles: structural design and applications. *J. Mater. Chem.* 20: 7924–7937.

237 Slowing, I.I., Vivero-Escoto, J.L., Wu, C.W., and Lin, V.S.Y. (2008). Mesoporous silica nanoparticles as controlled release drug delivery and gene transfection carriers. *Adv. Drug Deliv. Rev.* 60: 1278–1288.

238 Vallet-Regi, M., Balas, F., and Arcos, D. (2007). Mesoporous materials for drug delivery. *Angew. Chem. Int. Ed.* 46: 7548–7558.

239 García-Fernández, A., García-Laínez, G., Ferrándiz, M.L. et al. (2017). Targeting inflammasome by the inhibition of caspase-1 activity using capped mesoporous silica nanoparticles. *J. Control. Release* 248: 60–70.

240 Chan, W.C.W. and Nie, S.M. (1998). Quantum dot bioconjugates for ultrasensitive nonisotopic detection. *Science* 281: 2016–2018.

241 Bruchez, M. Jr.,, Moronne, M., Gin, P. et al. (1998). Semiconductor nanocrystals as fluorescent biological labels. *Science* 281: 2013–2016.

242 Medintz, I.L., Uyeda, H.T., Goldman, E.R., and Mattoussi, H. (2005). Quantum dot bioconjugates for imaging, labelling and sensing. *Nat. Mater.* 4: 435–446.

243 Chang, Y., Pinaud, F., Antelman, J., and Weiss, S. (2008). Tracking bio-molecules in live cells using quantum dots. *J. Biophotonics* 1: 287–298.
244 Clarke, S., Pinaud, F., Beutel, O. et al. (2010). Covalent monofunctionalization of peptide-coated quantum dots for single-molecule assays. *Nano Lett.* 10: 2147–2154.
245 Derfus, A.M., Chan, W.C.W., and Bhatia, S.N. (2004). Probing the cytotoxicity of semiconductor quantum dots. *Nano Lett.* 4: 11–18.
246 Kirchner, C., Liedl, T., Kudera, S. et al. (2005). Cytotoxicity of colloidal CdSe and CdSe/ZnS nanoparticles. *Nano Lett.* 5: 331–338.
247 Jakubek, Z.J., deVries, J., Lin, S. et al. (2008). Exciton recombination and upconverted photoluminescence in colloidal CdSe quantum dots. *J. Phys. Chem. C* 112: 8153–8158.
248 Wu, C., Hansen, S.J., Hou, Q. et al. (2011). Design of highly emissive polymer dot bioconjugates for *in vivo* tumor targeting. *Angew. Chem. Int. Ed.* 50: 3430–3434.
249 Mukthavaram, R., Wrasidlo, W., Hall, D. et al. (2011). Assembly and targeting of liposomal nanoparticles encapsulating quantum dots. *Bioconjug. Chem.* 22: 1638–1644.
250 Yang, K., Zhang, F.J., Tang, H. et al. (2011). *In-vivo* imaging of oral squamous cell carcinoma by EGFR monoclonal antibody conjugated near-infrared quantum dots in mice. *Int. J. Nanomedicine* 6: 1739–1745.
251 Yang, K., Cao, Y.A., Shi, C. et al. (2010). Quantum dot-based visual *in vivo* imaging for oral squamous cell carcinoma in mice. *Oral Oncol.* 46: 864–868.
252 Jung, K.H., Choe, Y.S., Paik, J.Y., and Lee, K.H. (2011). ^{99m}Tc-Hydrazinonicotinamide epidermal growth factor-polyethylene glycol-quantum dot imaging allows quantification of breast cancer epidermal growth factor receptor expression and monitors receptor downregulation in response to cetuximab therapy. *J. Nucl. Med.* 52: 1457–1464.
253 Biswas, P., Cella, L.N., Kang, S.H. et al. (2011). Quantum-dot based protein module for *in vivo* monitoring of protease activity through fluorescence resonance energy transfer. *Chem. Commun.* 47: 5259–5261.
254 Resch-Genger, U., Grabolle, M., Cavaliere-Jaricot, S. et al. (2008). Quantum dots versus organic dyes as fluorescent labels. *Nat. Methods* 5: 763–775.
255 Larson, D.R., Zipfel, W.R., Williams, R.M. et al. (2003). Water-soluble quantum dots for multiphoton fluorescence imaging *in vivo*. *Science* 300: 1434–1436.
256 Michalet, X., Pinaud, F.F., Bentolila, L.A. et al. (2005). Quantum dots for live cells, *in vivo* imaging, and diagnostics. *Science* 307: 538–544.
257 Escobedo, J.O., Rusin, O., Lim, S., and Strongin, R.M. (2010). NIR dyes for bioimaging applications. *Curr. Opin. Chem. Biol.* 14: 64–70.
258 Guo, Z., Park, S., Yoon, J., and Shin, I. (2014). Recent progress in the development of near-infrared fluorescent probes for bioimaging applications. *Chem. Soc. Rev.* 43: 16–29.
259 Gao, X., Yang, L., Petros, J.A. et al. (2005). *In vivo* molecular and cellular imaging with quantum dots. *Curr. Opin. Biotechnol.* 16: 63–72.

260 Yang, X., Zhao, D., Leck, K.S. et al. (2012). Full visible range covering InP/ZnS nanocrystals with high photometric performance and their application to white quantum dot light-emitting diodes. *Adv. Mater.* 24: 4180–4185.

261 Brunetti, V., Chibli, H., Fiammengo, R. et al. (2013). InP/ZnS as a safer alternative to Cdse/Zns core/shell quantum dots: *in vitro* and *in vivo* toxicity assessment. *Nanoscale* 5: 307–317.

262 Chibli, H., Carlini, L., Park, S. et al. (2011). Cytotoxicity of InP/ZnS quantum dots related to reactive oxygen species generation. *Nanoscale* 3: 2552–2559.

263 Bharali, D.J., Lucey, D.W., Jayakumar, H. et al. (2005). Folate-receptor-mediated delivery of InP quantum dots for bioimaging using confocal and two-photon microscopy. *J. Am. Chem. Soc.* 127: 11364–11371.

264 Wang, Y., Wang, Y., Chen, G. et al. (2017). Quantum-dot-based theranostic micelles conjugated with an anti-EGFR nanobody for triple-negative breast cancer therapy. *ACS Appl. Mater. Interfaces* 9: 30297–30305.

265 Ma, L., Tu, C., Le, P. et al. (2016). Multidentate polymer coatings for compact and homogeneous quantum dots with efficient bioconjugation. *J. Am. Chem. Soc.* 10: 3382–3394.

266 Fedosyuk, A., Radchanka, A., Antanovich, A. et al. (2016). Determination of concentration of amphiphilic polymer molecules on the surface of encapsulated semiconductor nanocrystals. *Langmuir* 8: 1955–1961.

267 Reisch, A. and Klymchenko, A.S. (2016). Fluorescent polymer nanoparticles based on dyes: seeking brighter tools for bioimaging. *Small* 12: 1968–1992.

268 Ye, D.X., Ma, Y.Y., Zhao, W. et al. (2016). ZnO-based nanoplatforms for labeling and treatment of mouse tumors without detectable toxic side effects. *ACS Nano* 10: 4294–4300.

269 Evans, C.M., Love, A.M., and Weiss, E.A. (2012). Surfactant-controlled polymerization of semiconductor clusters to quantum dots through competing step-growth and living chain-growth mechanisms. *J. Am. Chem. Soc.* 134: 17298–17305.

270 Zhao, M.X. and Zhu, B.J. (2016). The research and applications of quantum dots as nano-carriers for targeted drug delivery and cancer therapy. *Nanoscale Res. Lett.* 11: 207.

271 Zhang, Q., Shen, C., Zhao, N., and Xu, F.-J. (2017). Redox-responsive and drug-embedded silica nanoparticles with unique self-destruction features for efficient gene/drug codelivery. *Adv. Funct. Mater.* 27: 1606229.

272 Busseron, E., Ruff, Y., Moulin, E., and Giuseppone, N. (2013). Supramolecular self-assemblies as functional nanomaterials. *Nanoscale* 5: 7098–7140.

273 Deng, T., Wang, J., Li, Y. et al. (2018). Quantum dots-based multifunctional nano-prodrug fabricated by ingenious self-assembly strategies for tumor theranostic. *ACS Appl. Mater. Interfaces* 10: 27657–27668.

274 Moghimi, S.M., Hunter, A.C., and Murray, J.C. (2005). Nanomedicine: current status and future prospects. *FASEB J.* 19: 311–330.

275 Li, K.Y., Eres, G., Howe, J. et al. (2013). Self-assembly of graphene on carbon nanotube surfaces. *Sci. Rep.* 3: 2353.

276 Sinha, N. and Yeow, J.T. (2005). Carbon nanotubes for biomedical applications. *IEEE Trans. Nanobiosci.* 4: 180–195.

277 Mehra, N.K., Jain, K., and Jain, N.K. (2015). Pharmaceutical and biomedical applications of surface engineered carbon nanotubes. *Drug Discov. Today* 20: 750–759.

278 Iijima, S. (1991). Helical microtubules of graphitic carbon. *Nature* 354: 56–58.

279 Endo, M., Kim, Y.A., Hayashi, T. et al. (2001). Vapor-grown carbon fibers (VGCFs) basic properties and their battery applications. *Carbon* 39: 1287–1297.

280 Kang, S.J., Kocabas, C., Ozel, T. et al. (2007). High-performance electronics using dense, perfectly aligned arrays of single-walled carbon nanotubes. *Nat. Nanotechnol.* 2: 230–236.

281 Scrosati, B. (2007). Nanomaterials: paper powers battery breakthrough. *Nat. Nanotechnol.* 2: 598–599.

282 Koo, J.H., Jeong, S., Shim, H.J. et al. (2017). Wearable electrocardiogram monitor using carbon nanotube electronics and color-tunable organic light-emitting diodes. *ACS Nano* 11: 10032–10041.

283 El-Sayed, R., Eita, M., Barrefelt, A. et al. (2013). Thermostable luciferase from *Luciola cruciate* for imaging of carbon nanotubes and carbon nanotubes carrying doxorubicin using *in vivo* imaging system. *Nano Lett.* 13: 1393–1398.

284 Peña, B., Bosi, S., Aguado, B.A. et al. (2017). Injectable carbon nanotube-functionalized reverse thermal gel promotes cardiomyocytes survival and maturation. *ACS Appl. Mater. Interfaces* 9: 31645–31656.

285 Tasis, D., Tagmatarchis, N., Bianco, A., and Prato, M. (2006). Chemistry of carbon nanotubes. *Chem. Rev.* 106: 1105–1136.

286 Baker, S.N. and Baker, G.A. (2010). Luminescent carbon nanodots: emergent nanolights. *Angew. Chem. Int. Ed.* 49: 6726–6744.

287 Zheng, X.T., Ananthanarayanan, A., Luo, K.Q., and Chen, P. (2015). Glowing graphene quantum dots and carbon dots: properties, syntheses, and biological applications. *Small* 11: 1620–1636.

288 Li, Q., Zhang, S., Dai, L., and Li, L.-S. (2012). Nitrogen-doped colloidal graphene quantum dots and their size-dependent electrocatalytic activity for the oxygen reduction reaction. *J. Am. Chem. Soc.* 134: 18932–18935.

289 Jin, H., Huang, H., He, Y. et al. (2015). Graphene quantum dots supported by graphene nanoribbons with ultrahigh electrocatalytic performance for oxygen reduction. *J. Am. Chem. Soc.* 137: 7588–7591.

290 Dong, Y., Cai, J., You, X., and Chi, Y. (2015). Sensing applications of luminescent carbon based dots. *Analyst* 140: 7468–7486.

291 Zhuo, S., Shao, M., and Lee, S.T. (2012). Upconversion and downconversion fluorescent graphene quantum dots: ultrasonic preparation and photocatalysis. *ACS Nano* 6: 1059–1064.

292 Liu, J., Liu, Y., Liu, N. et al. (2015). Metal-free efficient photocatalyst for stable visible water splitting via a two-electron pathway. *Science* 347: 970–974.

293 Kong, B., Zhu, A., Ding, C. et al. (2012). Carbon dot-based inorganic–organic nanosystem for two-photon imaging and biosensing of pH variation in living cells and tissues. *Adv. Mater.* 24: 5844–5848.

294 Li, H., Kang, Z., Liu, Y., and Lee, S.-T. (2012). Carbon nanodots: synthesis, properties and applications. *J. Mater. Chem.* 22: 24230–24253.

295 Sharker, S.M., Kim, S.M., Kim, S.H. et al. (2015). Target delivery of β-cyclodextrin/paclitaxel complexed fluorescent carbon nanoparticles: externally NIR light and internally pH sensitive-mediated release of paclitaxel with bio-imaging. *J. Mater. Chem. B* 3: 5833–5841.

296 Ge, J., Lan, M., Zhou, B. et al. (2014). Graphene quantum dot photodynamic therapy agent with high singlet oxygen generation. *Nat. Commun.* 5: 4596.

297 Feng, T., Ai, X., An, G. et al. (2016). Charge-convertible carbon dots for imaging-guided drug delivery with enhanced *in vivo* cancer therapeutic efficiency. *ACS Nano* 10: 4410–4420.

298 Kaur, R. and Badea, I. (2013). Nanodiamonds as novel nanomaterials for biomedical applications: drug delivery and imaging systems. *Int. J. Nanomed.* 8: 203–220.

299 Puzyr, A.P., Baron, A.V., Purtov, K.V. et al. (2007). Nanodiamonds with novel properties: a biological study. *Diamond Relat. Mater.* 16: 2124–2128.

300 Mochalin, V.N., Shenderova, O., Ho, D., and Gogotsi, Y. (2012). The properties and applications of nanodiamonds. *Nat. Nanotechnol.* 7: 11–23.

301 Osawaa, E., Sasakib, S., and Yamanoia, R. (2012). Deagglomeration of detonation nanodiamond. In: *Ultananocrystalline Diamond* (ed. O.A. Shenderova and D.M. Gruen), 165–179. Elsevier.

302 Alhaddad, A., Adam, M.P., Botsoa, J. et al. (2011). Nanodiamond as a vector for siRNA delivery to ewing sarcoma cells. *Small* 7: 3087–3095.

303 Zhu, Y., Li, J., Li, W. et al. (2012). The biocompatibility of nanodiamonds and their application in drug delivery systems. *Theranostics* 2: 302–312.

304 Liu, K.K., Cheng, C.L., Chang, C.C., and Chao, J.I. (2007). Biocompatible and detectable carboxylated nanodiamond on human cell. *Nanotechnology* 18: 325102.

305 Wang, X., Low, X.C., Hou, W. et al. (2014). Epirubicin-adsorbed nanodiamonds kill chemoresistant hepatic cancer stem cells. *ACS Nano* 8: 12151–12166.

306 Chang, Y.R., Lee, H.Y., Chen, K. et al. (2008). Mass production and dynamic imaging of fluorescent nanodiamonds. *Nat. Nanotechnol.* 3: 284–288.

307 Wu, Y., Jelezko, F., Plenio, M.B., and Weil, T. (2016). Diamond quantum devices in biology. *Angew. Chem. Int. Ed.* 55: 6586–6598.

308 Wang, L., Wu, Y., Guo, B., and Ma, P.X. (2015). Nanofiber yarn/hydrogel core–shell scaffolds mimicking native skeletal muscle tissue for guiding 3D myoblast alignment, elongation, and differentiation. *ACS Nano* 9: 9167–9179.

309 Yuan, H., Qin, J., Xie, J. et al. (2016). Highly aligned core–shell structured nanofibers for promoting phenotypic expression of vSMCs for vascular regeneration. *Nanoscale* 8: 16307–16322.

310 Sun, Y., Li, W., Wu, X. et al. (2016). Functional self-assembling peptide nanofiber hydrogels designed for nerve degeneration. *ACS Appl. Mater. Interfaces* 8: 2348–2359.

311 Wu, H., Liu, S., Xiao, L. et al. (2016). Injectable and pH-responsive silk nanofiber hydrogels for sustained anticancer drug delivery. *ACS Appl. Mater. Interfaces* 8: 17118–17126.

312 Smith, L.A. and Ma, P.X. (2004). Nano-fibrous scaffolds for tissue engineering. *Colloids Surf., B* 10: 125–131.

313 Baker, B.M., Gee, A.O., Metter, R.B. et al. (2008). The potential to improve cell infiltration in composite fiber-aligned electrospun scaffolds by the selective removal of sacrificial fibers. *Biomaterials* 29: 2348–2358.

314 Cipitria, A., Skelton, A., Dargaville, T.R. et al. (2011). Design, fabrication and characterization of PCL electrospun scaffolds—a review. *J. Mater. Chem.* 21: 9419–9453.

315 Chua, K.N., Chai, C., Lee, P.C. et al. (2006). Surface-aminated electrospun nanofibers enhance adhesion and expansion of human umbilical cord blood hematopoietic stem/progenitor cells. *Biomaterials* 27: 6043–6051.

316 Nam, J., Johnson, J., Lannutti, J.J., and Agarwal, S. (2011). Modulation of embryonic mesenchymal progenitor cell differentiation via control over pure mechanical modulus in electrospun nanofibers. *Acta Biomater.* 7: 1516–1524.

317 Rim, N.G., Kim, S., Shin, Y.M. et al. (2012). Mussel-inspired surface modification of poly(L-lactide) electrospun fibers for modulation of osteogenic differentiation of human mesenchymal stem cells. *Colloids Surf., B* 91: 189–197.

318 Kim, I.L., Khetan, S., Baker, B.M. et al. (2013). Fibrous hyaluronic acid hydrogels that direct MSC chondrogenesis through mechanical and adhesive cues. *Biomaterials* 34: 5571–5580.

319 Cardwell, R.D., Dahlgren, L.A., and Goldstein, A.S. (2014). Electrospun fibre diameter, not alignment, affects mesenchymal stem cell differentiation into the tendon/ligament lineage. *J. Tissue Eng. Regen. Med.* 8: 937–945.

320 Wu, Y., Wang, L., Guo, B., and Ma, P.X. (2017). Interwoven aligned conductive nanofiber yarn/hydrogel composite scaffolds for engineered 3D cardiac anisotropy. *ACS Nano* 11: 5646–5659.

321 Hartgerink, J.D., Beniash, E., and Stupp, S.I. (2001). Self-assembly and mineralization of peptide-amphiphile nanofibers. *Science* 294: 1684–1688.

322 Silva, G.A., Czeisler, C., Niece, K.L. et al. (2001). Selective differentiation of neural progenitor cells by high-epitope density nanofibers. *Science* 303: 1352–1355.

323 Hong, Y., Legge, R.L., Zhang, S., and Chen, P. (2003). Effect of amino acid sequence and pH on nanofiber formation of self-assembling peptides EAK16-II and EAK16-IV. *Biomacromolecules* 4: 1433–1442.

324 Coutinho, D.F., Sant, S.V., Shin, H. et al. (2010). Modified Gellan Gum hydrogels with tunable physical and mechanical properties. *Biomaterials* 31: 7494–7502.

325 Hwang, C.M., Sant, S., Masaeli, M. et al. (2010). Fabrication of three-dimensional porous cell-laden hydrogel for tissue engineering. *Biofabrication* 2: 035003.

326 Yamanlar, S., Sant, S., Boudou, T. et al. (2011). Surface functionalization of hyaluronic acid hydrogels by polyelectrolyte multilayer films. *Biomaterials* 32: 5590–5599.

327 Wan, A.C.A., Liao, I.C., Yim, E.K.F., and Leong, K.W. (2004). Mechanism of fiber formation by interfacial polyelectrolyte complexation. *Macromolecules* 37: 7019–7025.

328 Tai, B.C.U., Wan, A.C.A., and Ying, J.Y. (2010). Modified polyelectrolyte complex fibrous scaffold as a matrix for 3D cell culture. *Biomaterials* 31: 5927–5935.

329 Sant, S., Coutinho, D.F., Gaharwar, A.K. et al. (2017). Self-assembled hydrogel fiber bundles from oppositely charged polyelectrolytes mimic micro-/nanoscale hierarchy of collagen. *Adv. Funct. Mater.* 27: 1606273.

330 Dasgupta, N.P., Sun, J., Liu, C. et al. (2014). 25th anniversary article: semiconductor nanowires–synthesis, characterization, and applications. *Adv. Mater.* 26: 2137–2184.

331 Christesen, J.D., Zhang, X., Pinion, C.W. et al. (2012). Design principles for photovoltaic devices based on Si nanowires with axial or radial p–n junctions. *Nano Lett.* 12: 6024–6029.

332 Chan, C.K., Peng, H.L., Liu, G. et al. (2008). High-performance lithium battery anodes using silicon nanowires. *Nat. Nanotechnol.* 3: 31–35.

333 Gudiksen, M.S., Lauhon, L.J., Wang, J. et al. (2002). Growth of nanowire superlattice structures for nanoscale photonics and electronics. *Nature* 415: 617–620.

334 Law, M., Sirbuly, D.J., Johnson, J.C. et al. (2004). Nanoribbon waveguides for subwavelength photonics integration. *Science* 305: 1269–1273.

335 Janissen, R., Sahoo, P.K., Santos, C.A. et al. (2017). InP nanowire biosensor with tailored biofunctionalization: ultrasensitive and highly selective disease biomarker detection. *Nano Lett.* 17: 5938–5949.

336 Kim, K.K., Hong, S., Cho, H.M. et al. (2015). Highly sensitive and stretchable multidimensional strain sensor with prestrained anisotropic metal nanowire percolation networks. *Nano Lett.* 15: 5240–5247.

337 Hong, W., Lee, S., Chang, H.J. et al. (2016). Multifunctional magnetic nanowires: a novel breakthrough for ultrasensitive detection and isolation of rare cancer cells from non-metastatic early breast cancer patients using small volumes of blood. *Biomaterials* 106: 78–86.

338 Chen, X.Z., Hoop, M., Shamsudhin, N. et al. (2017). Hybrid magnetoelectric nanowires for nanorobotic applications: fabrication, magnetoelectric coupling, and magnetically assisted *in vitro* targeted drug delivery. *Adv. Mater.* 29: 1605458.

339 Park, K., Kim, K.C., Lee, H. et al. (2017). Suppressing mosaicism by Au nanowire injector-driven direct delivery of plasmids into mouse embryos. *Biomaterials* 138: 169–178.

340 Lin, Z.L.C., Xie, C., Osakada, Y. et al. (2014). Iridium oxide nanotube electrodes for highly sensitive and prolonged intracellular measurement of action potentials. *Nat. Commun.* 5: 3206.

341 Berthing, T., Bonde, S., Sørensen, C.B. et al. (2011). Intact mammalian cell function on semiconductor nanowire arrays: new perspectives for cell-based biosensing. *Small* 7: 640–647.

342 Robinson, J.T., Jorgolli, M., Shalek, A.K. et al. (2012). Vertical nanowire electrode arrays as a scalable platform for intracellular interfacing to neuronal circuits. *Nat. Nanotechnol.* 7: 180–184.

343 Xie, C., Lin, Z., Hanson, L. et al. (2012). Intracellular recording of action potentials by nanopillar electroporation. *Nat. Nanotechnol.* 7: 185–190.

344 Shalek, A.K., Robinson, J.T., Karp, E.S. et al. (2010). Vertical silicon nanowires as a universal platform for delivering biomolecules into living cells. *Proc. Natl. Acad. Sci. U.S.A.* 107: 1870–1875.

345 Yan, R., Park, J.H., Choi, Y. et al. (2011). Nanowire-based single-cell endoscopy. *Nat. Nanotechnol.* 7: 191–196.

346 Chevrier, N., Mertins, P., Artyomov, M.N. et al. (2011). Systematic discovery of TLR signaling components delineates viral-sensing circuits. *Cell* 147: 853–867.

347 Shalek, A.K., Gaublomme, J.T., Wang, L. et al. (2012). Nanowire-mediated delivery enables functional interrogation of primary immune cells: application to the analysis of chronic lymphocytic leukemia. *Nano Lett.* 12: 6498–6504.

348 Wang, L.L., Shalek, A.K., Lawrence, M. et al. (2014). Somatic mutation as a mechanism of Wnt/β-catenin pathway activation in CLL. *Blood* 124: 1089–1098.

349 Na, Y.-R., Kim, S.Y., Gaublomme, J.T. et al. (2013). Probing enzymatic activity inside living cells using a nanowire–cell "sandwich" assay. *Nano Lett.* 13: 153–158.

350 Aalipour, A., Xu, A.M., Leal-Ortiz, S. et al. (2014). Plasma membrane and actin cytoskeleton as synergistic barriers to nanowire cell penetration. *Langmuir* 30: 12362–12367.

351 Xu, A.M., Aalipour, A., Leal-Ortiz, S. et al. (2014). Quantification of nanowire penetration into living cells. *Nat. Commun.* 5: 3613.

352 Lee, J.-H., Zhang, A., You, S., and Lieber, C.M. (2016). Spontaneous internalization of cell penetrating peptide-modified nanowires into primary neurons. *Nano Lett.* 16: 1509–1513.

353 Xie, X., Xu, A.M., Angle, M.R. et al. (2013). Mechanical model of vertical nanowire cell penetration. *Nano Lett.* 13: 6002–6008.

354 Xie, X., Aalipour, A., Gupta, S.V., and Melosh, N.A. (2015). Determining the time window for dynamic nanowire cell penetration processes. *ACS Nano* 9: 11667–11677.

355 Wang, S., Wang, H., Jiao, J. et al. (2009). Three-dimensional nanostructured substrates toward efficient capture of circulating tumor cells. *Angew. Chem. Int. Ed.* 48: 8970–8973.

356 Kim, S.T., Kim, D.J., Kim, T.J. et al. (2010). Novel streptavidin-functionalized silicon nanowire arrays for $CD4^+$ T lymphocyte separation. *Nano Lett.* 10: 2877–2883.

357 Lee, S.K., Kim, D.J., Lee, G. et al. (2014). Specific rare cell capture using micro-patterned silicon nanowire platform. *Biosens. Bioelectron.* 54: 181–188.

358 Kwak, M., Kim, D.-J., Lee, M.-R. et al. (2014). Nanowire array chips for molecular typing of rare trafficking leukocytes with application to neurodegenerative pathology. *Nanoscale* 6: 6537–6550.

359 Kim, G.-S., Kim, D.-J., Hyung, J.-H. et al. (2014). Dependence of filopodia morphology and the separation efficiency of primary $CD4^+$ T-lymphocytes on nanopillars. *Anal. Chem.* 86: 5330–5337.

360 Malvankar, N.S., Vargas, M., Nevin, K.P. et al. (2011). Tunable metallic-like conductivity in microbial nanowire networks. *Nat. Nanotechnol.* 6: 573–579.

361 Chen, S., Wang, M., Ye, J. et al. (2013). Kinetics-controlled growth of aligned mesocrystalline SnO_2 nanorod arrays for lithium-ion batteries with superior rate performance. *Nano Res.* 6: 243–252.

362 Bass, J.D., Schaper, C.D., Rettner, C.T. et al. (2011). Transfer molding of nanoscale oxides using water-soluble templates. *ACS Nano* 5: 4065–4072.

363 Xi, G.C. and Ye, J.H. (2010). Ultrathin SnO_2 nanorods: template- and surfactant-free solution phase synthesis, growth mechanism, optical, gas-sensing, and surface adsorption properties. *Inorg. Chem.* 49: 2302–2309.

364 Chen, Z., Pan, D., Li, Z. et al. (2014). Recent advances in tin dioxide materials: some developments in thin films, nanowires, and nanorods. *Chem. Rev.* 114: 7442–7486.

365 Zhao, R., Han, X., Li, Y. et al. (2017). Photothermal effect enhanced cascade-targeting strategy for improved pancreatic cancer therapy by gold nanoshell@mesoporous silica nanorod. *ACS Nano* 11: 8103–8113.

366 Lee, H., Lee, J.H., Kim, J. et al. (2016). Hyaluronate-gold nanorod/DR5 antibody complex for noninvasive theranosis of skin cancer. *ACS Appl. Mater. Interfaces* 8: 32202–32210.

367 Jung, B.-K., Lee, Y.K., Hong, J. et al. (2016). Mild hyperthermia induced by gold nanorod-mediated plasmonic photothermal therapy enhances transduction and replication of oncolytic adenoviral gene delivery. *ACS Nano* 10: 10533–10543.

368 Yin, J., Chen, D., Wu, S. et al. (2017). Tumor-targeted nanoprobes for enhanced multimodal imaging and synergistic photothermal therapy: core–shell and dumbbell Gd-tailored gold nanorods. *Nanoscale* 9: 16661–16673.

369 Zhu, C., Xue, J., Gilroy, K.D. et al. (2016). Micropatterned polymer nanorod forests and their use for dual drug loading and regulation of cell adhesion. *ACS Appl. Mater. Interfaces* 8: 34194–34197.

370 Muhammed, M.A., Doblinger, M., and Rodriguez-Fernandez, J. (2015). Switching plasmons: gold nanorod-copper chalcogenide core–shell nanoparticle clusters with selectable metal/semiconductor NIR plasmon resonances. *J. Am. Chem. Soc.* 137: 11666–11677.

371 Liu, L., Ouyang, S., and Ye, J. (2013). Gold-nanorod-photosensitized titanium dioxide with wide-range visible-light harvesting based on localized surface plasmon resonance. *Angew. Chem. Int. Ed.* 52: 6689–6693.

372 Arimoto, H. and Egawa, M. (2015). Imaging wavelength and light penetration depth for water content distribution measurement of skin. *Skin Res. Technol.* 21: 94–100.

373 Tsai, M.-F., Chang, S.-H.G., Cheng, F.-Y. et al. (2013). Au nanorod design as light-absorber in the first and second biological near-infrared windows for *in vivo* photothermal therapy. *ACS Nano* 7: 5330–5342.

374 Lee, H., Lee, M., Bhang, S.H. et al. (2014). Hyaluronate-gold nanoparticle/tocilizumab complex for the treatment of rheumatoid arthritis. *ACS Nano* 8: 4790–4798.

375 Lee, M.-Y., Yang, J.-A., Jung, H.S. et al. (2012). Hyaluronic acid-gold nanoparticle/interferon α complex for targeted treatment of hepatitis C virus infection. *ACS Nano* 6: 9522–9531.

376 Jazayeri, M.H., Amani, H., Pourfatollah, A.A. et al. (2016). Various methods of gold nanoparticles (GNPs) conjugation to antibodies. *Sens. Biosensing Res.* 9: 17–22.

377 Zhan, P., Dutta, P.K., Wang, P. et al. (2017). Reconfigurable three-dimensional gold nanorod plasmonic nanostructures organized on DNA origami tripod. *ACS Nano* 11: 1172–1179.

378 Paviolo, C., Haycock, J.W., Yong, J. et al. (2013). Laser exposure of gold nanorods can increase neuronal cell outgrowth. *Biotechnol. Bioeng.* 110: 2277–2291.

379 Locatelli, E., Monaco, I., and Franchini, M.C. (2015). Surface modifications of gold nanorods for applications in nanomedicine. *RSC Adv.* 5: 21681–21699.

380 Sun, Q., You, Q., Pang, X. et al. (2017). A photoresponsive and rod-shape nanocarrier: single wavelength of light triggered photothermal and photodynamic therapy based on AuNRs-capped & Ce6-doped mesoporous silica nanorods. *Biomaterials* 122: 188–200.

381 Liu, Z., Liu, X., Yuan, Q. et al. (2012). Hybrid mesoporous gadolinium oxide nanorods: a platform for multimodal imaging and enhanced insoluble anticancer drug delivery with low systemic toxicity. *J. Mater. Chem.* 22: 14982–14990.

382 Xiong, L.-H., Cui, R., Zhang, Z.-L. et al. (2015). Harnessing intracellular biochemical pathways for *in vitro* synthesis of designer tellurium nanorods. *Small* 11: 5416–5422.

383 Li, A., Li, X., Yu, X. et al. (2017). Synergistic thermoradiotherapy based on PEGylated Cu_3BiS_3 ternary semiconductor nanorods with strong absorption in the second near-infrared window. *Biomaterials* 112: 164–175.

384 Ding, X., Liow, C.H., Zhang, M. et al. (2014). Surface plasmon resonance enhanced light absorption and photothermal therapy in the second near-infrared window. *J. Am. Chem. Soc.* 136: 15684–15693.

385 Wu, Z., Li, W., Luo, C. et al. (2015). Rattle-type Fe_3O_4@CuS developed to conduct magnetically guided photoinduced hyperthermia at first and second NIR biological windows. *Adv. Funct. Mater.* 25: 6527–6537.

386 Guo, C., Yu, H., Feng, B. et al. (2015). Highly efficient ablation of metastatic breast cancer using ammonium-tungsten-bronze nanocube as a novel 1064 nm-laser-driven photothermal agent. *Biomaterials* 52: 407–416.

387 Lee, J., Macosko, C.W., and Urry, D.W. (2001). Elastomeric polypentapeptides cross-linked into matrixes and fibers. *Biomacromolecules* 2: 170–179.

388 Zhang, S. (2001). Molecular self-assembly. In: *Encyclopedia of Materials: Science and Technology* (ed. K.H.J. Buschow, R.W. Cahn, M.C. Hemings, et al.), 5822–5828. Oxford: Elsevier Science.

389 Hu, Y., Lin, R., Zhang, P.C. et al. (2016). Electrostatic-driven lamination and untwisting of β-sheet assemblies. *ACS Nano* 10: 880–888.

390 Ghosh, A., Haverick, M., Stump, K. et al. (2012). Fine-tuning the pH trigger of self-assembly. *J. Am. Chem. Soc.* 134: 3647–3650.

391 Hayes, R., Warr, G.G., and Atkin, R. (2015). Structure and nanostructure in ionic liquids. *Chem. Rev.* 115: 6357–6426.

392 Frederix, P.W.J.M., Scott, G.G., Abul-Haija, Y.M. et al. (2015). Exploring the sequence space for (tri-)peptide self-assembly to design and discover new hydrogels. *Nat. Chem.* 7: 30–37.

393 Wang, J., Liu, K., Xing, R., and Yan, X. (2016). Peptide self-assembly: thermodynamics and kinetics. *Chem. Soc. Rev.* 45: 5589–5604.

394 Lee, J.J., Kang, J.A., Ryu, Y. et al. (2017). Genetically engineered and self-assembled oncolytic protein nanoparticles for targeted cancer therapy. *Biomaterials* 120: 22–31.

395 Fung, S.Y., Yang, H., Bhola, P.T. et al. (2009). Self-assembling peptide as a potential carrier for hydrophobic anticancer drug ellipticine: complexation, release and *in vitro* delivery. *Adv. Funct. Mater.* 19: 74–83.

396 Rudra, S., Tian, Y.F., Jung, J.P., and Collier, J.H. (2010). A self-assembling peptide acting as an immune adjuvant. *Proc. Natl. Acad. Sci. U.S.A.* 107: 622–627.

397 French, K.M., Somasuntharam, I., and Davis, M.E. (2016). Self-assembling peptide-based delivery of therapeutics for myocardial infarction. *Adv. Drug Deliv. Rev.* 96: 40–53.

398 Fan, Z., Sun, L., Huang, Y. et al. (2016). Bioinspired fluorescent dipeptide nanoparticles for targeted cancer cell imaging and real-time monitoring of drug release. *Nat. Nanotechnol.* 11: 388–394.

399 Bellomo, E.G., Wyrsta, M.D., Pakstis, L. et al. (2004). Stimuli-responsive polypeptide vesicles by conformation-specific assembly. *Nat. Mater.* 3: 244–248.

400 Ortony, J.H., Newcomb, C.J., Matson, J.B. et al. (2014). Internal dynamics of a supramolecular nanofibre. *Nat. Mater.* 13: 812–816.

401 Yu, X., Zhang, Z., Yu, J. et al. (2018). Self-assembly of a ibuprofen-peptide conjugate to suppress ocular inflammation. *Nanomedicine* 14: 185–193.

402 Park, W.M. and Champion, J.A. (2014). Thermally triggered self-assembly of folded proteins into vesicles. *J. Am. Chem. Soc.* 136: 17906–17909.

403 Zhang, H., Park, J., Jiang, Y., and Woodrow, K.A. (2017). Rational design of charged peptides that self-assemble into robust nanofibers as immune-functional scaffolds. *Acta Biomater.* 55: 183–193.

404 Hosseinkhani, H. and Hosseinkhani, M. (2008). Suppression effect of basic fibroblast growth factor on mesenchymal stem cell proliferation activity. Part I: Release characteristics. *Chem. Today* 26: 30–32.

405 Hosseinkhani, H. and Hosseinkhani, M. (2008). Suppression effect of basic fibroblast growth factor on mesenchymal stem cell proliferation activity. Part II: Biological characteristics. *Chem. Today* 26: 35–37.

406 Wang, K., Keasling, J.D., and Muller, S.J. (2005). Effects of the sequence and size of non-polar residues on self-assembly of amphiphilic peptides. *Int. J. Biol. Macromol.* 36: 232–240.

407 Adams, D.J., Holtzmann, K., Schneider, C., and Butler, M.F. (2007). Self-assembly of surfactant-like peptides. *Langmuir* 25: 12729–12736.

408 McMillan, R., Meeks, B., Bensebaa, F. et al. (2001). Cell adhesion peptide modification of gold-coated polyurethanes for vascular endothelial cell adhesion. *J. Biomed. Mater. Res.* 54: 272–283.

409 Sun, X., Sheardown, H., Tengvall, P., and Brash, J.L. (2000). Peptide modified gold-coated polyurethanes as thrombin scavenging surfaces. *Biomed. Mater. Res.* 49: 66–78.

410 Hosseinkhani, H., Hosseinkhani, M., Hattori, S. et al. (2010). Micro and nanoscale *in vitro* 3D culture system for cardiac stem cells. *J. Biomed. Mater. Res. A* 94: 1–8.

411 Hosseinkhani, H., Hosseinkhani, M., and Kobayashi, H. (2006). Design of tissue engineered nanoscaffold through self assembly of peptide amphiphile. *J. Bioact. Compat. Polym.* 21: 277–296.

412 Hosseinkhani, H., Hosseinkhani, M., Khademhosseini, A., and Kobayashi, H. (2007). Bone regeneration through controlled release of bone morphogenetic protein-2 from 3-D tissue engineered nano-scaffold. *J. Control. Release* 117: 380–386.

413 Hosseinkhani, H., Hosseinkhani, M., Tian, F. et al. (2007). Bone regeneration on a collagen sponge-self assembled peptide-amphiphile nanofibers hybrid scaffold. *Tissue Eng.* 13: 1–9.

414 Hosseinkhani, H., Hosseinkhani, M., Tian, F. et al. (2006). Enhanced angiogenesis through controlled release of basic fibroblast growth factor from peptide amphiphile for tissue regeneration. *Biomaterials* 27: 5836–5844.

415 Hosseinkhani, H., Hosseinkhani, M., Tian, F. et al. (2006). Ectopic bone formation in collagen sponge-self assembled peptide amphiphile nanofibers hybrid scaffold in a perfusion culture bioreactor. *Biomaterials* 27: 5089–5098.

416 Hosseinkhani, H., Hosseinkhani, M., and Kobayashi, H. (2006). Proliferation and differentiation of mesenchymal stem cells by using self assembly of peptide-amphiphile nanofibers. *Biomed. Mater.* 1: 8–15.

10

3D *In Vitro* Systems

10.1 Introduction

Many three-dimensional (3D) models currently in practice, however, require expensive equipment, large sample volumes, long incubation times, and/or extensive expertise, and the greatest disadvantages is that they are too far from the nature of human organs. Because of these problems, research and development on drug discovery, regenerative medicine, biotech, and pharmaceutical industries is very costly and takes several years to bring a single drug/product to the marketing. The goal of 3D technology is to merge biomaterials science, nanotechnology, and biological principles to generate 3D *in vitro* living systems to mimic organ/tissues, thus partially reducing the amount of *in vitro* and *in vivo* animal testing and clinical trials and solving the above problems. These models are useful for *in vitro* diagnostics tools and drug screening applications as well as for transplantation.

10.2 Conventional 3D Technology

The proposed technique of cell culture in 3D artificial materials is based on the use of 3D fibrous scaffold to guide cell organization. In comparison with conventional 2D culture, cells maintained in 3D culture more closely resemble the *in vivo* situation with regard to cell shape and cellular environment that can influence the behavior of cells. It has been recognized that induction of tissue regeneration based on tissue engineering can be achieved by the following three key steps: the proliferation of cells, the seeding of cells and proliferation in a suitable scaffold, and the maintenance of the differentiation phenotype of the engineered tissues [1]. The property of scaffold material for cell attachment is one of the major factors contributing to their morphology, proliferation, functions, and the subsequent tissue organization [2]. At first, cells attach to the material surface of scaffold, then spread, and proliferate. The 3D scaffold can provide larger surface area available for cell attachment and spreading than 2D systems (i.e. tissue culture plate). Xie et al. have reported that the initial rate of cell growth was higher for the 2D culture, but once the cells reached confluence, their proliferation stopped [3]. However, the cell growth in the 3D scaffold was continued for longer time periods than that of 2D

Biomedical Engineering: Materials, Technology, and Applications, First Edition. Hossein Hosseinkhani.

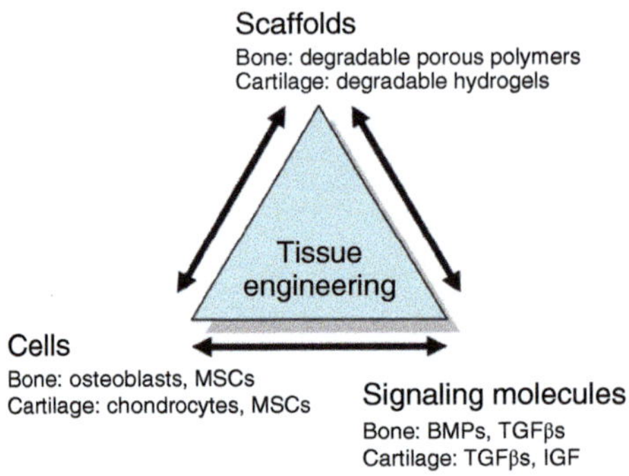

Figure 10.1 Schematic illustration of tissue engineering for regenerative medicine.

scaffold. Other reports have demonstrated that cell proliferation was superior in the 3D scaffold than the 2D one [4].

Regenerative medicine is an interdisciplinary field that combines engineering and live sciences in order to develop techniques that enable the restoration, maintenance, or enhancement of living tissues and organs. Its fundamental aim is the creation of natural tissue with the ability to restore missing organ or tissue function, which the organism has not been able to regenerate in physiological conditions. With that, it aspires to improve the health and quality of life for millions of people worldwide and to give solution to the present limitations: rejections, low quantity of donors, etc. [5]. Tissue engineering needs scaffolds to serve as a substrate for seeding cells and as a physical support in order to guide the formation of the new tissue [6]. The majority of the used techniques utilize 3D polymeric scaffolds, which are composed of natural or synthetic polymers. Synthetic materials are attractive because their chemical and physical properties (e.g. porosity and mechanical strength) can be specifically optimized for a particular application. The polymeric scaffolds structures are endowed with a complex internal architecture, channels, and porosity that provide sites for cell attachment and maintenance of differentiated function without hindering proliferation. Ideally, a polymeric scaffold for tissue engineering should have the following characteristics: (i) appropriate surface properties promoting cell adhesion, proliferation, and differentiation; (ii) biocompatible; (iii) highly porous, with a high surface area/volume ratio, with an interconnected pore network for cell growth and flow transport of nutrients and metabolic waste; and (iv) mechanical properties sufficient to whist and any *in vivo* stresses [7]. The last requisite is difficult to combine with the high porosity in volume of the material. That is why it is necessary to use polymeric matrices with special or reinforced properties, especially if the polymer is a hydrogel. Figure 10.1 illustrates the principle of tissue engineering for regenerative medicine.

Among many materials currently used as cell scaffolds, collagen has been widely used. The *in vivo* safety has been proven through the long-term applications to clinical medicine, cosmetics, and foods. The collagen sponge fabricated by freeze-drying method, followed by cross-linking of combined dehydrothermal, glutaraldehyde, and ultraviolent (UV), is highly porous with an interconnected pore structure, which is effective in the infiltration of cells and supplying oxygen and nutrients

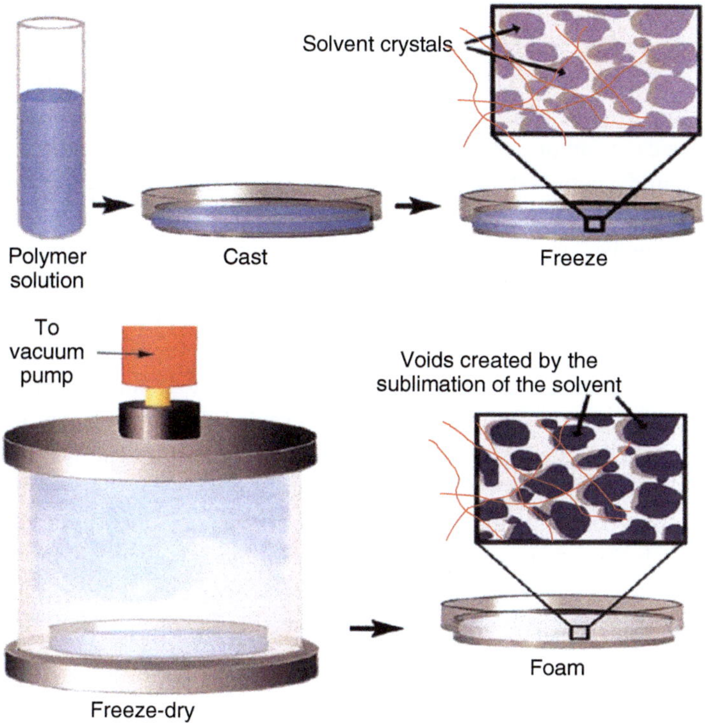

Figure 10.2 Schematic illustration of 3D scaffold fabrication.

to the cells or excluding the cells wastes, while the shape and bioresorbability can be readily regulated by changing the formulation conditions. Figure 10.2 shows schematic illustration of 3D scaffold fabrication.

However, as shown in Figure 10.3, the drawback of collagen sponge as a scaffold for cell proliferation and differentiation is its poor mechanical strength. To overcome the inherent material problem of sponge, combining it with other materials

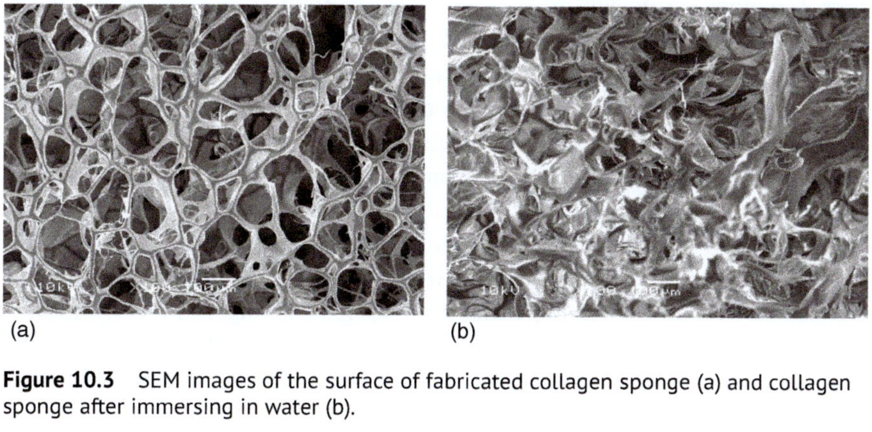

Figure 10.3 SEM images of the surface of fabricated collagen sponge (a) and collagen sponge after immersing in water (b).

has been attempted. Considering implantation, the materials to be combined should be also bioabsorbable. From the viewpoint of clinical application, it is preferable to select the material that has been clinically used. Several biodegradable synthetic polymers, such as poly(glycolic acid) (PGA) and its copolymers with L-lactic acid, DL-lactic acid, and ε-caprolactone, have been fabricated into the cell scaffolds of nonwoven fabric and sponge shapes for tissue engineering. The mechanical resistance of the scaffolds to compression is practically acceptable for the tissue engineering applications because of their hydrophobic nature. However, the cell attachment to the surface of synthetic polymer scaffolds is poor compared with that of collagen. PGA has been approved by US Food and Drug Administration for the clinical applications. Our previous study revealed that incorporation of PGA fiber enabled a collagen sponge to increase the resistance to compression *in vitro* and *in vivo* [8]. The *in vitro* culture experiment revealed that the number of MSC attached increased with the incorporation of PGA fiber to a significantly high extent compared with that of the original collagen sponge. It is a key for the present technology to fabricate mechanically strong collagen sponges by incorporating the PGA fiber of which the amount is as low as possible. Since collagen is more compatible to cells than PGA, at a higher amount of PGA fiber incorporated, the fiber may cause inflammation response to the sponge. Moreover, the collagen sponge does not become strong enough to resist the compressed deformation only by increasing the extent of cross-linking. Because the PGA fiber incorporation also suppressed the shrinkage of collagen sponge, it is possible that the volume available for cell attachment was larger, resulting in a higher number of cells attached. We have shown that mouse fibroblast L929 cells infiltrated into the collagen sponge incorporating PGA fiber more deeply than the collagen sponge. This phenomenon also can be explained in terms of suppressed shrinkage of sponge by PGA fiber incorporation. The collagen sponge mechanically reinforced by PGA fiber incorporation is a promising scaffold for tissue regeneration. The incorporation of PGA fiber enabled the sponge to increase the resistance to compression. On comparing *in vivo* degradability, the collagen scaffold is generally digested faster than the PGA fabric. This degradation profile greatly depends on the cross-linking extent of collagen sponge, the molecular weight of PGA, and the formulation shape. In our study, a combined cross-linking method of dehydrothermal, glutaraldehyde, and UV was used to prepare collagen sponges with or without PGA fiber incorporation. Weadock et al. have evaluated the physical, mechanical, and biological behaviors of collagen sponge cross-linked by physical (UV irradiation and dehydrothermal) and chemical (carbodiimide and glutaraldehyde) or combination of physical (dehydrothermal) and chemical (carbodiimide) [9]. The results revealed that combination of physical (dehydrothermal) and chemical (carbodiimide) cross-linking of collagen reduced significantly the swelling ratio, increased the collagenase resistance time, and low and high strain modulus compared with a single cross-linking of UV, dehydrothermal, and carbodiimide. The glutaraldehyde cross-linking itself showed the same physical and mechanical properties as combination of physical (dehydrothermal) and chemical (carbodiimide) cross-linking. Figure 10.4 shows scanning electron

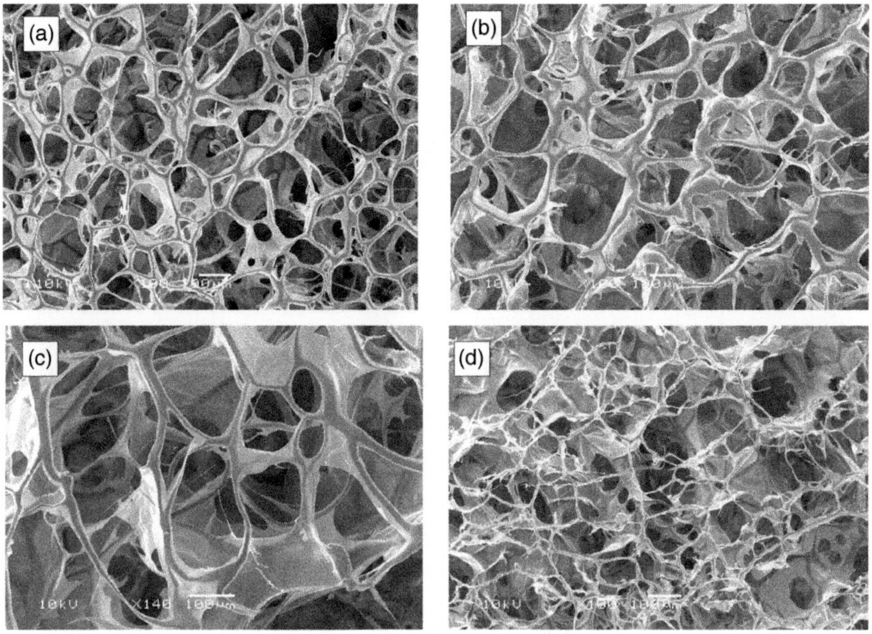

Figure 10.4 SEM images of the surface of fabricated collagen sponges containing 0 (a), 1.5 (b), 3 (c), or 6 mg of PGA nanofibers (d). Scale bar = 100 μm (NB: Figure 11c collected at ×140 magnification).

microscopy (SEM) images of collagen sponge reinforced with different of PGA nanofibers incorporation.

In vitro rabbit corneal (RC) cells on the previously mentioned scaffolds showed significant improvement of 3D cells culture on sponges reinforced with PGA nanofibers (Figure 10.5).

The polymeric scaffold design depends on the regarded applications, but in any case, it must achieve structures with the aforementioned characteristics, which are necessary to their correct function. To achieve it, two factors must be considered: (i) materials used, both the porogen and the reticulate polymer, which are infiltrated in the porogen to become a scaffold, and (ii) the structural architecture, both external and internal, basically shown by its porosity (high surface area/volume ratio), geometry, and pore size, as well as keeping in mind that the structures must be easily processed into 3D. Briefly, the sintered 3D mesh was placed between two glass plates covered with cellophane film and sealed by rubber spacers (i). Appropriate amounts of methylmethacrylate (MMA), cross-linking agent ethylenglycoldimethacrylate (EGDMA; 1% wt), and azobisisobutyronitrile (AIBN) as a thermal free-radical initiator (0.2% wt) in all solutions were mixed in a beaker. The homogeneous liquid mixture of monomer, initiator, and cross-linking agent was distributed in the glass mold (ii). Polymerization was carried out at 65 °C in a thermostatic bath during 30 minutes, in nitrogen atmosphere. Then, the mold was kept in an oven at 65 °C for four hours followed by an increase of temperature to 75 °C during 24 hours (iii). Once the polymerization process was completed,

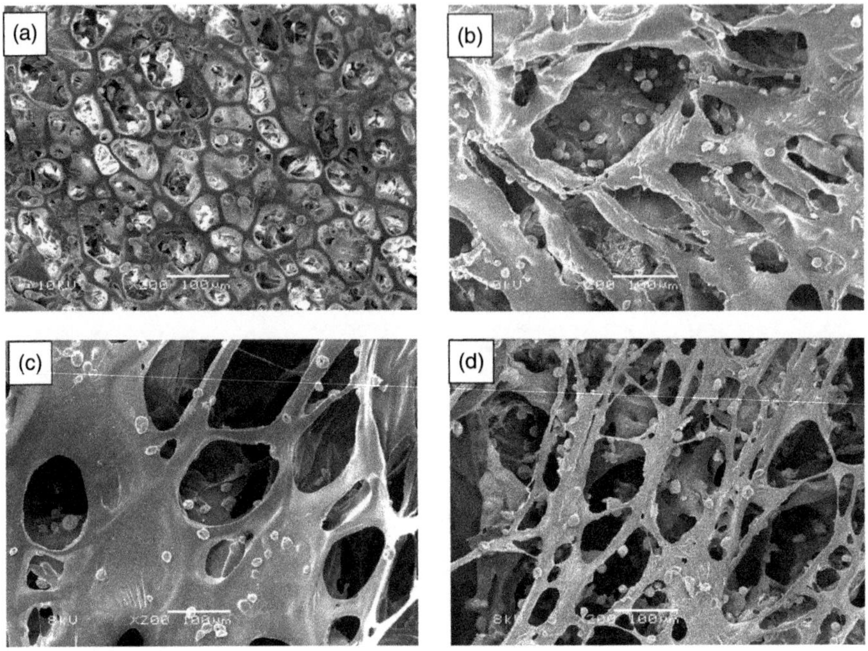

Figure 10.5 SEM images of RC cells cultured on collagen sponges reinforced with (a) 0, (b) 1.5, (c) 3, and (d) 6 mg PGA nanofibers. Images were collected at ×200 magnification. Scale bar = 100 μm.

the samples were removed from the glass molds (iv) and sanded in a LaboPol-21 (Struers) polishing machine. In order to remove the template, the samples were immersed in nitric acid to dissolve the nylon mesh at room temperature for five days (v), and then the material was washed in a Soxhlet extractor, using ethanol as a solvent for 48 hours, in order to extract low molecular weight substances and to remove the nitric acid remains. After that, the resulting scaffolds were dried at ambient conditions for a day, allowing them to shrink slowly, trying to avoid any possible bending or rolling of the scaffolds in the early stages of the drying, and finally, vacuum-dried at 120 °C for 48 hours (vi). The scaffolds were stored in a desiccator until characterization.

On basis of the extensive range of polymeric materials, different processing techniques have been developed to design and fabricate 3D scaffolds for tissue engineering implants [10]. They include (i) phase separation, (ii) gas foaming, (iii) fiber bonding, (iv) photolithography, (v) solid free-form (SFF), and (vi) solvent casting in combination with particle leaching.

However, none of these techniques have achieved a suitable model of 3D architecture so the scaffolds must be fulfilled using equipment with high cost even for the reasons that will be discussed next. So, using phase separation, a porous structure can be easily obtained by adjusting thermodynamic and kinetic parameters. However, because of the complexity of the processing variables involved in phase-separation technique, the pore structure cannot be easily controlled. Moreover, it is difficult to obtain large pores and may exhibit a lack of interconnectivity [11]. Gas foaming has the advantage of room temperature processing but produces a largely nonporous outer skin layer and a mixture of open and closed pores within the center leaving incomplete interconnectivity. The main disadvantage of the gas foaming method is that it often results in a non-connected cellular structure within the scaffold. Fiber bonding provides a large surface area for cell attachment and a rapid diffusion of nutrients in favor of cell survival and growth. However, these scaffolds, as the ones used to construct a network of bonded PGA, lacked the structural stability necessary for *in vivo* use. In addition, the technique does not lend itself to easy and independent control of porosity and pore size.

10.3 Advanced Techniques to Create 3D Systems

Photolithography has been employed for patterning, obtaining structures with high resolution, although this resolution may be unnecessary for many applications of patterning in cell biology. In any case, the disadvantage of this technique is the high cost of the equipment that limits their applicability [12]. SFF scaffold manufacturing methods provide excellent control over scaffold external shape, and internal pore interconnectivity and geometry but offer limited microscale resolution. Moreover, the minimum size of global pores is 100 μm. Additionally, SFF requires complex correction of scaffold design for anisotropic shrinkage during fabrication. Moreover, it needs high cost equipment. Solvent casting in combination with particulate leaching method, which involves the casting of a mixture of monomers and initiator solution and a porogen in a mold, polymerization, followed by leaching-out of the porogen with the proper solvent to generate the pores, is inexpensive but still has to overcome some disadvantages in order to find engineering applications, namely, the problem of residual porogen remains, irregular-shaped pores, and insufficient interconnectivity [13]. The proposed scaffolds may find applications as structures that facilitate either tissue regeneration or repair during reconstructive operations [14]. The new structure could also find applications in other areas in which the pore morphology may play an essential role, such as membranes and filters [15, 16]. In the United States alone, each year over 10 000 newly injured people are added to the total of more than 250 000 that are confined to their wheelchair [17]. A major limitation in treating nerve injury, central nervous system (CNS), and peripheral nervous system (PNS) is the failure of current therapies to induce nerve regeneration. Unfortunately, for CNS injury, and particularly spinal cord injury, there is currently no treatment available to restore nerve function [18]. One possible avenue for remedying this situation is to artificially engineer nerve tissue. It is commonly accepted that physical guidance of axons is a vital component of nerve repair. Many materials have been used in an attempt to physically guide the regeneration of damaged nerves. It has been shown that preferential alignment of channel pores may provide a unique advantage in certain medical applications, such as nerve regeneration [19]. Highly oriented polylactic acid (PLA) scaffold for spinal cord regeneration and highly oriented macroporous have shown to have efficiency in axonal regeneration both in the peripheral and CNS. Cell migration and angiogenesis were observed as well as expected orientation of axonal growth. The axons were perfectly aligned along the pore direction, which confirmed the crucial role of 3D polymer structures [20]. It has been demonstrated that 3D sponges of poly-hydroxy ethyl methacrylate (PHEMA) sponges are able to house a purified population of glial cells and provided a scaffold for regenerative growth of axons in the lesioned rat optic tract may be a candidate for use as prosthetic bridges in the repair of the damaged CNS. However, they deduce that further work is necessary to optimize their procedure, like providing a more oriented trabecular network within the hydrogel scaffold [21]. Macroporous foams with size of 100 μm were produced in the form of channels by the solid–liquid phase separation technique for nerve regeneration. Nerve regeneration can only occur through a structure of interconnected pores of ideal diameter in the range of 10–100 μm [22]. Poly(D,L-lactide) foams that have been developed with macroporous of 100 μm organized longitudinally were prepared by freeze-drying technique for spinal cord regeneration. They showed that the parallel assembly of rods of porous (diameter ~100 μm) containing an amphiphilic copolymer was a promising strategy to bridge a defect in the spinal cord of adult rats and they confirmed a high density of cells in the surface of porous interconnected structures as well [23, 24].

10.4 High-Throughput Screening (HTS) Technologies

Materials technology aims to develop a technique that draws from microscale engineering, novel biomaterials, and biological principles to overcome the limitations of the current approaches to generate 3D *in vitro* living organs. Fabrication technology such as photolithography is attractive to produce biodevices in a variety of shapes and sizes. For fabrication of 3D *in vitro* living systems, an advanced laboratory of microfabrication facilities based on lithography technology through collaboration with semiconductor industry on the future development of biochip technology is necessary. In addition, using microfluidic channels or micromolding techniques to deposit cells and materials on specific regions of a substrate, which may also be used for microfabrication, is important. To engineer 3D living organ model, soft lithography and photolithography techniques are followed. In this approach, the first layer of cells is cultured on hydrogels following photolithography. Next step is culturing the second layer of the cells inside the empty microwell. This process will be duplicated and continued to culture the third, fourth, and other layers of the cells, by which we are able to create an *in vitro* living organs to be used as intelligent diagnostic tools. High-throughput screening (HTS) technique offers great flexibility in testing of chemical libraries. For example, various doses of the same chemical can be easily tested from the hydrogel microarrays to assess the minimum toxic dosage. This can be used in arriving at toxic dose levels without expensive and ethically challenging animal experimentation. Also, the system can be modified to conduct other high-throughput experimentation. For example, various differentiation factors at different concentrations can be printed in the chip slide, and its effects on the stem cell differentiation can be studied in a multiplexed manner. One of the significant advantages of performing HTS studies in low volumes is the ability to highly concentrate the molecules that are secreted by the cells.

10.5 Development of 3D *In Vitro* Cell Culture Technology

Since several years, biomaterials were considered to be useful building blocks for materials engineering, and bioengineering is known as an important route to produce novel materials that can cover a wide range of biological applications.

Materials design for cell proliferation and differentiation is one of the key technologies for development of 3D *in vitro* culture systems. In conventional cell culture such as static tissue culture dish (2D), the initial rate of cell growth is higher, but the proliferation stops once the cells reached confluence. 3D structures have been investigated for the cell culture because they have larger surface for cell attachment and proliferation than 2D tissue culture dish and are preferable to assist the formation of 3D cell constructs that may resemble the structure and function of body tissues. In addition, 3D culture systems also play an important role in the substrate for *in vitro* cell culture to increase the number of cells as high as clinically applicable [24]. 3D materials able of regenerating or restoring tissue and/or organs have begun to revolutionize medicine and biomedical science and have been used to support and promote the regeneration of tissues. The proliferation of cells in the 3D structures needs oxygen and nutrition supply. In this circumstance, the 3D materials should provide such an environment for cells living in distance. Diffusion of nutrients, bioactive factors, and oxygen through 3D materials is sufficient for survival of large numbers of cells for extended periods of time. A major constraint of biodegradable polymeric materials for vascular tissue engineering is poor cell adhesion and lack of signals for new tissue generation. The presence of ECM within 3D structure is desirable for growth of stem cells and *in vitro* formation of remodeled vascular conduit. Figure 10.6 shows the process of cell culture in a typical 3D scaffolding materials.

10.6 Future Prospects

Developing 3D living systems will open many doors and lead to significant improvements in biological tools, drug discovery process, lead identification, and therapeutic approaches. The miniaturization of this approach allows one to perform many more experiments than previously possible in a simpler manner. 3D *in vitro* technology aims to develop sets of tools that are simple, inexpensive, portable, and robust that could be commercialized and used in various fields of biomedical sciences such as drug discovery, diagnostic tools, and therapeutic approaches in regenerative medicine. Materials technology aims at overcoming the fundamental limits to information processing. It will enhance the application of current technology of biology and scaffolding materials by creating new knowledge in selected areas of

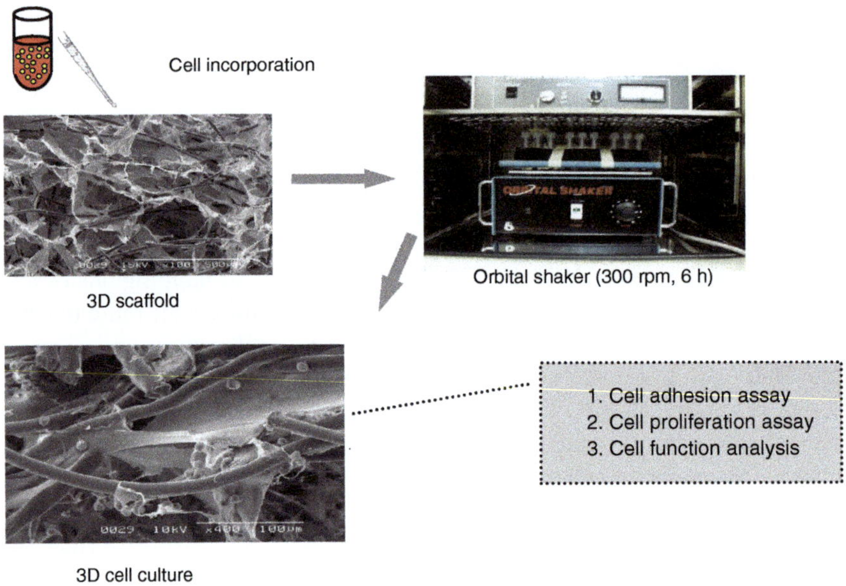

Figure 10.6 3D *in vitro* cell cultures.

focus. Also, it will consist of a top academic and high-tech objective to fabricate, investigate, and implement novel advanced microstructures, nanostructures, and superstructures based on ordered highly functionalized materials to meet the demands of maximum efficient, active hydrogel materials of high and sustained reactivity as well as long-term stability. It will also enable new fundamental research and development for the next generation of biomedical materials as well as exploit such novel structures to develop novel biomedical devices and transfer the knowledge to academia and industry for a future implementation of novel knowledge and technology to the world. In turn, it should increase the international competitiveness of the world into a knowledge-intensive micro- and bioengineering organ-based biosensor by focusing on high-impact research to generate new breakthroughs aimed at solving significant practical problems of biomedical sciences while seeking to extend the boundaries of understanding. It aims to raise research profile of bioengineering as a vibrant center for medical and technological applications through a bottom-up approach that embraces both elements of basic and applied research to enhance the competencies in existing technologies, to seek out promising new areas, and to develop an integrated, cutting-edge research program by growing a pool of top research talent and developing the platforms on which local universities could create research breakthroughs of importance to the world. Since advancement in device development technologies is a significant indicator of the developed societies, rapidly growing market for biomedical devices provide competitive advantage in R&D and commercialization for this field. Such advancement can be reached by means of setting up and helping research institutes to enhance the development of biodevices to be used as diagnostic tools.

References

1 Hosseinkhani, H., Inatsugu, Y., Inoue, S. et al. (2005). Perfusion culture enhances the osteogenic differentiation of rat mesenchymal stem cells in collagen sponge rein forced with poly (glycolic acid) fiber. *Tissue Eng.* 11: 1476–1488.

2 Tian, F., Hosseinkhani, H., Hosseinkhani, M. et al. (2008). Quantitative analysis of cell adhesion on aligned micro- and nanofibers. *J. Biomed. Mater. Res. Part A* 84: 291–299.

3 Xie, Y., Yang, S.T., and Kniss, D.A. (2010). Three-dimensional cell-scaffold constructs promote efficient gene transfection: implications for cell-based gene therapy. *Tissue Eng.* 7: 585–598.

4 Hosseinkhani, H., Azzam, T., Kobayashi, H. et al. (2006). Combination of 3-D tissue engineered scaffold and non-viral gene enhance *in vitro* DNA expression of mesenchymal stem cells. *Biomaterials* 27: 4269–4278.

5 Hosseinkhani, H., Yamamoto, M., Inatsugu, Y. et al. (2006). Enhanced ectopic bone formation using a combination of plasmid DNA impregnation into 3-D scaffold and bioreactor perfusion culture. *Biomaterials* 27: 1387–1398.

6 Hutmatcher, D.W. (2000). Scaffolds in tissue engineering bone and cartilage. *Biomaterials* 21: 2529–2543.

7 Hosseinkhani, H., Hosseinkhani, M., Khademhosseini, A. et al. (2008). DNA nanoparticles encapsulated in 3-D tissue engineered scaffold enhance osteogenic differentiation of mesenchymal stem cells. *J. Biomed. Mater. Res. Part A* 85: 47–60.

8 Khalaji, S., Ebrahimi, N.G., and Hosseinkhani, H. (2021). Enhancement of biocompatibility of PVA/HTCC blend polymer with collagen for skin care application. *Int. J. Polym. Mater. Polym. Biomater.* 70: 459–468.

9 Weadock, K.S., Miller, E.J., Bellincampi, L.D. et al. (1995). Physical crosslinking of collagen fibers: comparison of ultraviolet irradiation and dehydrothermal treatment. *J. Biomed. Mater. Res.* 29: 1373–1379.

10 Cai, Z. and Cheng, G. (2003). Novel method to produce poly(3-hydroxybutyrate) scaffolds with controlled multi-pore size. *J. Mater. Sci. Lett.* 22: 153–155.

11 Zhang, X., Jiang, X.N., and Sun, C. (1997). Micro-stereolithography of polymeric and ceramic microstructures. *Sens. Actuators* 77: 149–156.

12 Kane, R.S., Takayama, S., Ostuni, E. et al. (1999). Patterning proteins and cells using soft lithography. *Biomaterials* 20: 2363–2376.

13 Mikos AG, Sakarinos GJ, Vacanti P, et al. (1996). Biocompatible polymer membranes and methods of preparation of three dimensional membrane structures. US Patent 5514378.

14 Hosseinkhani, H., Hosseinkhani, M., Khademhosseini, A. et al. (2006). Enhanced angiogenesis through controlled release of basic fibroblast growth factor from peptide amphiphile for tissue regeneration. *Biomaterials* 27: 5836–5844.

15 Lindstrom, S., Iles, A., Persson, J. et al. (2010). Nanoporous titania coating of microwell chips for stem cell culture and analysis. *J. Biomech. Sci. Eng.* 5: 272–279.

16 Mohageri, S., Hosseinkhani, H., Ebrahimi, N.G. et al. (2010). Proliferation and differentiation of mesenchymal stem cell on collagen sponge reinforced with polypropylene/polyethylene terephathalate blend fibers. *Tissue Eng. Part A* 16: 3821–3830.

17 Hosseinkhani, H., Hosseinkhani, M., Hattori, S. et al. (2010). Micro and nanoscale *in vitro* 3D culture system for cardiac stem cells. *J. Biomed. Mater. Res. Part A* 94: 1–8.

18 Schmidt, C.E. and Leach, J.B. (2003). Neural tissue engineering: strategies for repair and regeneration. *Ann. Rev. Biomed. Eng.* 5: 293–347.

19 Hosseinkhani, H., Hosseinkhani, M., Tian, F. et al. (2006). Osteogenic differentiation of mesenchymal stem cells in self assembled-peptide amphiphile nanofibers. *Biomaterials* 27: 4079–4086.

20 Hosseinkhani, H., Hosseinkhani, M., and Kobayashi, H. (2006). Design of tissue engineered nanoscaffold through selfassembly of peptide amphiphile. *J. Bioact. Compat. Polym.* 21: 277–296.

21 Hosseinkhani, H., Hosseinkhani, M., and Khademhosseini, A. (2006). Emerging applications of hydrogels and microscale technologies in drug discovery. *Drug Discovery* 1: 32–34.

22 Hosseinkhani, H., Hosseinkhani, M., and Kobayashi, H. (2006). Proliferation and differentiation of mesenchymal stem cells by using self assembly of peptide-amphiphile nanofibers. *Biomed. Mater.* 1: 8–15.

23 Maquet, V., Martin, D., Scholtes, F. et al. (2001). Poly(D,L-lactide) foams modified by poly(ethylene oxide)-block-poly(D,L-lactide) copolymers and a-FGF: *in vitro* and *in vivo* evaluation for spinal cord regeneration. *Biomaterials* 22: 1137–1146.

24 Cavallini, A., Notarnicola, M., Berloco, P. et al. (2000). Use of macroporous polypropylene filter to allow identification of bacteria by PCR in human fecal simples. *J. Microbiol. Methods* 39: 265–270.

11

Frontiers in Biomedical Engineering

11.1 Introduction

Current developments in the technological fields of biomedical and tissue engineering, bioengineering, biomechanics, microfabrication, and microfluidics have led to highly complex and pertinent new tools for *in vitro* and *in vivo* applications [1–5]. The purpose of biomaterials technology is to mimic organ tissues *in vitro* in order to partially reduce the amount of *in vivo* testing. These types of systems can enhance the functionality of cells by mimicking the tissue architecture complexities when compared to *in vitro* analysis but at the same time present a more rapid and simple process when compared to *in vivo* testing procedures. The development of new technology for analysis of engineered tissues can be achieved through the combination of these research domains. Combining these advanced research domains, we then present new area of technology that allows *in vitro* analysis on engineered tissues [6–8]. An extension of the biomaterials technology has also allowed tissue and organ development, which can be considered as a first step toward the replacement of animal testing using a combined organ model. From the viewpoint of designing a biomaterial, understanding materials' characteristics is critical for predicting what function the material can exhibit as a biomaterial [9–12]. For example, the design of a ceramic or metal implant emphasizes the mechanical functions, such as the geometric shape, surface, and mechanical properties, to bear and transmit loading. The metal stent serves as mechanical supports to prevent localized flow constriction by its rigid structure. The contact lenses made by polymer provide optical function to allow visible light to precisely transmit and refract into pupil. Cochlear implant using electrical impulses that stimulate cochlear nerves can create a perception of sound. The electric function such as cardiac pacemaker stimulates muscle to induce rhythmical pulses. Also, piezoelectric implant transfers mechanical stress into electric current to improve bone growth. From the viewpoint of safety, in general, toxic reaction is directly associated with chemical or electrochemical properties of biomaterials interacting with body fluid. The passive characteristic is the basic and important requirement for invaded biomaterials. Accordingly, the physical interactions between biomaterials and living body rather than chemical interactions should be particularly concerned. From several clinical cases, failures also occur when utilizing biomaterials with unfavorable physical

Biomedical Engineering: Materials, Technology, and Applications, First Edition. Hossein Hosseinkhani.

properties. For instance, comparing with bone, a relatively higher elastic modulus of titanium alloy that served as an artificial hip joint may induce the stress shielding effect and may result in implant loosening after a long-term implantation. The main reason for stress shielding effect is an unsymmetrical loading-stress transfer from implant to neighboring bone due to their mismatch stiffness.

11.2 Personalized Medicine

Personalized medicine is a new and better approach that uses an individual's genetic profile to health care based on the prevention, diagnosis, and treatment of disease. Novel biomaterials are useful for *in vitro* diagnostics and drug screening applications at early stage cancer formation. Furthermore, development of new technology based on biomaterials significantly contributes in developing microscale biosensors that will be more portable and scalable for point-of-care sample analysis and real-time diagnosis. Modern biosensors based on micro- and nanoscale techniques have the potential to greatly enhance methods of detecting foreign and potentially dangerous toxins and may result in cheaper, faster, and easier-to-use analytical tools. Building micro/nano interfaces to cells and organisms and exploring bio-derived fabrication methods based on biomaterials technology for developing microfabricated interfaces for diagnosis devices are in central goal of personalized medicine. There is enormous need in the market for new tools for the detection of DNA, proteins, bacteria, and viruses, which can provide a fast and reliable response, allowing at the same time *in situ* analysis. A biosensor is the ideal tool that can fulfill all the aforementioned requirements. In the development of the latter, a key issue is represented by the choice of both the sensing platform and the detection technique. Personalized medicine is based on engineering novel *in vitro* biomaterials technology to develop set of tools that are simple, inexpensive, portable, and robust that could be commercialized and used in various fields of biomedical sciences such as drug discovery, diagnostic tools, and therapeutic approaches in medicine.

11.3 Intelligent Biomaterials

Intelligent materials are based on microenvironment interaction and have properties that are changeable by external factors, such as stress, moisture, electric or magnetic fields, light, temperature, pH, or chemical compounds. The extracellular matrix (ECM) is the extracellular part of animal tissue that usually provides structural support to the animal cells in addition to performing various other important functions. The cellular microenvironment/niche plays a significant role in the regulation of a host of physiological and pathophysiological processes. The ability to control the cellular microenvironment is the key to controlling

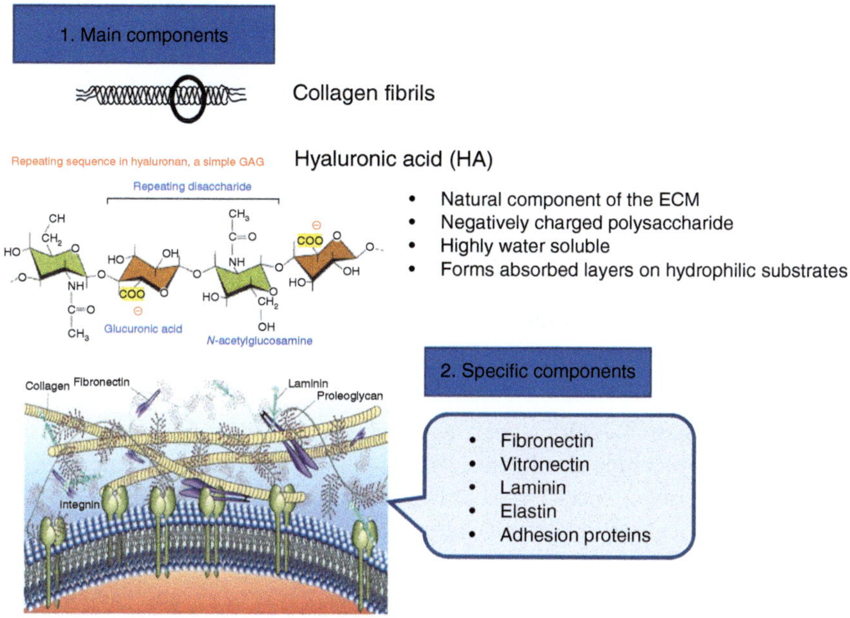

Figure 11.1 Cellular microenvironment showing extracellular matrix (ECM) main and specific components.

cell viability, growth, migration, apoptosis, and differentiation. Niche refers to a microenvironment in which a cell is situated. Thus, a stem cell niche is a microenvironment surrounding a stem cell. In addition to possible contributors to the stem cell, niche, the microenvironments also comprised many ECMs, and signaling molecules that are entrapped inside the ECM. Adult human mammary stem and progenitor cells exhibit impressive plasticity in response to hundreds of unique combinatorial microenvironments. The rational modulation of the microenvironmental milieu can impose specific differentiation phenotypes on normal stem or progenitor cells and perhaps even impose phenotypically normal behavior on malignant cells during tissue genesis. All of this points to the rational manipulation of adult stem and progenitor cells as a promising pathway for beneficial therapies. The ability of adult stem cells to self-maintain, as well as to give rise to progenitor cells that are targeted to become a specific tissue cell, indicates an ability to respond to changing microenvironmental demands, which would mean that a stem or progenitor cell is receiving instructional information from its surroundings. Figure 11.1 provides a simple illustration to the readers to understand the basic knowledge of ECM. The main components of ECM are proteoglycans, heparan sulfate, chondroitin sulfate, keratan sulfate. Also, non-proteoglycan polysaccharides are involved in ECM, which include collagen, elastin, fibronectin, and laminin.

11.4 Artificial Intelligence (AI)

Artificial intelligence (AI) is a broad field of study dedicated to complex problem solving. AI is based on various algorithms that can finally make a robot behave as a human. There are many types of algorithm for a machine to learn something. This process is generally based on another machine (or software) or human that prepares huge set of training data (pairs of example data and label for the data). Next, the data (training example) is fed to the learning algorithm (neural net). The machine learning (ML) algorithm evaluates the training data and produce a result. The machine compares the produced result with the desired result. The machine adjusts the internal parameters of the ML based on the result of the comparison. AI is divided in three categories:

1. **Narrow Intelligence**: Artificial narrow intelligence (ANI) is AI that specializes in one area. It is machine intelligence that equals or exceeds human intelligence or efficiency but in one specific area only. Smartphone apps, spam filters, Google translate, and Google search are all examples of ANI.
2. **General Intelligence**: Artificial general intelligence refers to a computer that is as smart as a human across the board and that can perform any intellectual task that a human being can.
3. **Super Intelligence**: Artificial super-intelligence is an intellect that is much smarter than the best human brains in practically every field, including scientific creativity, general wisdom, and social skills. Artificial super-intelligence ranges from a computer that is just a little smarter than a human to one that is trillions of times smarter across the board.

The differences between AI and human intelligence are summarized as follows:

Artificial intelligence	Human intelligence
1. Ability to stimulate human behavior and cognitive processes 2. Capture and preserve human expertise 3. Fast responses. The ability to comprehend large amounts of data quickly	1. Intuition, common sense, judgment, creativity, and beliefs 2. The ability to demonstrate their intelligence by communicating effectively 3. Reasoning and critical thinking

Potential use cases of AI in medicine are very wide in biomedical engineering such as early detection, diagnosis, treatment, training, and research. The following are emerging fields of AI in biomedical engineering:

AI Surgeon: This technique can reduce the treatment time and the scope of error.

Virtual Assistants for Nursing: AI is able to keep healthcare providers and patients in touch with each other all the time.

AI Assistant for Administration and Workflow: This application can help keep a record of all treatments and patients to allow predictive diagnosis.

Automated Image Diagnosis: This technology can identify anomalies and diseases based on medical images better than doctors.

Drug Creation: By scanning the existing medicines, the application will be able to redesign them so that they can fight the disease.

Clinical Trial Participation: Clinical trials collect a plethora of data that needs to be organized in order to get the correct theory. The AI applications facilitate the outcome-driven approach of clinical trials.

Dosage Error Reduction: The application is designed to reduce the margin of medical errors that may occur when giving medicines to patients.

Treatment Design: The application will help you to narrow down the correct and customized treatment path for a patient.

Pharma Industry: The future of medicine is data-driven and AI-enabled. AI will impact diagnosis, treatment decisions, and patient care. AI and ML have huge potential in the pharma industry, identifying the right candidates at a faster rate and accelerating R&D timelines.

AI-Driven Drug Discovery: Drug metabolism research plays a key role in the discovery and development of drug. Based on the discovery of drug metabolites, new chemical entities can be identified, and potential safety hazards caused by reactive or toxic metabolites can be minimized. AI can model diseased human cells by varying the levels of sugar and oxygen the cells were exposed to and then tracked their lipid, metabolite, enzyme, and protein profiles.

On the other hand, only 12% of drug development programs ended in success in a 2000–2019 study, according to recent research. Some of the problems with drug development are inability to demonstrate efficacy or safety, flawed study design, participant dropouts, and unsuccessful recruitment. Therefore, diversity of AI solutions is as the following:

1. **Information Engines:** Information engines often work across drug discovery and clinical trials, providing valuable information to both sides of the drug development process. IBM's Watson is using AI to improve clinical trial matching and enrollment, particularly for oncology trials. Aetion analyzes medical and pharmacy claims to understand which treatments work best for which patients and at which times. Concerto HealthAI extracts insights from oncology patients' experiences with drugs to generate evidence for new therapeutic approaches.
2. **AI Models of Adverse Drug Reactions:** Identifying adverse drug reactions (ADRs) of new pharmaceuticals is a huge challenge for drug developers today. It is estimated that 100 000 people in the United States die each year as a result of ADRs, which are often not apparent until after the drug has hit the market. By using databases containing known ADRs for FDA-approved drugs, and statistical probabilities of drug-to-protein binding, the AI is able to develop predictive models for theoretical drugs before any *in vitro* research on those drug candidates had been completed. This computer model can be used early on in the drug development process to decrease the chances of a drug failing in the final phases of development due to an ADR.
3. **Patient Stratification:** The success of clinical trials often relies on having the right group of individuals and the ability of researchers to track their appropriate responses. WuXi NextCODE and Tempus Labs use DNA sequencing and genetic

biomarkers to subtype patients and associate them with treatment responses. nQ Medical and WinterLight Labs are using speech biomarkers and patient interactions with electronic devices and touchscreens to quantitatively detect and track progression of neurological disorders for more automated classification.

4. **Operations**: Unlearn.AI is trying to reduce the number of subjects required for clinical trials by using digital twins as synthetic control arms for a proportion of the placebo controls. Trials.ai, Antidote.me, and Deep 6 AI develop AI applications that accelerate the recruitment process by making it easier for patients to enroll and engage or by analyzing medical records and relevant information to identify more patients suitable for a specific trial.
5. **Ecosystem in Clinical Trials:** AI is changing the paradigm of biomedical research, particularly in the clinical trial space, by increasing success rates and decreasing costs. AI technology vendors are most commonly working directly with the pharmaceutical company to sponsor trials, optimizing trials using their own identified data collated directly from hospitals and academic centers and in many cases additional enterprise clinical trial data from the sponsor. Data management vendors such as InterSystems, LifeImage, Medexprim, and Segmed offer faster and easier ways to obtain training data sets and connect with pharmaceutical companies' enterprise data. Microsoft and Google are also working directly with pharmaceutical companies to build AI capabilities around the massive amounts of enterprise and clinical trial data often stored at these companies.

11.4.1 AI in Document Processing

Large companies deal with mountain of documents: everything from contracts, invoices, purchase orders, and policies and procedures, to reports, activity logs, and emails. Document processing is a manual, human-intensive effort and is one of the most time- and resource-consuming tasks; AI and ML can help companies reduce document review and processing time by 90%, decrease costs by 80%, and cut risk by 20%. Automated document process is not much different from the old-fashioned, human-driven way of doing it, but it offers guaranteed accuracy, can run 24/7 without getting sick or being late, and never gets bored.

11.5 Quality Assurance Statement of Biomaterials

Quality assurance statements are necessary for analytical and quantitative measurements of nanomaterials. They must address the following issues:

- The activities to be performed or hypotheses to be tested, and criteria for determining acceptable data quality in terms of precision, accuracy, representativeness, completeness, and comparability or in terms of data quality objectives or acceptance criteria. These criteria must also be applied. Secondary data may be defined as data collected for other purposes or from other sources, including the literature, compilations from computerized data bases, or results from mathematical models of environmental processes and conditions.

- The study design, including sample type and location requirements, all statistical analyses used to estimate the types and numbers of samples required for physical samples, or equivalent information for studies using survey and interview techniques.
- The procedures for the handling and custody of samples, including collection, identification, preservation, transportation, and storage.
- The procedures that will be used in the calibration and performance evaluation of all analytical instrumentation and all methods of analysis used. Explain how the effectiveness of any new technology will be measured and how it will be benchmarked to improve an existing process, such as those used by industry.
- The procedures for data reduction and reporting, including a description of all statistical methods with reference to any statistical software to be used to make inferences and conclusions; discuss any computer models to be designed or utilized with associated verification and validation techniques.
- Describe quantitative and/or qualitative procedures including plans, study design, or analytical methods prior to data collection.

11.6 Environmental, Health, and Safety Issues of Materials

The most important issue of using materials is safety to the people who handle them, to people using the products, and to the environment. Nanoscience is often referred to as key, horizontal, or enabling, since it can virtually encompass all technological areas. Today, more than 800 new and "value-added" products have come to the market as a result of cutting-edge nanotechnology [13–20]. In 2006 sales of products incorporating nanotechnology reached US$50 billion, and nano-formulated drugs alone accounted for sales of US$3 billion. According to predictions, worldwide sales of nanobiotechnology, more specifically for medical, pharmaceutical, forest products, materials, coatings, catalysts, food, and food processing industries, as well as green energy organizations, will invest significantly in developing products based on nanobiotechnology [21–30]. While development and sales of nanotechnology-based products and materials are growing exponentially, the real or perceived risks to human health and the environment could be the limit public acceptance and successful commercialization [31–40]. This poses both a challenge and an opportunity for many industries including the insurance industry.

There are several issues that need to be clearly considered, such as the following:

- Lack of meaningful specifications for scaffold and matrix materials, including failure to differentiate between scaffolds and matrices.
- Pragmatic use of biomaterials with prior FDA approval for medical devices is not a sensible scientific specification.
- Nanomaterials and nanotopography may have significant rules in scaffold material applications.
- Little knowledge about the toxicology of nanoscale of materials.
- The nanoscale is defined as ranging from atomic dimensions to around 100 nm, when very high surface area-to-volume ratios and quantum effect become significant.
- Effects are increasingly being used in positive ways to benefit industries and the society, including medical applications such as drug delivery systems, nonviral vectors for gene therapy, imaging systems (including quantum dots), and tissue engineering.
- Little is known about the potential risks of the nanoscale. The control of adverse effects is a high political priority in the United States and elsewhere.
- It is not believed that there are inherent health risks with the nanoscale per se, since nanomaterials have been with us forever. Nor is it considered that the bottom-up approach to nanomaterials through self-assembly carries any risks of uncontrolled proliferation.
- Nor is it considered that nanotopography on monolithic materials carries any additional risks.
- There are strict concerns regarding biomaterials, especially poor soluble or non-degradable materials, both for humans and to the environment, as well as adverse effects of nanomaterials on cell morphology.
- Biomaterials are already used in sunscreens, cosmetics, and drug delivery systems. The disposal of nanoscale-based consumer products is already giving concern regarding the accumulation of biopersistent particles in the environment.
- Most evidence comes from the inhalation route.
- There is evidence of absorption of materials in nanoscale and translocation to many organs, including the brain and the liver. Passage of nanomaterials along the olfactory nerve is possible. Trans-placental transport cannot be ruled out.
- Transdermal absorption cannot be ignored.
- The fate of insoluble nanomaterials within cells is not clearly understood. Toxicity of nanomaterials is likely to be associated with oxidative stress.
- Consider the use of insoluble nanomaterials within a biodegradable matrix in a tissue engineering scaffold. What will be the ultimate fate of those particles?
- What test systems are required to determine the level of risk?

11.7 Biohazard of Materials

Biohazards are infectious agents or hazardous biological materials that present a risk or potential risk to the health of humans, animals, or the environment.

The risk can be either direct through infection or indirect through damage to the environment. Biohazardous materials include certain types of recombinant DNA; organisms and viruses infectious to humans, animals, or plants (e.g. parasites, viruses, bacteria, fungi, prions, and rickettsia); and biologically active agents (e.g. toxins, allergens, and venoms) that may cause disease in other living organisms or cause significant impact to the environment or community. Biological materials not considered to be biohazardous may be controlled by regulations and guidelines as biohazardous materials. Many research studies emphasize the necessity of robust experimental procedures and multimodal biohazard assessment of products incorporating nanomaterials. Relevant information about biohazard of nanomaterials should be provided to facilitate safer development of nanotechnology. Rapid progress in nanotechnology has been accompanied by insufficient data on biohazard identification. Nanomaterials are biointeractive in that their size and existence themselves induce bioreaction to cells and tissue [41–50]. They are also bioreactive, since they induce the intrinsic functions of biological organisms that are both merit (bioactive) and demerit (toxic) for human beings. The bilateral potential for both high functional performance (nanotechnology) and unintentionally occurring biohazard (nanotoxicology) is the essential property of nano/micro-materials. It is critical to note that the applied approaches in the evaluation of ENM-related risks and hazards are difficult, since the physicochemical properties of ENMs may differ markedly. It is challenging to determine conclusions about the risks of ENMs in the current regulatory format, unlike the huge worldwide investment in nanosafety research [51–60]. It has become increasingly clear that nanosafety research has predominantly employed by a material-by-material method using regular low-throughput assays to evaluate toxicity end points. Each of these established assays has its own advantages and disadvantages that vary for the nanoparticles (NPs). It is necessary to establish highly accurate assays for prompt screening of different types of NPs in various contexts and under different conditions [60–70]. The development of HTS methods and high-content assays (HCAs) can lead to notable savings in research costs and time [71–75]. HTS is a potent assay for screening of the hazard potential of many NPs. HTS can be used for assessment of cytotoxicity, immunotoxicity, and genotoxicity of ENMs. HTS is a suitable method in drug discovery, and it is progressively used to validate predictive testing of chemicals [76–80]. Both HTS and HCA assays can yield substantial information regarding the risk assessment and potential toxicity of NPs, and they may be further utilized with other advanced approaches enabling accurate, fast, and affordable production of data essential for NP hazard prediction.

11.8 Limitations of Materials in Medicine

The science and knowledge that the scientific community has about nanotechnology and its potential versatile applications is only based on laboratory research. This research is being conducted to understand how matter behaves at the nanoscale level. Factors and conditions governing the behavior of macrosystems do not really apply to nanosystems. The major limitations and technological hurdles faced by nanotechnology and its applications in nanobiotechnology are discussed as follows:

- The scientific community has not yet completely understood how the human body reacts to these nanomaterials and nanosystems acting as carriers.
- Nanomaterials have larger surface areas relative to their volume. Friction and clumping of the nanomaterials into a larger structure is inevitable, which may affect their function as a delivery system.
- Due to their minute size, these drug carriers can be cleared from the body by excretory pathways.
- When not excreted, larger nanomaterials can accumulate in vital organs causing toxicity, leading to organ failure.
- Polymeric micelles reportedly cause acute hypersensitivity reactions in animal tests.
- Liposomes have certain drawbacks like being captured by the human body's defense system. The drug loading capacity of liposomes is being tested by researchers and still remains inconclusive.
- All previous studies resulted in post-treatment accumulation of the nanomaterials in the skin and eyes.
- Accumulation of nanomaterials in internal organs.
- Once nanomaterials are administered into the human body, they should be controlled by an external monitor preventing them from causing adverse effects.
- These new technologies are in various stages of research and development. It is expected that these discoveries will later come into practical use within the next 15–20 years.

There is a need to create new and alternative methods to face problems using current technology in delivering biomolecule drugs, such as protein, growth factors, and DNA. Such systems have previously been created as an alternative to enhance *in vitro* and *in vivo* localization of biomolecule drugs [80–90]. Nanotechnology promises numerous nanoscale carriers that can act as drug delivery systems. Research will hopefully result in discovery of newer and smarter drug carrier systems to treat various diseases. Even though there are certain technological hurdles, these can be overcome by understanding the drawbacks of a drug delivery system and finding an alternative to overcome drawbacks [90–100].

Nanotechnology and its versatile applications have kindled the interest of scientists and researchers to find newer applications in pharmacology and drug delivery. Nanotechnology is evolving rapidly with more potential impacts in the treatment of diseases like cancer, diabetes, respiratory diseases like asthma, ocular diseases, and gene therapy [101–110]. Results in nanotechnology are inspiring the scientific community to discover new innovative non-invasive tools at the nanoscale level. NP technology has been enhanced by major advances in genomics with identification of the structure and function of thousands of genes in the human genome as well as advances in controlling the delivery and expression of plasmid-based gene expression systems *in vivo*. Recent advances in the application *in vitro* as well as *in vivo* use of DNA NPs for gene delivery to cells and tissue have been considered [111–120]. While improvements to transfection efficiencies have been made, and research toward understanding the underlying mechanism behind several gene delivery vehicles has been carried out, issues of toxicity, transfection efficiency, and host response still remain problems for some gene delivery applications. There are many aspects regarding the application of nonviral gene delivery vehicles for human treatment still to be clarified [120–130].

11.9 Clinically Approved Biomaterials

Of all nanomaterials, those reaching preclinical or clinical trials are of great importance in nanomedicine, since they can be employed as efficient materials for treatment of many diseases. Currently, the US FDA controls the approval process for NP-based medicines. The FDA is also responsible for the assessment of additional materials/systems in addition to nanostructures such as biologic, drug, or devices that have potential applications in medicine or related fields. It is predicted that the entire approval process requires almost US$1 billion and approximately 10–15 years for each new nanomedicine to be approved. About 50 nanopharmaceuticals currently available for clinical use have been approved by the FDA since 1995 [131–135]. These nanopharmaceuticals are generally administrated orally or intravenously, while transdermal administration is less common [135]. It is necessary to run preclinical testing phase prior to clinical use. These tests usually comprised *in vivo* trials to determine safety, efficacy, toxicity, and suitable dosage of NPs. A comprehensive knowledge of fabrication processes as well as physicochemical parameters (comprehensively explained in Section 10.5) that affect the final properties of NPs are vital. To gain information about these processes, researchers can read publications of the Nanotechnology Characterization Laboratory (NCL), founded by the National Cancer Institute. They can send NPs to NCL for testing and validation based on a number of protocols [136, 137]. Comprehensive *in vivo* studies regarding the precise behavior of NPs within real biological environments along with accurate assays that can simply distinguish multiple effects of NPs are required. The FDA also provides information about characterization of NPs through diverse guidance documents. The physicochemical properties, efficacy, and toxicity of NPs can be submitted in an Investigational New Drug (IND) application for FDA review and approval [10–69]. When IND approves new nanomedicines, human trials can clearly specify the efficacy and safety of the particular NP. Human/clinical trials can be classified into Phase I, Phase II, Phase III, and Phase IV. Phase I is the first in-human trial, and it is defined as observation of dosage, toxicity, and excretion in a small group of patients. Phase II evaluates effectiveness for specific diseases and overall safety in a larger group. Phase III represents the final verification of efficacy and safety of proposed materials or devices. Lastly, Phase IV (post-marketing) examines safety evaluations during marketing mainly at the request of the FDA, other groups, and clinicians [134]. FDA-approved NPs reached a peak from 2001

Table 11.1 Mechanism of toxicity of nanomaterials.

Mechanism	NPs	Application	Remark	References
Oxidative stress	SWCNT	Cell research	SWCNT-mediated oxidative stress resulted in loss of multicellular chirality	[139]
	ZnO	Nanomedicine	ZnO NP-triggered oxidative stress notably influenced the BJ fibroblast p53 pathway. At low concentration of NPs caused ROS, p53 induced expression of antioxidant genes, whereas at high level, the elevated amount of intracellular ROS activated the apoptotic pathway via p53	[140]
	Fe_3O_4@carbon	Prostate cancer	NPs enhanced ascorbic acid-triggered oxidative stress in cancer cells and adjusted the ROS-mediated antineoplastic efficacy of exogenous ascorbic acid	[141]
Apoptosis	TMO, REO	Liver research	Pro-oxidative transition-metal oxides (TMOs) triggered the activation of caspases 3 and 7, leading to apoptosis in both hepatocytes and KCs, while rare-earth oxide (REOs) caused caspase 1 activation, NLRP3 inflammasome activation, lysosomal damage, and pyroptosis in KCs	[142]
	Micelle		PEG–phosphoethanolamine (PE) micelles accumulated in the endoplasmic reticulum (ER) and induced apoptosis through the activation of the unfolded protein response (UPR) signaling, and high expression of cell death effector CHOP, as well as proapoptotic Bax/Bak	[143]
	Fab′-MORF1 and P-MORF2	Chemotherapy and radiotherapy	Two nanoconjugates including *N*-(2 hydroxypropyl)methacrylamide (HPMA) polymer and an anti-CD20 Fab′ fragment were self-assembled through oligonucleotide hybridization and initiated apoptosis. Both NPs destroyed cancer cells and generated long-term survivors	[144]
Autophagy	IONPs	Cancer	Photothermal effect of IONPs caused autophagy induction in cancerous MCF-7 cells in a laser dose-dependent manner. Laser treatment of IONPs and autophagy inhibitor improved the rate of tumor inhibition from 43.26% to 68.56%	[145]
	Polymer	Breast cancer	The self-assembled polymer-Beclin-1 NPs promoted the stability and cellular uptake of Beclin-1 *in vitro*. The disassociated NPs induced autophagy, inhibited the growth of tumors, and reduced tumor development *in vivo*	[146]
	Ag	Tumor therapy	The PVP-coated AgNPs induced autophagy through ROS-mediated mTOR signaling pathway. Also, autophagy was inhibited via introduction of ATG5 or 3-methyladenine highly reduced the cytotoxicity of AgNPs	[147]
	Ag	Cancer	AgNPs induced nucleus transcription factor EB (TFEB) translocation in HeLa cells, thus improved expression of autophagy-required genes. Knocking down the expression of TFEB weakened the autophagy response, and enhanced cell killing	[148]

Necrosis	ZnO	Cell transformation and cancer	ZnO NPs resulted in acute cytoskeletal collapse that induced fast necrotic process, and a ROS-dependent apoptotic process	[149]
	CaP	Biomimetic nanomaterials	The internalized calcium phosphate (CaP) NPs caused lysosomal ruptures due to the fast dissolution of CaP in lysosomes. The osmotic pressure balance across the lysosomal membranes was disturbed by the released ions, causing lysosomal ruptures and then cell necrosis	[150]
Genotoxicity	Ag	Cancer	AgNPs reduced ATP content of the cell resulted in damage to mitochondria and improved generation of ROS in a dose-dependent manner. The DNA damage was also observed to be dose-dependent. The presence of AgNPs within the nucleus and mitochondria revealing their impact in the DNA damage and the mitochondrial toxicity	[151]
	Silicate	Biomedical areas	No DNA damage was observed after 24 hours of incubation with nanosilicate platelets (NSPs) using the comet assay, and the micronucleus assay showed no important micronucleus induction in the Chinese hamster ovary (CHO) cells. Similarly, the *Salmonella* gene mutation assay on five strains indicated no mutations	[152]
	Graphene	Tissue engineering	GoNPs induced genotoxicity on the stem cells via chromosomal aberrations and DNA fragmentations even at low concentration of 0.1 mg ml^{-1}	[153]
	USPIONs	Cancer	As an important feature of USPIONs, iron redox state influenced their induction of DNA damage	[154]
Immune response	Silica	Cancer	Nano-enabled method for delivery of indoximod (IND) and oxaliplatin (OX) to the pancreatic ductal adenocarcinoma (PDAC) could be applied in a synergistic immunotherapy response	[155]
	Micelle	Cancer	The micelles enhanced cytokine generation, dendritic cells, and the higher cross-presentation of antigen to $CD8^+$ T cells which resulted in induction of immune responses *in vivo*	[156]
	Polymer	Nanomedicine	To study the effect of hydrophilicity in maintaining the equilibrium of the immune system, three different nanogels, namely, PEG, poly(sulfobetaine) (PSB), and poly(carboxybetaine) (PCB) were synthesized. Among them, PCB nanogels exhibited the strongest effect in reducing immune responses with the highest degree of hydrophilicity, followed by PSB and PEG, showing that the hydrophilicity of NPs has significant impact of immune system	[157]
	$[Gd@C_{82}(OH)_{22}]_n$	Tumor	The $[Gd@C_{82}(OH)_{22}]_n$ NPs activated dendritic cells and Th1 immune responses	[158]

Table 11.2 The common cell viability and cytotoxicity *in vitro* assays.

Endpoint	Cellular process	Assays	Advantages	Disadvantages	References
Cell viability	Metabolic activity	MTT	Water soluble and readily enters viable cells	SWCNTs bind to insoluble MTT-formazan and prevent its resolubilization	[159]
		MTS	Allows the sequential analysis of treated cells with other assays	Overestimation of the cell viability due to the presence of iron oxide NPs. The false increase of cell viability due to changes in redox metabolism in the presence of NPs	[160]
		WST-1	More accurate than MTT for evaluation of CNTs	Do not penetrate cells since they are negatively charged	[161]
		XTT	Solubilization of the XTT formazan product	Overestimate cell viability caused by O_2^- induced by TiO_2 NPs	[162]
	ATP content	ATP	Sensitive and fast. No possible interference with the particles' intrinsic light absorbance	Susceptible to temperature flux	[163]
Cytotoxicity	Necrosis	LDH	LDH enzyme is naturally available in the body	Released cytosolic enzyme cannot be a definite parameter	[164]
		Propidium iodide	Provides a more consistent assessment and studies many cells each second	Requires expensive, complex instrument, and long testing time	[165]
		Trypan blue	Estimation of the actual number of viable cells is easy	Manual counting of alive/dead cells	[166]
		NRU	Very sensitive, readily quantifiable, and relatively cheap	Once initiated it must be completed immediately	[165]
	Apoptosis	Caspase-3/7 activation	Reliable marker for cytotoxicity	Requires several wash steps	[167]
		Annexin V	Provide quick and reliable detection methods	It is usually used with other assays	[167]
		ELISA	Antibody/antigen recognition	The procedure of antigen immobilization is not special	[168]
	Autophagy	Electron microscopy	Produce good quality images	Various stabilization methods are required for almost all biological samples	[169]

to 2005, and then there was a considerable reduction after 2006, possibly related to lower investments related to the economic crisis of 2008 [138]. In comparison with conventional formulations, the majority of approved nanopharmaceuticals decreases toxicity. Most nanomedicines rejected in clinical development do not achieve a notable enhancement in efficacy [135]. Hence, the currently clinically used nanomedicines have exhibited better efficacy and promising results. Such FDA-approved NPs available for clinical use can be classified into polymers (i.e. ADYNOVATE®, Cimzia®, Copaxone®, Krystexxa®, and Neulasta®); liposomes (i.e. AmBisome®, Doxil®, Marqibo®, and Onivyde®); crystals (i.e. Avinza®, Emend®, Invega®, and Ryanodex®); polymeric micelle (i.e. Estrasorb®); inorganic NPs (i.e. INFeD®, Venofer®, and Nanotherm®); and proteins (i.e. Abraxane®). The complete list of approved nanomedicines with detailed information regarding their advantage as well as material description can be found elsewhere [134]. These NPs have been extensively employed in treatment of diseases including cancer, rheumatoid arthritis, neutropenia, hepatitis, psychostimulant, and schizophrenia. Tables 11.1 and 11.2 summarized some of the recent biomaterials that are currently used in the market and their potential health risk.

11.10 Future Prospects

The increasing interest in biomedical engineering has stimulated the researchers to scrutinize biological elements and learn from nature. The topic discussed earlier on frontiers technology gives us a wide knowledge about the basic principle underlying biomedical engineering systems. It gives a general overview of different kinds of new systems. The elucidation of the aforementioned goals will open many doors and lead to significant improvements in biological tools, drug discovery process, lead identification, and therapeutic approaches. The miniaturization of this approach allows one to perform many more experiments than previously possible in a simpler manner. These new technologies aim to develop set of tools that are simple, inexpensive, portable, and robust that could be commercialized and used in various fields of biomedical sciences such as drug discovery, diagnostic tools, and therapeutic approaches in regenerative medicine. Throughout the topics discussed earlier, it clearly guides the reader about the potential applications of new *in vitro* systems that can be applied for fabricating wider range of novel biomaterials for the use in biotechnology. With appropriate references and examples, it opens up the reader's mind incorporating a wider range of knowledge about new *in vitro* systems. These specific structures have inspired the researchers to use them in various areas of science like biotechnology, nanotechnology, and medicine. Discussed in this chapter are a few applications of new technologies and the advantages of new materials in medicine. These new technologies promise numerous micro- and nanoscale biomaterials that can act as bio-mimicking systems.

References

1 Khalaji, S., Ebrahimi, N.G., and Hosseinkhani, H. (2021). Enhancement of biocompatibility of PVA/HTCC blend polymer with collagen for skin care application. *Int. J. Polym. Mater. Polym. Biomater.* 70: 459–468.

2 Domb, A.J., Sharifzadeh, G., Nahum, V., and Hosseinkhani, H. (2021). Safety evaluation of nanotechnology products. *Pharmaceutics* 13: 1615–1662.

3 Steinman, N.Y., Campos, L.M., Feng, Y. et al. (2020). Cyclopropenium nanoparticles and gene transfection in cells. *Pharmaceutics* 12: 768–779.

4 Hosseinkhani, H. (2020). siRNA delivery systems in cancer therapy. *Org. Med. Chem. Int. J.* 10 (1): 1–3.

5 Hosseinkhani, H. and Domb, A.J. (2019). Biodegradable polymers in gene-silencing technology. *Polym. Adv. Technol.* 30: 2647–2655.

6 Hosseini, V., Maroufi, N.F., Saghati, S. et al. (2019). Current progress in hepatic tissue regeneration by tissue engineering. *J. Transl. Med.* 17 (1): 383.

7 Toosi, S., Naderi-Meshkin, H., Kalalinia, F. et al. (2019). Bone defect healing is induced by collagen sponge/polyglycolic acid. *J. Mater. Sci. - Mater. Med.* 30 (3): 33.

8 Ebrahimi, N., Soleimani, A.A., Rashidiani, J. et al. (2019). Chitosan/fluoride nanoparticles for preventing dental caries. *Curr. Dentistry* 1: 61–67.

9 Abedini, F., Ebrahimi, M., Roozbehani, A.H. et al. (2018). Overview on natural hydrophilic polysaccharide polymers in drug delivery. *Polym. Adv. Technol.* 29: 2564–2573.

10 Saberianpour, S., Heidarzadeh, M., Geranmayeh, M.H. et al. (2018). Tissue engineering strategies for the induction of angiogenesis using biomaterials. *J. Biol. Eng.* 12: 1–15.

11 Abedini, F., Ebrahimi, M., and Hosseinkhani, H. (2018). Technology of RNA interference in advanced medicine. *MicroRNA* 7: 74–84.

12 Hosseinkhani, H. (2018). 3-D culture systems for cell culture technology. *J. Cell Res.* 1: 1–4.

13 Sharifzadeh, G. and Hosseinkhani, H. (2017). Biomolecule-responsive hydrogels in medicine. *Adv. Healthcare Mater.* 6: 1–35.

14 Mottaghitalab, F., Rastegari, A., Farokhi, M. et al. (2017). Prospects of siRNA applications in regenerative medicine. *Int. J. Pharm.* 524: 312–329.

15 Ghadiri, M., Vasheghani-Farahani, E., Atyabi, F. et al. (2017). Transferrin-conjugated magnetic dextran-spermine nanoparticles for targeted drug transport across blood-brain barrier. *J. Biomed. Mater. Res. A* 105: 2851–2864.

16 Shahrezaei, M., Habibzadeh, S., Babaluo, A.A. et al. (2017). Study of synthesis parameters and photocatalytic activity of TiO_2 nanostructures. *J. Exp. Nanosci.* 12: 45–61.

17 Toosi, S., Naderi-Meshkin, H., Kalalinia, F. et al. (2017). Long bone mesenchymal stem cells (Lb-MSCs): clinically reliable cells for osteo-diseases. *Cell Tissue Banking* 18: 489–500.

18 Ghadiri, M., Vasheghani-Farahani, E., Atyabi, F. et al. (2017). *In-vitro* assessment of magnetic dextran-spermine nanoparticles for capecitabine delivery to cancerous cells. *Iran. J. Pharm. Res.* 16: 1320–1334.

19 Abassi, N., Hashemei, S.M., Salehi, M. et al. (2016). Influence of oriented nanofibrous PCL scaffolds on quantitative gene expression during neural differentiation of mouse embryonic stem cells. *J. Biomed. Mater. Res. A* 104: 155–164.

20 Farokhi, M., Mottaghitalab, F., Shokrghozar, M.A. et al. (2016). Importance of dual delivery systems for bone tissue engineering. *J. Controlled Release* 225: 152–169.

21 Gholivand, M.B., Mohammadi-Behzad, L., and Hosseinkhani, H. (2016). Application of Cu-chitosan/multi-walled carbon nanotube film modified electrode for the sensitive determination of rutin. *Anal. Biochem.* 493: 35–43.

22 Alibolandi, M., Abnous, K., Sadeghi, F. et al. (2016). Folate receptor-targeted multimodal polymersomes for delivery of quantum dots and doxorubicin to breast adenocarcinoma: *in vitro* and *in vivo* evaluation. *Int. J. Pharm.* 500: 162–178.

23 Toosi, S., Naderi-Meshkin, H., Kalalinia, F. et al. (2016). PLGA-incorporated collagen: toward a biodegradable composite scaffold for bone-tissue engineering. *J. Biomed. Mater. Res. A* 104: 2020–2028.

24 Toosi, S., Naderi-Meshkin, H., Kalalinia, F. et al. (2016). Comparative characteristics of mesenchymal stem cells derived from reamer–irrigator–aspirator, iliac crest bone marrow, and adipose tissue. *Cell. Mol. Biol.* 62: 68–74.

25 Mottaghitalab, F., Shokrghozar, M.A., Farokhi, M., and Hosseinkhani, H. (2015). Silk fibrin nanoparticles as novel drug delivery systems. *J. Controlled Release* 206: 161–176.

26 Mottaghitalab, F., Hosseinkhani, H., Shokrghozar, M.A. et al. (2015). Silk as potential candidate for bone tissue engineering. *J. Controlled Release* 215: 112–128.

27 Seyedpour, S.M., Pachenari, M., Janmaleki, M. et al. (2015). Effects of antimitotic drug on mechanical behaviors of cytoskeleton in distinct grades of colon cancer cell lines. *J. Biomech.* 48: 1172–1178.

28 Han, H.C., Lo, H.C., Chen, K.H. et al. (2015). Nano-textured fluidic biochip as biological filter for selective survival of cells and biological microorganisms. *J. Biomed. Mater. Res. A* 103: 2015–2023.

29 Jahani, H., Azizi Jalilian, F., Kaviani, S. et al. (2015). Controlled surface morphology and hydrophilicity of polycaprolactone towards differentiation of mesenchymal stem cells into neural cells. *J. Biomed. Mater. Res. A* 103: 1875–1881.

30 Hosseinkhani, H., Abedini, F., Ou, K.L., and Domb, A.J. (2015). Polymers in gene therapy technology. *Polym. Adv. Technol.* 26: 198–211.
31 Baheiraei, N., Azami, M., and Hosseinkhani, H. (2015). Investigation of magnesium incorporation within gelatin/calcium phosphate nanocomposite scaffold for bone tissue engineering. *Int. J. Appl. Ceram. Technol.* 12: 245–253.
32 Jain, A., Hosseinkhani, H., Domb, A.J., and Khan, W. (2015). Cationic polymers for the delivery of therapeutic nucleotides. *Polysaccharides: Bioactivity Biotechnol.* 1969–1990.
33 Pachenari, M., Seyedpour, M., Babazadeh Shayan, S. et al. (2014). Mechanical properties of cancer cytoskeleton depend on actin filaments to microtubules content: investigating different grades of colon cancer cell lines. *J. Biomech.* 47: 373–379.
34 Rahbarghazi, R., Nassiri, S.M., Ahmadi, S.H. et al. (2014). Dynamic induction of pro-angiogenic milieu after transplantation of marrow-derived mesenchymal stem cells in experimental myocardial infarction. *Int. J. Cardiol.* 173: 453–466.
35 Yeo, W.Y., Hosseinkhani, H., Rahman, S.A. et al. (2014). Safety profile of dextran-spermine gene delivery vector in mouse lungs. *J. Nanosci. Nanotechnol.* 14: 3328–3336.
36 Alibolandi, M., Abnous, K., Ramezani, M. et al. (2014). Synthesis of AS1411-aptamer-conjugated CdTe quantum dots with high fluorescence strength for probe labeling tumor cells. *J. Fluoresc.* 24: 1519–1529.
37 Ghodsizadeh, A., Hosseinkhani, H., Piryaei, A. et al. (2014). Galactosylated collagen matrix enhanced *in vitro* maturation of human embryonic stem cell-derived hepatocyte-like cells. *Biotechnol. Lett.* 36: 1095–1106.
38 Shi, D., Tatu, R., Liu, Q., and Hosseinkhani, H. (2014). Stem cells based tissue engineering for regenerative medicine. *Nano LIFE* 4 (2): 1–13.
39 He, W., Hosseinkhani, H., Mohammadinejad, R. et al. (2014). Polymeric nanoparticles for therapy and imaging. *Polym. Adv. Technol.* 25: 1216–1225.
40 Hu, C.S., Tang, S.L., Chiang, C.H. et al. (2014). Characterization and antitumor effects of chondroitin sulfate-chitosan nanoparticles delivery system. *J. Nanopart. Res.* 16: 1–15.
41 Ou, K.L. and Hosseinkhani, H. (2014). Development of 3D *in vitro* technology for medical applications. *Int. J. Mol. Sci.* 15: 17938–17962.
42 Ou, K.L., Chu, J.S., Hosseinkhani, H. et al. (2014). Biomedical nanostructured coating for minimally invasive surgery devices applications: characterization, cell cytotoxicity evaluation and an animal study in rat. *Surg. Endoscopy* 28: 2174–2188.
43 Huang, C.F., Huang, C.F., Chiang, H.J. et al. (2014). Comparison of cell response and surface characteristics on titanium implant with SLA and SLAffinity functionalization. *J. Electrochem. Soc.* 161: 15–20.
44 Hosseinkhani, H., Hong, P.D., and Yu, D.S. (2013). Self-assembled proteins and peptides for regenerative medicine. *Chem. Rev.* 113: 4837–4861.
45 Hosseinkhani, H., Chen, Y.R., He, W. et al. (2013). Engineering of magnetic DNA nanoparticles for tumor-targeted therapy. *J. Nanopart. Res.* 15: 1345–1355.

46 Hosseinkhani, H., Hiraoka, Y., Li, C.H. et al. (2013). Engineering three-dimensional collagen-IKVAV matrix to mimic neural microenvironment. *ACS Chem. Neurosci.* 4: 1229–1235.

47 Hosseinkhani, H., He, W.J., Chiang, C.H. et al. (2013). Biodegradable nanoparticles for gene therapy technology. *J. Nanopart. Res.* 15: 1794–1809.

48 Amini, R., Azizi Jalilian, F., Abdullah, S. et al. (2013). Dynamics of PEGylated-dextran-spermine nanoparticles for gene delivery to leukemic cells. *Appl. Biochem. Biotechnol.* 170: 841–853.

49 He, W.-J., Hosseinkhani, H., Hong, P.-D. et al. (2013). Magnetic nanoparticles for imaging technology. *Int. J. Nanotechnol.* 10: 930–944.

50 Chiang, C.-H., Hosseinkhani, H., Cheng, W.-S. et al. (2013). Improving drug loading efficiency and delivery performance of micro- and nanoparticle preparations through optimizing formulation variables. *Int. J. Nanotechnol.* 10: 996–1006.

51 Ou, S.-F., Chen, C.-S., Hosseinkhani, H. et al. (2013). Surface properties of nano-structural silicon-doped carbon films for biomedical applications. *Int. J. Nanotechnol.* 10: 945–958.

52 Hosseinkhani, H. and Chen, K.H. (2013). Editorial: nanotechnology research in Taiwan. *Int. J. Nanotechnol.* 10: 837–839.

53 Hosseinkhani, H., Hong, P.D., Yu, D.S. et al. (2012). Development of 3D *in vitro* platform technology to engineer mesenchymal stem cells. *Int. J. Nanomed.* 7: 3035–3043.

54 Khan, W., Hosseinkhani, H., Ickowicz, D. et al. (2012). Polysaccharide gene transfection agents. *Acta Biomater.* 8: 4224–4232.

55 Abedini, F., Hosseinkhani, H., Ismail, M. et al. (2012). Cationized dextran nanoparticles-encapsulated CXCR4-siRNA enhanced correlation between CXCR4 expression and serum ALP in colorectal cancer. *Int. J. Nanomed.* 7: 4159–4168.

56 Subramani, K., Pathak, S., and Hosseinkhani, H. (2012). Recent trend in diabetes treatment using nanotechnology. *Digest J. Nanomater. Biostruct.* 7: 85–95.

57 Hosseinkhani, H. (2012). 3D *in vitro* technology for drug discovery. *Curr. Drug Safety* 7: 37–43.

58 Mohammad-Taheri, M., Vasheghani-Farahani, E., Hosseinkhani, H. et al. (2012). Fabrication and characterization of a new MRI contrast agent based on a magnetic dextran-spermine nanoparticle system. *Iran. Polym. J.* 21: 239–251.

59 Amini, R., Hosseinkhani, H., Abduamir, A.S. et al. (2012). Engineered smart biomaterials for gene delivery. *Gene Ther. Mol. Biol.* 14: 72–86.

60 Mahmoudi, M., Hosseinkhani, H., Hosseinkhani, M. et al. (2011). Magnetic resonance imaging tracking of stem cells *in vivo* using iron oxide nanoparticles as a tool for the advancement of clinical regenerative medicine. *Chem. Rev.* 111: 253–280.

61 Hosseinkhani, H. (2011). Editorial: on nanomedicine. *Int. J. Nanotechnol.* 8: 615–617.

62 Hosseinkhani, H., Hosseinkhani, M., Chen, Y.R. et al. (2011). Innovative technology of engineering magnetic DNA nanoparticles for gene therapy. *Int. J. Nanotechnol.* 8: 724–735.

63 Hosseinkhani, M., Hosseinkhani, H., Chen, Y.R., and Subramani, K. (2011). *In vitro* physicochemical evaluation of DNA nanoparticles. *Int. J. Nanotechnol.* 8: 736–748.

64 Abedini, F., Hosseinkhani, H., Ismail, M. et al. (2011). *In vitro* intracellular trafficking of biodegradable nanoparticles of dextran-spermine in cancer cell lines. *Int. J. Nanotechnol.* 8: 712–723.

65 Sarabi, R.S., Sadeghi, E., Hosseinkhani, H. et al. (2011). Polyrotaxane capped quantum dots as new candidates for cancer diagnosis and therapy. *J. Nanostruct. Polym. Nanocomposit.* 7: 18–31.

66 Subramani, K., Mathew, R., Hosseinkhani, H., and Hosseinkhani, M. (2011). Bone regeneration around dental implants as a treatment for peri-implantitis: a review of the literature. *J. Biomimetics Biomater. Tissue Eng.* 11: 21–33.

67 Kalhor, H.R., Shahin, F.V., Fouani, M.H., and Hosseinkhani, H. (2011). Self-assembly of tissue transglutaminase into amyloid-like fibrils using physiological concentration of Ca^{2+}. *Langmuir* 27: 10766–10784.

68 Abedini, F., Ismail, M., Hosseinkhani, H. et al. (2011). Effects of CXCR4 siRNAs/dextran-spermine nanoparticles on CXCR4 expression and serum LDH levels in a mouse model of colorectal cancer metastasis to the liver. *Cancer Manag. Res.* 3: 301–309.

69 Hosseinkhani, H., Hosseinkhani, M., Hattori, S. et al. (2010). Micro and nanoscale *in vitro* 3D culture system for cardiac stem cells. *J. Biomed. Mater. Res. A* 94: 1–8.

70 Mohageri, S., Hosseinkhani, H., Ebrahimi, N.G. et al. (2010). Proliferation and differentiation of mesenchymal stem cell on collagen sponge reinforced with polypropylene/polyethylene terephathalate blend fibers. *Tissue Eng. Part A* 16: 3821–3830.

71 Lindstrom, S., Iles, A., Persson, J. et al. (2010). Nanoporous titania coating of microwell chips for stem cell culture and analysis. *J. Biomech. Sci. Eng.* 5: 272–279.

72 Abdullah, S., Yeo, W.Y., Hosseinkhani, H. et al. (2010). Gene transfer into the lung by nanoparticle dextran-spermine/plasmid DNA complexes. *BioMed Res. Int.* 2010: 1–10.

73 Abedini, F., Ismail, M., Hosseinkhani, H. et al. (2010). Toxicity evaluation of dextran-spermine polycation as a tool for gene therapy *in vitro*. *J. Cell Anim. Biol.* 4: 170–176.

74 Amini, R., Rosli, R., Abdullah, S. et al. (2010). Delivery of plasmid expressing green fluorescent protein by PEGylated dextran-spermine to acute myeloid leukemic cells. *Myeloid Leukemia* 65: 1–10.

75 Balbarini, A., Magera, A., Barsotti, M.C. et al. (2010). Self assembling peptide amphiphile nanofibers as a scaffold for endothelial progenitor cells. *J. Am. Coll. Cardiol.* 55: E1652.
76 Subramani, K., Hosseinkhani, H., Khraisat, A. et al. (2009). Targeting nanoparticles as drug delivery systems for cancer treatment. *Curr. Nanosci.* 5: 134–140.
77 Hosseinkhani, H. and Hosseinkhani, M. (2009). Biodegradable polymer–metal complexes for gene and drug delivery. *Curr. Drug Safety* 4: 79–83.
78 Hosseinkhani, H., Hosseinkhani, M., Vasheghani, E., and Nekoomanesh, M. (2009). *In vitro* sustained release and degradation study of biodegradable poly(D,L-lactic acid) microspheres loading theophylline. *Adv. Sci. Lett.* 2: 70–77.
79 Di Stefano, R., Barsotti, M.C., Magera, A. et al. (2009). A biological self assembling amphiphilic peptide enhances endothelial progenitor cells growth and paracrine release. *Eur. Heart J.* 30: 177–177.
80 Tian, F., Hosseinkhani, H., Hosseinkhani, M. et al. (2008). Quantitative analytical of cell adhesion on aligned micro- and nanofibers. *J. Biomed. Mater. Res. A* 84: 291–299.
81 Hosseinkhani, H., Hosseinkhani, M., Khademhosseini, A. et al. (2008). DNA nanoparticles encapsulated in 3-D tissue engineered scaffold enhance osteogenic differentiation of mesenchymal stem cells. *J. Biomed. Mater. Res. A* 85: 47–60.
82 Hosseinkhani, H. and Hosseinkhani, M. (2008). Suppression effect of basic fibroblast growth factor on mesenchymal stem cell proliferation activity; part I: release characteristics. *Chem. Today* 26: 30–32.
83 Hosseinkhani, H. and Hosseinkhani, M. (2008). Suppression effect of basic fibroblast growth factor on mesenchymal stem cell proliferation activity; part II: biological characteristics. *Chem. Today* 26: 35–37.
84 Hosseinkhani, M., Hosseinkhani, H., Khademhosseini, A. et al. (2007). Bone morphogenetic protein-4 enhances cardiomyocytes differentiation of cynomolgus monkey ES cells in knockout serum replacement medium. *Stem Cells* 25: 571–580.
85 Furong, T., Hosseinkhani, H., Estrada, G., and Kobayashi, H. (2007). Quantitative method for the analysis of cell attachment using the aligned scaffold structure. *J. Phys.* 61: 587–590.
86 Hosseinkhani, H., Yamamoto, M., Inatsugu, Y. et al. (2006). Enhanced ectopic bone formation using combination of impregnation of plasmid DNA into 3-D scaffold and bioreactor perfusion culture. *Biomaterials* 27: 1387–1398.
87 Hosseinkhani, H., Hosseinkhani, M., Tian, F. et al. (2006). Osteogenic differentiation of mesenchymal stem cells in self-assembled-peptide amphiphile nanofibers. *Biomaterials* 27: 4079–4086.

88 Hosseinkhani, H., Azzam, T., Kobayashi, H. et al. (2006). Combination of 3-D tissue engineered scaffold and non-viral gene carrier enhance in vitro DNA expression of mesenchymal stem cells. *Biomaterials* 27: 4269–4278.

89 Hosseinkhani, H., Hosseinkhani, M., Tian, F. et al. (2006). Ectopic bone formation in collagen sponge-self assembled peptide amphiphile nanofibers hybrid scaffold in a perfusion culture bioreactor. *Biomaterials* 27: 5089–5098.

90 Hosseinkhani, H., Hosseinkhani, M., Khademhosseini, A. et al. (2006). Enhanced angiogenesis through controlled release of basic fibroblast growth factor from peptide amphiphile for tissue regeneration. *Biomaterials* 27: 5836–5844.

91 Hosseinkhani, H., Hosseinkhani, M., and Kobayashi, H. (2006). Proliferation and differentiation of mesenchymal stem cells by using self-assembly of peptide-amphiphile nanofibers. *Biomed. Mater.* 1: 8–15.

92 Hosseinkhani, H., Hosseinkhani, M., and Kobayashi, H. (2006). Design of tissue-engineered nanoscaffold through self-assembly of peptide amphiphile. *J. Bioact. Compat. Polym.* 21: 277–296.

93 Hosseinkhani, H. and Tabata, Y. (2006). Self-assembly of DNA nanoparticles with polycations for the delivery of genetic materials into cells. *J. Nanosci. Nanotechnol.* 6: 2320–2328.

94 Hosseinkhani, H. (2006). DNA nanoparticles for gene delivery to cells and tissue. *Int. J. Nanotechnol.* 3: 416–461.

95 Hosseinkhani, H., Hosseinkhani, M., and Khademhosseini, A. (2006). Emerging applications of hydrogels and microscale technologies in drug discovery. *Drug Discovery* 1: 32–34.

96 Hosseinkhani, H., Hosseinkhani, M., and Khademhosseini, A. (2006). Tissue regeneration through self-assembled peptide amphiphile nanofibers. *Yakhte Med. J.* 8: 204–209.

97 Hosseinkhani, H., Kobayashi, H., and Tabata, Y. (2006). Design of tissue-engineered nano-scaffold using peptide-amphiphile for regenerative medicine. *Pept. Sci.* 2005: 341–344.

98 Hosseinkhani, H., Kobayashi, H., and Tabata, Y. (2006). Selective differentiation cardiomyocyte cells by using peptide-amphiphile nanofibers. *Pept. Sci.* 2005: 63–66.

99 Konishi, M., Tabata, Y., Kariya, M. et al. (2005). *In vivo* anti-tumor effect of dual release of cisplatin and adriamycin from biodegradable gelatin hydrogel. *J. Controlled Release* 103: 7–19.

100 Hosseinkhani, H., Inatsugu, Y., Hiraoka, Y. et al. (2005). Impregnation of plasmid DNA into 3-D scaffold and medium perfusion enhance *in vitro* DNA expression of mesenchymal stem cells. *Tissue Eng.* 11: 1459–1475.

101 Hosseinkhani, H., Inatsugu, Y., Hiraoka, Y. et al. Perfusion culture enhances osteogenic differentiation of rat mesenchymal stem cells in collagen sponge reinforced with poly (glycolic acid) fiber. *Tissue Eng.* 11: 1476–1488.

102 Hosseinkhani, H., Azzam, T., Tabata, Y., and Domb, A.J. (2004). Dextran-spermine polycation: an efficient non-viral vector for *in vitro* and *in vivo* gene transfection. *Gene Ther.* 11: 194–203.

103 Hosseinkhani, H., Aoyama, T., Ogawa, O., and Tabata, Y. (2003). Ultrasound enhances the transfection of plasmid DNA by non-viral vector. *Curr. Pharm. Biotechnol.* 4: 109–122.

104 Hosseinkhani, H., Aoyama, T., Ogawa, O., and Tabata, Y. (2002). Liver targeting of plasmid DNA by pullulan conjugation based on metal coordination. *J. Controlled Release* 83: 287–302.

105 Jain, A., Domb, A.J., Hosseinkhani, H., and Khan, W. (2015). Cationic polymers for the delivery of therapeutic nucleotides. In: *Polysaccharides, Bioactivity and Biotechnology* (ed. K.G. Ramawat and J.-M. Mérillon), 1969–1990. Springer Publication.

106 Hosseinkhani, H. (2013). Innovation technology to engineer 3D living organs as intelligent diagnostic tools. In: *Characterization and Development of Biosystems and Biomaterials* (ed. A. Öchsner, L.F.M. Silva and H. Altenbach), 183–192. Springer Publication.

107 Hosseinkhani, H. (2013). Controlled release systems for bone regeneration. In: *Polymeric Biomaterials, Third Edition, Volume II: Medicinal and Pharmaceutical Applications of Polymers and Technology* (ed. S. Dumitriu and V. Popa), 643–654. USA: CRC Press/Taylor and Francis.

108 Hosseinkhani, H., Hosseinkhani, M., and Subramani, K. (2012). Bone regeneration using self-assembled nanoparticles-based scaffolds. In: *Emerging Nanotechnologies in Dentistry* (ed. K. Subramani and W. Ahmed), 225–237. UK: Elsevier.

109 Hosseinkhani, H. and Hosseinkhani, M. (2009). Tissue engineered scaffolds for stem cells and regenerative medicine. In: *Trend in Stem Cells Biology and Technology* (ed. H. Baharvand), 367–387. USA: HUMANANA Press Inc.

110 Hosseinkhani, H., Hosseinkhani, M., Zhang, S., and Subramani, K. (2008). Self-assembly of nanomaterials for engineering cell microenvironment. In: *Micro and Nanoenginernig of the Cell Microenvironment: Applications and Technologies* (ed. A. Khademhosseini, J. Borenstein, M. Toner and S. Takayama), 275–290. USA: Artech House Publishers.

111 Hosseinkhani, H., Hosseinkhani, M., and Khademhosseini, A. (2007). Emerging technology of hydrogels in drug discovery. In: *Topics in Multifunctional Biomaterials and Devices* (ed. N. Ashammakhi), 110–119. London, UK.

112 Hosseinkhani, H., Aoyama, T., Ogawa, O., and Tabata, Y. (2005). Tumor targeting of plasmid DNA by dextran conjugation based on metal coordination. In: *Key Engineering Materials*, vol. 288–289 (ed. X. Zhang, J. Tanaka, Y. Yu and Y. Tabata), 109–112. Switzerland: Trans Tech Publications.

113 Yamamoto, M. and Hosseinkhani, H. (2004). *Liver Targeting of Plasmid DNA by Polymer–Metal Conjugation*, 119–132. Japan: Yodosha Tech Publications, Japanese version, Med Tech Publications.

114 Hosseinkhani, H. and Hosseinkhani, M. (2009). Design of 3D culture systems to enhance *in vitro* gene expression of mesenchymal stem cells. *Proc. Mol. Ther.* 17: S270–S270.

115 Hosseinkhani, H., Hosseinkhani, M., and Khademhosseini, A. (2007). A new injectable tissue engineered scaffold induces angiogenesis. *Proc. AIChE Annu. Meeting*.

116 Hosseinkhani, M. and Hosseinkhani, H. (2007). Post-translational modification of GATA-4 involved in the differentiation of monkey ES cell into cardiac myocytes. *Circulation* 116 (16): 202–203.

117 Hosseinkhani, M. and Hosseinkhani, H. (2007). Bone morphogenetic protein-4 enhances cardiomyocyte differentiation of cynomolgus monkey ES cells in knockout serum replacement medium. *Eur. Heart J.* 28: 230–231.

118 Hosseinkhani, H., Kobayashi, H., and Tabata, Y. (2006). Design of a nano-vessel-like network for controlled proliferation and differentiation of mesenchymal stem cells for regenerative medicine. *Tissue Eng.* 12 (4): 993–994.

119 Hosseinkhani, H., Hosseinkhani, M., and Khademhosseini, A. (2006). A new injectable tissue engineered scaffold for regenerative medicine. *Proc. Int. Conf. Microtechnol. Med. Biol.* 4281294: 10–11.

120 Hosseinkhani, H. (2005). Selective differentiation cardiomyocyte cells by using peptide-amphiphile nanofibers, *Proceedings of the 42nd Japanese Peptide Symposium*, Osaka, 30: 25–30.

121 Hosseinkhani, H. (2005). Design of tissue-engineered nano-scaffold using peptide-amphiphile for regenerative medicine, *Proceedings of the 42nd Japanese Peptide Symposium*, Osaka, 30: 21–34.

122 Hosseinkhani, H. and Tabata, Y. (2004). PEGylation enhances tumor targeting of plasmid DNA by an artificial cationized protein with repeated RGD sequences, Pronectin® cationized, *Proceedings 7th World Biomaterials Congress*, Sydney, Australia.

123 Hosseinkhani, H. and Tabata, Y. (2004). Tumor targeting of plasmid DNA by spermine derivative of dextran combined with ultrasound. *Polym. Preprints* 53 (2): 2PE179.

124 Hosseinkhani, H. and Tabata, Y. (2004). Ultrasound enhances expression level of plasmid DNA by PEGylation of cationized dextran in tumor. *Journal Code: X0225A* 19 (3): 299.

125 Hosseinkhani, H., Aoyama, T., Ogawa, O., and Tabata, Y. (2002). Tumor targeting of plasmid DNA by dextran conjugation based on metal coordination. *Drug Delivery* 17 (3): 260.

126 Vasheghani, E., Hosseinkhani, H., and Nekomanesh, M. (2001). Effect of preparation conditions on theophylline release from biodegradable poly (DL-lactic acid) microspheres, Proceedings of the JCR Symposium. *J. Controlled Release* 72: 287–291.

127 Hosseinkhani, H. (2019). *Nanomaterials in Advanced Medicine*. Wiley. ISBN: 978-3-527-34549-6.

128 Wang, H., Hosseinkhani, H., Chaswal, V. et al. (2016). *The World Scientific Encyclopedia of Nanomedicine and Bioengineering I: Nanotechnology for Translational Medicine: Tissue Engineering, Biological Sensing, Medical Imaging, and Therapeutics (a 5-Volume Set)*. World Scientific Publishing, Inc. ISBN: ISBN-13: 978-9814667654.

129 Hosseinkhani, H. and Ou, K.L. (2014). *The Nanoscience in Translational Medicine*. Wu-Nan Book Inc. ISBN: 978-957-11-7839-4.

130 Hosseinkhani, H. and Ou, K.L. (2013). *Advanced Biomaterials for Biomedical Engineering*. Wu-Nan Book Inc. ISBN: 978-957-11-7169-2.

131 Fadeel, B., Farcal, L., Hardy, B. et al. (2018). Advanced tools for the safety assessment of nanomaterials. *Nat. Nanotechnol.* 13: 537–543.

132 Collins, A.R., Annangi, B., Rubio, L. et al. (2017). High throughput toxicity screening and intracellular detection of nanomaterials. *Wiley Interdiscip. Rev. Nanomed. Nanobiotechnol.* 9: e1413.

133 Richard, A.M., Judson, R.S., Houck, K.A. et al. (2016). ToxCast chemical landscape: paving the road to 21st century toxicology. *Chem. Res. Toxicol.* 29: 1225–1251.

134 Bobo, D., Robinson, K.J., Islam, J. et al. (2016). Nanoparticle-based medicines: a review of FDA-approved materials and clinical trials to date. *Pharm. Res.* 33: 2373–2387.

135 Caster, J.M., Patel, A.N., Zhang, T., and Wang, A. (2017). Investigational nanomedicines in 2016: a review of nanotherapeutics currently undergoing clinical trials. *Wiley Interdiscip. Rev. Nanomed. Nanobiotechnol.* 9: e1416.

136 Tinkle, S., McNeil, S.E., Mühlebach, S. et al. (2014). Nanomedicines: addressing the scientific and regulatory gap. *Annu. Rep.* 1313: 35–56.

137 Dobrovolskaia, M.A. (2015). Pre-clinical immunotoxicity studies of nanotechnology-formulated drugs: challenges, considerations and strategy. *J. Controlled Release* 220: 571–583.

138 Havel, H.A. (2016). Where are the nanodrugs? An industry perspective on development of drug products containing nanomaterials. *AAPS J.* 18: 1351–1353.

139 Kotov, N.A. (2010). Inorganic nanoparticles as protein mimics. *Science* 330: 188–189.

140 Wilhelm, S., Kaiser, M., Wuerth, C. et al. (2015). Water dispersible upconverting nanoparticles: effects of surface modification on their luminescence and colloidal stability. *Nanoscale* 7: 1403–1410.

141 Colombo, M., Carregal-Romero, S., Casula, M.F. et al. (2012). Biological applications of magnetic nanoparticles. *Chem. Soc. Rev.* 41: 4306–4334.

142 Lin, C.-A.J., Yang, T.-Y., Lee, C.-H. et al. (2009). Synthesis, characterization, and bioconjugation of fluorescent gold nanoclusters toward biological labeling applications. *ACS Nano* 3: 395–401.

143 Parak, W.J., Pellegrino, T., and Plank, C. (2005). Labelling of cells with quantum dots. *Nanotechnology* 16: R9–R25.

144 Zhang, R., Yang, J., Sima, M. et al. (2014). Sequential combination therapy of ovarian cancer with degradable *N*-(2-hydroxypropyl)methacrylamide copolymer paclitaxel and gemcitabine conjugates. *Proc. Natl. Acad. Sci. USA* 111: 12181–12186.

145 Sharifzadeh, G., Wahit, M.U., Soheilmoghaddam, M. et al. (2016). Kappa-carrageenan/halloysite nanocomposite hydrogels as potential drug delivery systems. *J. Taiwan Inst. Chem. Eng.* 67: 426–434.

146 Guo, P. (2010). The emerging field of RNA nanotechnology. *Nat. Nanotechnol.* 5: 833–842.

147 Nurunnabi, M., Khatun, Z., Reeck, G.R. et al. (2014). Photoluminescent graphene nanoparticles for cancer phototherapy and imaging. *ACS Appl. Mater. Interfaces* 6: 12413–12421.

148 Yu, S.J., Kang, M.W., Chang, H.C. et al. (2005). Bright fluorescent nanodiamonds: no photobleaching and low cytotoxicity. *J. Am. Chem. Soc.* 127: 17604–17605.

149 Lee, Y., Fukushima, S., Bae, Y. et al. (2007). A protein nanocarrier from charge-conversion polymer in response to endosomal pH. *J. Am. Chem. Soc.* 129: 5362–5363.

150 Sapsford, K.E., Tyner, K.M., Dair, B.J. et al. (2011). Analyzing nanomaterial bioconjugates: a review of current and emerging purification and characterization techniques. *Anal. Chem.* 83: 4453–4488.

151 Schaak, R.E. and Williams, M.E. (2012). Full disclosure: the practical side of nanoscale total synthesis. *ACS Nano* 6: 8492–8497.

152 Wolf, L.K. (2012). Sweating the small stuff. *Chem. Eng. News* 90: 48–50.

153 Kovalenko, M.V., Manna, L., Cabot, A. et al. (2015). Prospects of nanoscience with nanocrystals. *ACS Nano* 9: 1012–1057.

154 Chou, L.Y.T., Ming, K., and Chan, W.C.W. (2011). Strategies for the intracellular delivery of nanoparticles. *Chem. Soc. Rev.* 40: 233–245.

155 Yildiz, I., Shukla, S., and Steinmetz, N.F. (2011). Applications of viral nanoparticles in medicine. *Curr. Opin. Biotechnol.* 22: 901–908.

156 Pokorski, J.K. and Steinmetz, N.F. (2011). The art of engineering viral nanoparticles. *Mol. Pharmaceutics* 8: 29–43.

157 Voigt, N., Henrich-Noack, P., Kockentiedt, S. et al. (2014). Surfactants, not size or zeta-potential influence blood-brain barrier passage of polymeric nanoparticles. *Eur. J. Pharm. Biopharm.* 87: 19–29.

158 Rosenholm, J.M., Meinander, A., Peuhu, E. et al. (2009). Targeting of porous hybrid silica nanoparticles to cancer cells. *ACS Nano* 3: 197–206.

159 Deng, T., Wang, J., Li, Y. et al. (2018). Quantum dots-based multifunctional nano-prodrug fabricated by ingenious self-assembly strategies for tumor theranostic. *ACS Appl. Mater. Interfaces* 10: 27657–27668.

160 Moghimi, S.M., Hunter, A.C., and Murray, J.C. (2005). Nanomedicine: current status and future prospects. *FASEB J.* 19: 311–330.

161 Li, K.Y., Eres, G., Howe, J. et al. (2013). Self-assembly of graphene on carbon nanotube surfaces. *Sci. Rep.* 3: 2353.

162 Sinha, N. and Yeow, J.T. (2005). Carbon nanotubes for biomedical applications. *IEEE Trans. Nanobiosci.* 4: 180–195.

163 Mehra, N.K., Jain, K., and Jain, N.K. (2015). Pharmaceutical and biomedical applications of surface engineered carbon nanotubes. *Drug Discovery Today* 20: 750–759.

164 Iijima, S. (1991). Helical microtubules of graphitic carbon. *Nature* 354: 56–58.

165 Endo, M., Kim, Y.A., Hayashi, T. et al. (2001). Vapor-grown carbon fibers (VGCFs) basic properties and their battery applications. *Carbon* 39: 1287–1297.

166 Kang, S.J., Kocabas, C., Ozel, T. et al. (2007). High-performance electronics using dense, perfectly aligned arrays of single-walled carbon nanotubes. *Nat. Nanotechnol.* 2: 230–236.

167 Hu, Q.D., Tang, G.P., and Chu, P.K. (2014). Cyclodextrin-based host-guest supramolecular nanoparticles for delivery: from design to applications. *Acc. Chem. Res.* 47: 2017–2025.

168 Koo, J.H., Jeong, S., Shim, H.J. et al. (2017). Wearable electrocardiogram monitor using carbon nanotube electronics and color-tunable organic light-emitting diodes. *ACS Nano* 11: 10032–10041.

169 Behzadi, S., Serpooshan, V., Tao, W. et al. (2017). Cellular uptake of nanoparticles: journey inside the cell. *Chem. Soc. Rev.* 46: 4218–4244.

Index